AF776100

Weed management in Conservation Agriculture systems

It is widely recognised that agriculture is a significant contributor to global warming and climate change. Agriculture needs to reduce its environmental impact and adapt to current climate change whilst still feeding a growing population, i.e. become more 'climate-smart'. Burleigh Dodds Science Publishing is playing its part in achieving this by bringing together key research on making the production of the world's most important crops and livestock products more sustainable.

Based on extensive research, our publications specifically target the challenge of climate-smart agriculture. In this way we are using 'smart publishing' to help achieve climate-smart agriculture.

Burleigh Dodds Science Publishing is an independent and innovative publisher delivering high quality customer-focused agricultural science content in both print and online formats for the academic and research communities. Our aim is to build a foundation of knowledge on which researchers can build to meet the challenge of climate-smart agriculture.

For more information about Burleigh Dodds Science Publishing simply call us on +44 (0) 1223 839365, email info@bdspublishing.com or alternatively please visit our website at www.bdspublishing.com.

Related titles:

Advances in Conservation Agriculture Volume 1
Print (ISBN 978-1-78676-264-1); Online (ISBN 978-1-78676-266-5, 978-1-78676-267-2)

Advances in Conservation Agriculture Volume 2
Print (ISBN 978-1-78676-268-9); Online (ISBN 978-1-78676-270-2, 978-1-78676-271-9)

Advances in Conservation Agriculture Volume 3
Print (ISBN 978-1-78676-475-1); Online (ISBN 978-1-78676-477-5, 978-1-78676-478-2)

Advances in integrated weed management
Print (ISBN 978-1-78676-745-5); Online (ISBN 978-1-78676-747-9, 978-1-78676-748-6)

Improving organic crop cultivation
Print (ISBN 978-1-78676-184-2); Online (ISBN 978-1-78676-186-6, 978-1-78676-187-3)

Integrated weed management for sustainable agriculture
Print (ISBN 978-1-78676-164-4); Online (ISBN 978-1-78676-166-8, 978-1-78676-167-5)

Chapters are available individually from our online bookshop:
https://shop.bdspublishing.com

climate•SMART•publishing
IN AGRICULTURAL SCIENCE

BURLEIGH DODDS SERIES IN AGRICULTURAL SCIENCE
NUMBER 160

Weed management in Conservation Agriculture systems

Edited by Professor Gottlieb Basch, University of Evora, Portugal, Professor Emilio González-Sánchez, University of Cordoba, Spain, Mr John Geraghty, South East Technological University, Ireland, Dr Seyed Vahid Eslami, University of Birjand, Iran, Professor Sjoerd W. Duiker, Penn State University, USA, Dr Saidi Mkomwa, African Conservation Tillage Network, Kenya and Dr Marie Bartz, Municipal Center for Culture and Development of Idanha-a-Nova and University of Coimbra, Portugal

burleigh dodds
SCIENCE PUBLISHING

Published by Burleigh Dodds Science Publishing Limited
82 High Street, Sawston, Cambridge CB22 3HJ, UK
www.bdspublishing.com

Burleigh Dodds Science Publishing, 1518 Walnut Street, Suite 900, Philadelphia, PA 19102-3406, USA

First published 2025 by Burleigh Dodds Science Publishing Limited
© Burleigh Dodds Science Publishing, 2025. All rights reserved.

This book contains information obtained from authentic and highly regarded sources. Reprinted material is quoted with permission and sources are indicated. Reasonable efforts have been made to publish reliable data and information but the authors and the publisher cannot assume responsibility for the validity of all materials. Neither the authors nor the publisher, nor anyone else associated with this publication shall be liable for any loss, damage or liability directly or indirectly caused or alleged to be caused by this book.

No part of this publication may be reproduced, stored in a retrieval system or transmitted in any form or by any means electronic, mechanical, photocopying, recording or otherwise without the prior written permission of the publisher.

The consent of Burleigh Dodds Science Publishing Limited does not extend to copying for general distribution, for promotion, for creating new works, or for resale. Specific permission must be obtained in writing from Burleigh Dodds Science Publishing Limited for such copying.

Permissions may be sought directly from Burleigh Dodds Science Publishing at the above address. Alternatively, please email: info@bdspublishing.com or telephone (+44) (0) 1223 839365.

Trademark notice: Product or corporate names may be trademarks or registered trademarks and are used only for identification and explanation, without intent to infringe.

Notice
No responsibility is assumed by the publisher for any injury and/or damage to persons or property as a matter of product liability, negligence or otherwise, or from any use or operation of any methods, products, instructions or ideas contained in the material herein.

Library of Congress Control Number: 2025931068

British Library Cataloguing in Publication Data
A catalogue record for this book is available from the British Library

ISBN 978-1-80146-784-1 (print)
ISBN 978-1-80146-785-8 6 (ePub)
ISBN 978-1-80146-786-5 (PDF)
ISSN 2059-6936 (print)
ISSN 2059-6944 (online)

DOI: 10.19103/AS.2024.0145

Typeset by Deanta Global Publishing Services, Dublin, Ireland

Contents

© Burleigh Dodds Science Publishing Limited, 2025. All rights reserved.

Part 1 Cultural, physical and chemical weed management techniques

© Burleigh Dodds Science Publishing Limited, 2025. All rights reserved.

© Burleigh Dodds Science Publishing Limited, 2025. All rights reserved.

Part 2 Weed management in specific production systems

© Burleigh Dodds Science Publishing Limited, 2025. All rights reserved.

© Burleigh Dodds Science Publishing Limited, 2025. All rights reserved.

© Burleigh Dodds Science Publishing Limited, 2025. All rights reserved.

Series list

© Burleigh Dodds Science Publishing Limited, 2025. All rights reserved.

© Burleigh Dodds Science Publishing Limited, 2025. All rights reserved.

© Burleigh Dodds Science Publishing Limited, 2025. All rights reserved.

© Burleigh Dodds Science Publishing Limited, 2025. All rights reserved.

© Burleigh Dodds Science Publishing Limited, 2025. All rights reserved.

© Burleigh Dodds Science Publishing Limited, 2025. All rights reserved.

© Burleigh Dodds Science Publishing Limited, 2025. All rights reserved.

© Burleigh Dodds Science Publishing Limited, 2025. All rights reserved.

© Burleigh Dodds Science Publishing Limited, 2025. All rights reserved.

© Burleigh Dodds Science Publishing Limited, 2025. All rights reserved.

© Burleigh Dodds Science Publishing Limited, 2025. All rights reserved.

Acknowledgements

Cover image copyright:

- Cover image: Picture of no-till maize emerging through rolled/crimped hairy vetch/oat cover crop residue. 2021. © Sjoerd W. Duiker.

© Burleigh Dodds Science Publishing Limited, 2025. All rights reserved.

Introduction

Since the beginning of sedentary agriculture, soil tillage has played a primary role in managing weeds. However, an important lesson learnt from the Dust Bowl in the United States in the 1930s is that tillage is also a root cause of soil degradation. This realisation led to the implementation of conservation tillage practices and, in the late 1990s, to the approach known as Conservation Agriculture (CA) based on:

- eliminating soil tillage altogether and replacing it with no-tillage systems
- permanent soil cover with crop residues and other organic materials
- diversified cropping systems with wide and diverse crop rotations or associations

However, the absence of soil tillage has made weed management more challenging. The continued use and reliance on herbicides such as glyphosate, as well as their misuse, has led to problems of weed resistance and unwanted herbicide residues. This requires Conservation Agriculture to develop approaches other than the use of herbicides to manage weeds. This collection reviews research on ways to optimise weed management in Conservation Agriculture systems without the current reliance on herbicides.

Chapter 1 helps to set the scene by exploring the dynamics of weed communities within Conservation Agriculture systems. It reviews current research on patterns of weed density, shifts in species and seedbank dynamics and uses this understanding to assess the efficacy of the main Conservation Agriculture pillars (minimum soil disturbance, diversified crop rotations and cover cropping) in mitigating weed pressure. The chapter also considers the challenges of changes in weed abundance, shifts in weed diversity, and invasive species. Identifying weed vulnerabilities helps to optimise management practices for their control.

Chapter 2 introduces weed population modelling as an essential tool for combining existing knowledge of weed growth with an assessment of the effectiveness of weed control methods. This chapter provides tools to support the development of Conservation Agriculture through decision support systems which integrate population dynamic models in order to improve the effectiveness and outcomes of CA weed control practices. It concludes by considering challenges in modelling and how they can be addressed to support its use within Conservation Agriculture.

© Burleigh Dodds Science Publishing Limited, 2025. All rights reserved.

Part 1 Cultural, physical and chemical weed management techniques

The first group of chapters provides an overview of cultural, physical and chemical techniques used to control weeds in CA systems. Chapter 3 provides an overview of cultural techniques used to control weeds, including selection of appropriate varieties, crop plant spacing and density of cover to occupy space and outcompete weeds. The chapter discusses the use of mulches, cover crops and rotations (including ley pastures) to control lower-level weed populations, making it harder for weeds to become established as well as promoting soil health. These methods are explored further through a case study from southern Brazil, which explores the dynamics of weeds in the region, how cover crops can be used to disrupt weeds and practical applications for farmers.

Chapter 4 focuses on allelopathic weed control, which involves release of chemicals from plant roots that can inhibit growth of other species and which can be exploited to control weed growth. It considers how components of a Conservation Agriculture system, such as minimum tillage, soil cover, and diversified crop rotation, provide an unfavourable habitat for weeds and the allelopathic strategies that can support overcoming these challenges. The chapter details how using allelopathic methods such as weed-suppressive crop varieties and natural allelopathic extracts can be used in combination with other techniques such as mulching, crop rotation, intercropping and cover crops.

Chapter 5 considers the impact of thermal weed control in Conservation Agriculture. Thermal weed control techniques include flaming, electrical weed control, hot water steam treatments and laser technology. The chapter discusses their respective strengths in terms of proven weed control efficacy, energy consumption and impact on soil health and crop safety. Research shows the effectiveness of these methods on younger, smaller weeds and discusses how advancements in selective targeting such as drones and AI will need to be introduced to optimise efficacy, avoid damage to crops and minimise risk of fire.

Chapter 6 reviews the state-of-the-art in the use of herbicides in Conservation Agriculture. The chapter provides an overview of current herbicide use in Conservation Agriculture, as well as an analysis of herbicide performance and effects on soil. This chapter discusses the particular reliance on glyphosate and how it has become a key herbicide in Conservation Agriculture through the introduction of glyphosate-tolerant crop varieties. The chapter assesses how herbicide efficacy can potentially be reduced by higher levels of soil organic matter and use of crop residues, including variables such as precipitation and irrigation, herbicide formulation and application method and amount and type of residue. The chapter concludes by discussing resistance and responses to this challenge such as occasional tillage or planting green systems.

© Burleigh Dodds Science Publishing Limited, 2025. All rights reserved.

Complementing the previous chapter, Chapter 7 considers advances in the reduction of herbicide use in Conservation Agriculture. It focusses on research carried out on experimental stations and farmer networks to synthesize knowledge on how herbicide-free no-till agriculture can best be achieved. The chapter identifies two proven pathways: use of crop rotation and cover cropping where tillage is not an option, or using occasional tillage or techniques such as strip till with minimal soil disturbance that allows quick crop establishment and growth to outcompete weeds.

Part 2 Weed management in specific production systems

The second group of chapters utilises the ideas introduced in Part 1 and puts them into context of specific production systems in order to illustrate their effectiveness in supporting Conservation Agriculture systems. These chapters consider weed management in perennial, horticultural and organic systems. Chapter 8 highlights the use of ground covers as a method for weed management in perennial Conservation Agriculture systems. This chapter considers the ways of selecting different types of ground cover species, including selected or spontaneous plants (including sown grass, legume or cruciferous species). The chapter concludes by explaining techniques for ground cover management as well as the equipment required.

Chapter 9 describes cultivation of vegetables using the no-tillage CA cultivation system developed in Brazil (SPDH+: Sistema de Plantio Direto de Hortaliças). Based on agroecological principles, this allows more sustainable cultivation of vegetables without reliance on synthetic fertilizers and pesticides. The SPDH+ system is based on a holistic approach which combines vegetable cultivation with green soil cover, combining planted species with spontaneous plant growth. The chapter describes both underpinning principles and practical implementation with case studies, including cultivation of Solanaceae and Fabaceae, Brassicae and Cucurbits, leafy vegetables and Liliaceae.

Chapter 10 discusses weed management in organic Conservation Agriculture. The chapter highlights the similarities and differences between organic CA and the emerging idea of organic regenerative agriculture. It offers solutions for weed management within these systems, particularly through crop diversification. This chapter combines the theoretical and practical issues involved in implementing these practices successfully and provides case studies on how they can be used in practice. The chapter concludes by presenting future opportunities and challenges of organic conservation and regenerative systems.

Chapter 11 discusses experience with CA-based cultivation of annual crops such as maize, sorghum and cassava. It emphasises the particularly important, multifunctional role of cover crops not just in direct weed control but

© Burleigh Dodds Science Publishing Limited, 2025. All rights reserved.

in promoting broader soil health and crop vigour. This can create an integrated resilient system that can successfully overcome both biotic pressures (such as weeds) as well as abiotic stresses and deliver a range of ecosystem services including food production. As it indicates, weed control by cover crops is mainly associated with the quantity and quality of biomass input, as well as the weed species and mulch permanence (allelopathic and shade effects) that provide consistent soil cover.

Chapter 12 completes the book by focusing on farmers' testimonies on their practical experience on weed management in Conservation Agriculture. The testimonies include farmers from Denmark, Spain, Brazil and Canada and their experiences of weed management under Conservation Agriculture systems. The farms operate across a range of climates, and crops, including cereals, oilseeds and pulses. This chapter provides key insights from those involved directly in weed management in Conservation Agriculture systems.

© Burleigh Dodds Science Publishing Limited, 2025. All rights reserved.

Chapter 1

Weed ecology in Conservation Agriculture systems: an overview

Seyed Vahid Eslami, University of Birjand, Iran; Carol C. Baskin and Jerry M. Baskin, University of Kentucky, USA

1 Introduction

Weeds are a major challenge in agriculture, and their management is crucial for the success of crop production. Traditional or conventional agriculture uses tillage equipment to achieve weed control, and it also increases the fineness of tilth and soil smoothness, aeration, artificial porosity, friability, and optimum moisture content. Further, well-cultivated soil facilitates the subsequent sowing and covering of the seeds (Baker et al., 2007). This intensive tillage-based agricultural system, however, causes several problems such as the deterioration of soil's physical, chemical, and biological properties that lead to soil erosion, compaction, and reduced microbial fauna (Johansen et al., 2012). The problems caused by conventional agriculture have convinced farmers to consider adopting Conservation Agriculture (CA), which involves abandoning or reducing plowing practices and leaving residues on the soil surface (Eslami, 2014). CA is a set of farming practices that minimizes soil disturbance, maintains a permanent soil cover, and includes diversified crop rotations. In other words, it is a farming system that aims to improve soil health, reduce soil erosion, and

http://dx.doi.org/10.19103/AS.2024.0145.01
© Burleigh Dodds Science Publishing Limited, 2025. All rights reserved.

enhance crop productivity while minimizing the use of external inputs. CA is based on enhancing natural biological processes above and below ground, and it has several benefits such as decreased soil erosion, a positive impact on soil microbes, and enhancement of water infiltration through the soil profile. Overall, CA reduces the use of diesel fuel and other production costs, while it supports sustainability of crop production and weed suppression over time (Friedrich et al., 2009).

Despite its advantages, CA has some disadvantages, especially in the first years of conversion to the new system. Weeds and their management are among the most significant challenges in CA, and farmers complain about the increased abundance of weeds when they convert from conventional to CA (Cordeau, 2022). There are many challenges to weed management in CA. One challenge is that weeds are more likely to germinate and establish in CA than in traditional agriculture systems because CA practices, such as no-till (NT) and cover cropping, help create a favorable environment for weed seed germination and seedling establishment. This system relies on reduced tillage or NT practices, which leaves crop residues on the soil surface, thereby creating a favorable environment for the growth of some weed species, especially in the absence of herbicides (Derrouch et al., 2020).

Eliminating or reducing soil disturbance forces farmers to rely more on herbicides for weed control; however, crop residues left on the soil surface can intercept herbicides, preventing them from reaching the weeds and reducing their efficacy. Thus, there is a great likelihood of weed escapes if crop residue is not managed properly or herbicide application timings or rates are not adjusted, thereby resulting in enhanced densities of weed species (Lee and Thierfelder, 2017; Khalil, 2018). Furthermore, herbicide resistance is a growing problem in CA, as the repeated use of herbicides can select for herbicide-resistant weeds. This poses a significant challenge to CA since there are limited options for controlling weeds (Nichols et al., 2015).

In fact, reducing soil disturbance under CA would restructure the physical nature of the soil seedbank and change the relationship between seedbank and weed seedling density. Farmers lose their mechanical tools for burying seeds in CA, so weed seeds remain primarily on the soil surface or in the top layer of the soil where they may germinate. If seeds are buried in the soil and have a light requirement for germination, they will not germinate. However, under traditional agriculture seeds buried in the soil could be moved to the surface where light may promote germination. Increased densities of weed seeds in the soil seedbank with the simplification of tillage have been documented. The success of the CA system depends largely on a good understanding of the dynamics of the weed seedbank in soil (Schwartz-Lazaro, and Cope, 2019).

© Burleigh Dodds Science Publishing Limited, 2025. All rights reserved.

In addition, the transition to CA and low soil-disturbance systems can affect weed density as well as diversity, i.e. new species of weeds may appear under CA. Weeds species tolerant to shade, wet conditions, and reduced temperatures may increase in fields with reduced tillage (Martin et al., 2002). Further, the reduction or elimination of soil disturbance in CA creates a more favorable environment for perennial weed species compared to systems with frequent soil disturbances. Another challenge with CA systems is the introduction/proliferation of invasive species in crops, and these may be native or introduced species. In particular, global warming due to climate change may enhance the movement of weedy species of plants and insects into new regions, and fields under CA are likely to promote survival and increase of these species if management is lacking (Sorte et al., 2013).

Growers lose the predictability and regularity of seed behavior in NT and reduced tillage systems, thus great variability in seed germination behavior is experienced over seasons and years. Species composition of the weed flora depends upon the type of agroecosystem, e.g. the weed flora in rainfed systems differs from that of irrigated systems. Likewise, annual crops will have different weeds than perennial cropping systems. Altogether dominance of the weed flora depends on the type of crop, soil type, input level, microclimate, and cropping system (monocrop/diversified crop) (Shrestha et al., 2002; Naeem et al., 2021a). CA in many sites has been selected for large-seeded dicots and/or grassy weeds as well as shade-tolerant ones, especially due to covering the soil surface with residues (Sosnoskie et al., 2006).

Since most seeds in CA are concentrated on the soil surface or in the soil's upper layers, we need to think more about seeds on the soil surface. Further, it is essential to focus on factors that drive changes in the weed flora, the level of change, and the responses of weed species to environmental factors. Here, we will focus on management practices that modify environmental factors on the soil surface, which may lead to a reduction in weed pressure in crops. In this chapter, we evaluate the ecological principles that affect the behavior of weeds in CA farming systems with the aim of identifying the vulnerabilities of weeds to help combat the challenges associated with weed management in this system. Some of the questions addressed in this chapter are as follows:

- (i) How will the density and composition of weeds change when converting to CA, and what new challenges do these changes cause?
- (ii) What is the status of the soil seedbank under CA, and what factors affect it?
- (iii) How can the findings from population dynamics of weeds be used in the management of weeds in CA cropping systems?

© Burleigh Dodds Science Publishing Limited, 2025. All rights reserved.

2 Weed density and species shifts under Conservation Agriculture

2.1 Changes due to lowered soil disturbance in CA

Widespread adoption of CA has been impeded by major obstacles, including shifts in weed species and increased weed density in fields leading to decreased crop yield. Minimum tillage or NT practices under CA disturb the soil less compared to conventional tillage methods, which helps improve soil health and reduce erosion. However, CA results in farmers losing their mechanical tools against weeds and creates a favorable environment for certain weed species that thrive in undisturbed soil. Moreover, CA often involves leaving crop residues on top of the soil as mulch or cover crops. This mulch layer can provide shelter and protection for some weed seeds and seedlings, allowing them to germinate/survive while evading herbicides. Failure to properly manage soil surface residues or neglect to adjust herbicide application timing or rate in CA can increase the likelihood of weed escapes. This is because the residues on the soil surface can interfere with herbicide application, resulting in higher weed densities in CA.

Increased weed densities have been reported for farms devoted to CA for a long period of time (Baker et al., 2018; Govindasamy et al., 2020; Légère et al., 2011b; Mishra et al., 2012; Steponaviciene et al., 2021). Mishra et al. (2012) observed that over the course of time, ZT (zero-till) favored relatively higher weed growth over CT (conventional-till) in a rice–wheat system. While weed growth in the initial year was not higher under continuous ZT than under CT, in the third year total weed dry mass was significantly higher under the ZT than the CT. Steponaviciene et al. (2021) observed an increase in the abundance of weed species in the reduced and ZT systems compared with deep ploughing. Şimon et al. (2019) and Chetan et al. (2022) found that reducing the intensity and depth of soil disturbance caused an increase in weed density, especially perennial weeds. In a 3-year study, Minhas et al. (2023) found that levels of weed infestation and weed biomass in a wheat crop under three different tillage systems were: ZT > MT (minimum tillage) > CT.

Since the type of agroecosystem determines the composition of the weed flora, converting to CA changes the weed species composition in the field. Long-term changes in the weed flora in an agroecosystem are driven by interactions of several factors, including tillage practices, environment and climatic conditions, crop rotation, crop type, timing and type of weed management practices, and existing weed species in the area (Shrestha et al., 2002). Thus, studies that monitor changes in weed communities under various environmental and agricultural conditions yield valuable information about the alterations in the weed flora. Owen (2008) found that the use of herbicides with a single mode of action in CA can accelerate shifts in weed populations. In a 6-year field study,

© Burleigh Dodds Science Publishing Limited, 2025. All rights reserved.

Woźniak and Soroka (2015) found that reduced tillage significantly enhanced the biomass and diversity of weed species in pea crops. In another experiment, these researchers observed that reduced tillage significantly enhanced weed diversity in spring wheat (Woźniak and Soroka, 2018). Growers who shift to CA systems should be cautious about possible changes in the prevalence of weed species and devise suitable strategies for weed control.

Different strategies employed in CA create a changed environment compared to the conventional system, especially at the soil surface and in the top layer of soil. When tillage is eliminated, the growing environment selects for different types of weeds, e.g. weed seeds remain on the soil surface where they may accumulate. Annual broadleaved species are favored in the high-disturbance environments of CT (conventional tillage) systems, whereas perennial weeds and species that can germinate on the soil surface, such as annual grasses, are favored in low-disturbance NT (no tillage) systems. This shift in species has been demonstrated by various studies such as those by Hume et al. (1991), Moyer et al. (1994), Barberi et al. (2001), Menalled et al. (2001), Tuesca et al. (2001), and Taa et al. (2004). The failure to bury weed seeds in the NT system may promote an increase in small-seeded annual weeds (Moyer et al., 1994; Swanton et al., 1999) whose seeds require light for germination. Increased densities of small-seeded species, including *Amaranthus tuberculatus, Chamaesyce humistrata,* and *Lamium amplexicaule* were recorded in the NT system compared with the CT system (Govindasamy et al., 2020), which is most likely due to the fact that these seeds of these species germinate to higher percentages on the soil surface or in shallow soil. Barberi and Lo Cascio (2001) found that winter annual weeds were more prevalent in the NT system (60%) than in the CT system (≤43%). Zhang and Wu (2021) conducted a study on the weed composition in winter wheat fields using different tillage systems (NT and CT) and crop residue management (CS and NS = retention and removal of crop residues, respectively). They discovered that in the NT system with crop residue retention (NT-CS) graminoid weeds made up more than 85% of the total weed biomass, whereas in other treatments they accounted for less than one-third of the total weed biomass. These researchers attributed the significantly higher aboveground dry biomass of graminoids in NT-CS to increased soil water content and reduced soil disturbance associated with NT and crop residue retention. The shallow and fibrous roots of graminoid weeds may restrict their access to deeper soil water, while tap-rooted forbs are able to obtain water from deeper soil layers, enabling them to survive drought conditions.

There has been a significant shift from easily controlled annual weeds to perennials that are difficult to control in croplands transitioned to CA. Cropping systems that do not involve tillage may lead to increased persistence of perennial weeds due to a lack of disturbance to their underground structures

© Burleigh Dodds Science Publishing Limited, 2025. All rights reserved.

(Armengot et al., 2016; Zhang and Wu, 2021). Kumar et al. (2022) reported that in CT annual weed species were dominant, whereas perennial weeds dominated in NT. In this study, a shift in weed species with greater dominance of monocots and a marginal decrease in dicots was observed with NT. Trichard et al. (2013) studied weed shifts following the adoption of direct drilling (DD) with a cover crop in 52 winter wheat fields based on the number of years under NT management (1 to 13 years) in France. Under DD, monocots and perennial weed species were favored, while dicots and annuals were not, resulting in their lower representation in fields that had been converted to DD for several years. Monocots thrive in DD systems due to two factors commonly found in such systems: lack of mechanical weed control through superficial soil disturbance (Zanin et al., 1997; Holland, 2004) and a shift toward the use of foliar-applied herbicides, which are less effective on monocots than dicots (Délye et al., 2008). The authors' findings suggest that as time since conversion increased, weeds invested more energy in maintaining their root system in the soil than in seed production, which is consistent with previous research (Fried et al., 2012). The decline in germination success could be attributed to the discontinuation of burying weed seeds in the soil by DD systems, leading to a notable reduction in their ability to germinate (Yenish et al., 1992; Tourdonnet and Nozières, 2007) and sometimes leading to a depletion of the soil seedbank after several years (Murphy et al., 2006). Furthermore, weeds tended to flower earlier in DD systems, likely because crops are sown earlier compared to CT practices (Chauhan et al., 2012).

Campiglia et al. (2018) studied the impact of different soil tillage methods (inversion and non-inversion) on weed community composition in conventional and organic systems in a wheat-tomato-chickpea rotation over 3 years. These researchers found that the non-inversion system had the highest weed biomass, weed density, and weed community richness. The increase in weed biomass in the non-inversion tillage was primarily due to an abundance of perennial weeds. Non-inversion tillage led to a significant change in the weed community and an increase in the biomass of harmful perennial weed species. In a study on the effects of long-term (36-year) NT and CT systems on weed species dynamics, the NT system had higher densities of johnsongrass (*Sorghum halepense*) compared to the CT system (Govindasamy et al., 2020). The greater density of johnsongrass as a perennial weed under NT was attributed to a lack of tillage and improved availability of soil moisture. In the study by Thomas et al. (2017), perennial species such as *Cirsium arvense* and *Sonchus arvensis* were associated with reduced- and no-tillage systems.

Although an NT system may lead to an increase in perennial weeds, it does not consistently favor perennial weeds (Nichols et al., 2015). Surprisingly, it has been observed in some studies that minimum soil disturbance systems are likely to be less favorable for some perennials such as *Asclepias syriaca*,

© Burleigh Dodds Science Publishing Limited, 2025. All rights reserved.

Cyperus esculentus, and *Helianthus tuberosa*, which tend to multiply when their underground parts are disturbed (Sharma et al., 2023; Swanton et al., 1993). Sharma et al. (2023) evaluated weed shifts in a wheat field in a long-term (12-year) study and found that the CA-based NT permanent beds reduced the occurrence of the perennial weed *Cyperus esculentus* compared to CT. They attributed this opposite result to the ecological differences in this perennial weed. *C. esculentus* differs from *C. rotundus* despite being a perennial sedge. It grows quickly and can germinate in crowded conditions. Its small tubers resemble seeds and can sprout immediately when the soil is disturbed. These tubers differ in shape and size compared to those of *C. rotundus*. The frequent germination and abundant tuber formation of *C. esculentus* may be due to regular tilling operations in CT practice. Reduced tillage can increase the selection of weed species that have life cycles and resource requirements that complement those of agronomic crops, regardless of whether they are annual or perennial (Chancellor, 1985; Dorado et al., 1999). Nevertheless, in certain cases transitioning to NT practices within crop rotations consisting of two or more crops did not result in an increase in perennial weed growth (Derksen et al., 1993; Blackshaw et al., 2001; Tuesca et al., 2001). This highlights the significance of crop rotation in managing weeds, particularly in reduced tillage systems.

Some authors emphasize the risks associated with the presence of invasive species under minimum soil tillage systems (Mayor and Maillard, 1995; Tørresen et al., 2003; Légère et al., 2005; Feledyn-Szewczyk et al., 2020). That is, if tillage is discontinued invasive weeds have a tendency to increase rapidly (Travlos et al., 2018). Feledyn-Szewczyk et al. (2020) assessed the qualitative and quantitative changes in weed flora under reduced tillage and NT (direct sowing) in comparison with traditional ploughing. They discovered that the presence of invasive alien weed species like *Conyza canadensis* was higher in reduced tillage systems as opposed to ploughing. In Woźniak's research (2018), when harrowing was used as a substitute for ploughing two invasive dominant weed species, *Echinochloa crus-galli* and *Viola arvensis,* increased significantly. Cussans (2022) also observed the presence of invasive species like *Echinochloa crus-galli* after the adoption of CA in the UK. According to Grey et al. (2017), tillage alone has proven to be a successful method in managing the potentially invasive *Pennisetum purpureum*.

In evaluating floristic changes under minimum soil tillage, it is useful to consider ecological succession. Also, we need to distinguish between simple vegetational fluctuations and permanent changes in the flora. Changes in floral composition are reversible in vegetational fluctuations, but they are unidirectional and continuous in succession. Soil disturbance is a crucial factor that selects for species, depending on how regularly it occurs. Therefore, the composition and density of weed floras are expected to be significantly affected by any cultivation practices, and a decrease in mechanical soil

© Burleigh Dodds Science Publishing Limited, 2025. All rights reserved.

disturbance may lead to a systematic replacement of species and biological groups. This process can be compared to old-field succession, which occurs on abandoned cropland (Zanin et al., 1997). Zanin et al. (1997) evaluated the influence of tillage system on floral changes of weeds over 8 years and found that undisturbed soils favored the geophyte and hemicryptophyte species, *Digitaria sanguinalis*, *Conyza canadensis,* and *Kickxia elatine,* as well as wind-dispersed weeds. After 8 years, floristic changes in response to reduced tillage indicated ecological succession toward a pioneer community dominated by annual species (therophytes) and wind-dispersed plants but with a potential for a more mature community similar to the woodland edge with more perennial species, shrubs, and bird-dispersed plants.

In no-tillage or minimum soil disturbance systems, perennial and biennial species thrive due to less soil disturbance, resulting in more diverse weed communities. Conversely, annual weeds are more abundant in conventional tillage than in minimum soil tillage (Gaba et al., 2014). Despite general predictions based on previous research regarding weed flora changes under CA, predicting these shifts is often difficult. Soil tillage, for instance, hinders the growth of phanerophytes, chamephytes, and a majority of hemicryptophytes; however, its impact on perennial species with bulbs, tubers, or rhizomes is difficult to determine. Depending on the specific conditions during tillage, perennial weeds with rhizomes or bulbs can either benefit (division and dispersal of vegetative propagules) or suffer (drying out of the propagules) (Lemieux et al., 1993). Part of the reason that it is difficult to predict and characterize weed species shifts is that weeds associated with minimum soil disturbance systems may rely on multiple regeneration strategies, and two important regenerative strategies in relation to minimum soil disturbance systems are vegetative regeneration and wind dispersal (Swanton et al., 1993).

2.2 Changes due to cover cropping in CA

In CA, permanent soil cover is maintained by the use of cover crops or crop residues. Weisberger et al. (2023) in a meta-analysis of cover crop usage in southeastern USA confirmed the negative impact of cover crops on weed density. By increasing soil coverage, farmers aim to increase the competition between sown species (crops and cover crops) and weeds, which results in the selection of more competitive and shade-tolerant weeds (Derrouch et al., 2021). Cover crops have been shown to limit seed germination and seedling emergence, and growth by altering light quantity and quality due to the presence of a physical barrier, and they may increase seed predation (Teasdale 2018). Surface residues conserve soil moisture, reduce light levels, and lower temperatures at the soil surface. High residue accumulation in NT reflects solar radiation and alters the albedo of the soil surface, leading to a reduction in

© Burleigh Dodds Science Publishing Limited, 2025. All rights reserved.

surface soil temperature (Cox et al., 1990). There is a delay in the emergence of some weed species under CA systems with residues present on the soil surface due to reduced temperatures at the soil surface despite increased soil moisture. A delay in seedling emergence of *Amaranthus tuberculatus* and *Sorghum halopense* in NT, when compared to CT systems, was observed by Govindasamy et al. (2020). Refsell and Hartzler (2009) found that 50% of *A. tuberculatus* seedling emergence was achieved 10 days after planting in CT, whereas it took 35 days in NT.

On the other hand, changed conditions at the soil surface provided by cover crops and crop residues may create a favorable environment for weed species that can tolerate shade, wet conditions, and cool temperatures. Therefore, these species are more likely to thrive in CA and can cause problems in reduced tillage systems (Martin et al., 2002). In other words, problematic species like *Amaranthus palmeri, Caperonia palustris, Poa annua* with the ability to acclimate to great light reductions are expected to thrive under CA conditions (Yasin et al., 2017). A higher density of wind-dispersed species has been observed in NT systems (Derksen et al., 1994; Gill and Arshad, 1995; Tuesca et al., 2001). This increased density could be due to the vulnerability of the species to soil disturbance or the accumulation of crop residue on the soil surface in minimum soil disturbance systems, which can catch wind-borne seeds and/or promote weed growth. Certain cover crops can increase parasitic weed infestation by providing a host for the weedy species. For example, Cussans (2022) stated that the holoparasite *Orobanche minor* was observed growing on a vetch cover crop, while the holoparasite *Cuscuta campestris* was found in a crimson clover cover crop in the UK.

In addition to factors such as climate and environmental factors, the life history of the cover crop and of the weed species as well as the type of cover crop management also may have an impact on weed density and their species composition. In a Spanish vineyard, five different cover crops were evaluated over three seasons: *Festuca arundinacea* (the only perennial cover crop species studied), *Hordeum vulgare*, *Vulpia ciliata*, a mixture of *Bromus* species, and *Medicago rugosa* (Recasens et al., 2023). Weed community composition was determined by the different structure and management of the cover crops over the course of 3 years, and *F. arundinacea* and *H. vulgare* were the most effective at controlling *Cynodon dactylon*. Weeds associated with *F. arundinacea* differed from those associated with the annual cover crop species, with *F. arundinacea* being the cover crop with the lowest number of weed species. *F. arundinacea* effectively created a weed community consisting of only a few species, including *Malva sylvestris* and *Picris echioides*. The presence of these species can be related to the lack of tillage needed to manage this perennial cover crop, and to their tolerance to shredding, i.e. mechanical chopping or cutting cover crops into small fragments, which can hinder weed growth. However, certain species,

© Burleigh Dodds Science Publishing Limited, 2025. All rights reserved.

such as *Malva sylvestris* and *Picris echioides*, exhibit tolerance to this method of control. Soil disturbance prior to planting annual cover crops resulted in a greater weed pressure compared to the perennial cover crop *F. arundinacea*, which did not require any disturbance.

While most studies indicate that cover crops have a suppressive effect on weeds, there are cases in which cover crops have led to an increase in weed densities. To determine whether cover crops definitively suppress weeds and can be relied upon as the sole method of weed management, various factors need to be considered. These factors include the type of cover crop used, timing of its planting, density, and biomass of plants, when it is terminated, subsequent cash crop in the rotation, and various climatic conditions. Additionally, differences in weed suppression also may exist between annual and perennial cropping systems. Some studies also have shown that using a mixture of cover crop species is more effective in suppressing weeds compared to using a single species (Fernando and Shrestha, 2023).

2.3 Changes due to diversified crop rotation in CA

Monoculture cropping impoverishes both organic matter and microbes in the soil (Balota et al., 2004; Maillard et al., 2016), increases weed infestation, and worsens the phytosanitary condition of a crop (Hernandez-Restrepo et al., 2016; Tillmann et al., 2017; Calzarano et al., 2018). These negative consequences ultimately lead to a decrease in grain yield and a deterioration in grain quality (Woźniak and Stępniowska, 2017; Woźniak, 2020). Crop monocultures often lead to weed simplification with only a few dominant weeds (Moyer et al., 1994; Anderson et al., 1998; Blackshaw et al., 2001; Cardina et al., 2002), potentially simplifying the choice of herbicide but potentially increasing selection pressure for herbicide-resistant weeds. Diversified crop rotation is another principle of CA that contributes to variations in density and diversity of weeds in the system. Crop rotation helps interrupt the weed-crop relationship and break the weed cycle (Minhas et al., 2023). In a 30-year study, Woźniak (2020) found that weed density and biomass were significantly higher where durum wheat was planted in monoculture compared to a crop rotation system (pea- winter wheat-durum wheat). Minhas et al. (2023) compared six rotation systems, including cotton-wheat, sorghum-wheat, mung bean-wheat, rice-wheat, sunflower-wheat, and fallow wheat (control), for weed infestation at three locations and concluded that the sorghum-wheat cropping system had the lowest weed infestation, while the fallow-wheat cropping system had the highest weed infestation and biomass at all locations. The inclusion of allelopathic crops such as sorghum and sunflower in the rotation decreased the establishment of weed flora. Shahzad et al. (2016b) and Naeem et al. (2021a) observed similar trends in wheat and barley crops, respectively.

© Burleigh Dodds Science Publishing Limited, 2025. All rights reserved.

Over an extended period, implementing appropriate crop rotations potentially can benefit the preservation of weed flora biodiversity in the environment, which likely prevents the dominance of noxious weed species. That is, agricultural land serves as a natural habitat for specific segetal (grain-field) weeds that are increasingly at risk of extinction due to intensive farming methods. Mhlanga et al. (2015) reported that rotations with cover crops can result in decreased numbers of dominant weeds, leading to lower levels of weed infestation and an increase in the diversity of weed species. However, these outcomes are influenced by various factors including seasonal conditions and other agricultural practices. Therefore, crop rotations have the potential to enhance the diversity of weed species (Baker et al., 2018). The 19-year study conducted by Nikolic et al. (2018) indicates that the cultivation of winter wheat in long-term crop rotations has a notable impact on the floristic composition and community structure of weeds, but it does not lead to a significant increase in a number of invasive plant species. Therefore, it can be inferred that the cultivation of winter wheat in long-term crop rotations affects weed growth without promoting invasive plant species.

Weisberger et al. (2019) conducted a meta-analysis of 54 studies across six continents, using 298 paired observations to compare the effects of simple and diverse crop rotations on weed responses. Their findings showed that diversifying from simple rotations resulted in a 49% reduction in weed density, but it did not significantly affect weed biomass. The study suggests that increasing variance around crop planting dates through diversification is more effective in suppressing weeds than simply increasing crop species richness. Additionally, increasing rotational diversity had a greater impact on reducing weed density under zero-tillage conditions (65%) compared to tilled conditions (41%), regardless of environmental context and auxiliary herbicide use. In fact, crop diversification can reduce weed abundance by diversifying selection pressure.

Steponaviciene et al. (2021) found that the crop rotation applied in their experimental field favored the spread of weed species from various communities. The diversification of crops can create new temporal niches, which in turn may result in a variation of weed life cycle periods (Derrouch et al., 2021). Naeem et al. (2021a) in their study on the impact of different crop sequences on weed infestation and productivity of barley found that every crop promoted and suppressed specific weed species. This might be due to competition, crop-specific allelochemical properties (Latif et al., 2017), soil disturbance, and mechanical damage to the weeds. All these factors eventually changed the macro- and microenvironment for the development of weeds, leading to the promotion of one species and suppression of others. For example, cotton has a deep root system and can take up most of the nutrients from deep soils (Riaz Marral et al., 2020). Therefore, weeds with roots present in the upper soil layer can benefit

© Burleigh Dodds Science Publishing Limited, 2025. All rights reserved.

from the nutrients available in the upper layers or applied fertilizers. Similarly, soils of mung bean-barley rotation had increased nutrients due to N-fixation by mung beans, which promoted weed diversity (Shahzad et al., 2016b; Kadziene et al., 2020). Naeem et al. (2021b) compared weed composition in different barley-based cropping systems under various tillage practices and concluded that the weed flora of all cropping systems consisted of broadleaved weeds (>65 %) and grassy weeds (> 30 %) during both years of the study. They found that the inclusion of diverse crops in the rotation affected weed composition, and the sorghum-barley and maize-barley cropping systems suppressed various broadleaved and grassy weed species due to their allelopathic potential.

2.4 Variations in diversity indices under CA

Changes made in the cropping system in CA usually lead to changes in weed diversity indices. In a 3-year field study, minimum soil disturbance systems (no-tillage and chisel plough) allowed the build-up of a more diverse community, whereas the most disturbed system (moldboard ploughing) prevented high diversity in the weed community (Feldman et al., 1998). Several studies have found a higher Shannon-Wiener index (*H*) and species richness (S) values in NT than in CT systems (Dorado et al., 1999; Sosnoskie et al., 2006; Légère et al., 2011b; Feledyn-Szewczyk et al., 2020; Govindasamy et al., 2020), indicating higher weed densities where tillage was not practiced. Govindasamy et al. (2020) attributed the greater weed species diversity (*H*) and species richness (S) in the NT system to a relatively stable environment, longer persistence of weed seeds owing to lack of incorporation into the soil, and higher soil moisture levels compared with the CT system. The structure of weed communities was analyzed using diversity indices under reduced tillage, NT (direct sowing) and traditional ploughing (Feledyn-Szewczyk et al., 2020). These authors found that the greatest level of weed biodiversity, as measured by Shannon's index, was observed under reduced tillage (2.56), which can be attributed to a more even distribution of weed species within the community. On the other hand, the direct sowing system had the lowest value of Shannon's diversity index (2.28) and a high value of Simpson's dominance index (0.15), mainly due to dominance of two species: *Galium aparine* and *Fallopia convolvulus*. Sorensen's qualitative and quantitative similarity indices showed that weed communities in the NT (the direct sowing) and ploughing systems were the least similar. In a 10-year study comparing weed diversity between CT and various CA practices in a maize-wheat-mung bean rotation, Ghosh et al. (2023) concluded that CA practices increased weed diversity (as evident from higher richness and evenness indices) compared to CT.

Increased weed diversity is thought to be ecologically beneficial both agronomically and environmentally, regardless of the specific weed species

© Burleigh Dodds Science Publishing Limited, 2025. All rights reserved.

(Storkey and Neve, 2018). The presence of diversified weed communities may help mitigate the negative effects of competitive and dominant species on productivity while also promoting agricultural sustainability (Storkey and Neve, 2018). In CT, a decrease in species richness could lead to increased selection pressure on a small number of weeds and encourage the evolution of biotypes that are better adapted to management practices (Neve et al., 2014). Zhang et al. (2021) concluded that residue retention on the soil surface could impact weed diversity, leading to a decrease in the relative abundance of annual weeds, while the relative abundance of perennial weeds increased with crop residue retention. The species richness was higher with no residue retention, while the Pielou evenness index was lower when no residue was maintained on the soil surface.

It is worth considering that evaluating the competitive potential of a weed community should not rely solely on species richness. Weed communities with a high number of species, including highly competitive ones or those with high densities may pose more competition to crops compared to communities with low species richness but consisting of less competitive species or lower densities. Hence, it is crucial to supplement species richness data with information regarding the relative abundance and frequency of species within the weed community as well as the type of weed species present in the area (Moonen and Barberi, 2004).

According to Travlos et al.'s (2018) review, minimum soil disturbance systems generally promote higher weed richness and diversity due to the creation of more favorable conditions for certain weed species. However, there are cases in which minimum soil disturbance systems result in less diverse weed communities compared to more intensive tillage systems due to the dominance of specific weed species. In contrast, Légère et al. (2005) found that diversity indices indicated a minimal influence of the soil cultivation method on weed diversity, but the soil cultivation method significantly influenced the species composition of the weed community. Similarly, Plaza et al. (2010), based on a 23-year experiment, discovered no major differences between tillage systems concerning weed diversity, except that NT seemed to support more species than the other two systems. Additionally, species richness and Shannon diversity index varied greatly over the years in all tillage systems. These results suggest that studies on weed diversity and its changes due to different tillage methods should be conducted over an extended period of time to avoid conflicting results found in existing literature (Feledyn-Szewczyk et al., 2020). In their study, Sharma et al. (2023) examined the diversity of weeds in CT (conventional tillage) and CA systems from 2010 to 2022. They discovered that the Shannon-Weiner index and Simpson dominance index were higher in the CA system during 2021–2022 compared to 2010–2011 and 2015–2016, suggesting a positive change in weed diversity over time. Additionally, the Sorensen similarity index

© Burleigh Dodds Science Publishing Limited, 2025. All rights reserved.

revealed that there was little variation in weed diversity during 2010–2011. However, in 2015–2016 and 2021–2022, only 89% (0.89) and 62% (0.62) of weed species were shared between CT and CA systems, indicating a shift in weed species within the long-term CA system during 2021–2022.

Research has shown that the length of time elapsed since adopting CA affects the trend of weed species shifts in the field. Trichard et al. (2013) examined how weed communities responded to the implementation of all three principles of CA over a period of 5 years after adoption. They found that during this initial period the effects observed on minimum soil disturbance systems persisted, including an increase in perennial and graminoid species. Derrouch et al. (2021) investigated the functional composition of weed communities in 100 winter wheat fields implementing CA continuously from 1 to 21 years in France. Over time, the adoption of CA led to an increase in phanerophyte and geophyte life forms as well as to summer-germinating species and spring-and summer-flowering species. The proportion of therophytes in the weed communities in CA was not affected. Most of the common (*Alopecurus myosuroides* and *Chenopodium album*) and abundant (*Amaranthus hybridus*, *Digitaria sanguinalis*, *Setaria pumila*, and *Veronica hederifolia*) species were therophytes. Shade-tolerant species were not initially present in recently converted fields, but they were established later. Although there was an increase in shade-tolerant weed species, the majority of weed species that remained in the CA weed community were heliophytes. Seed mass and percentage of graminoids tended to increase with time under CA. Trichard et al. (2013) attributed this change to the fact that large seeds usually exhibited a higher ability to germinate/emerge under a cover of litter or soil than small seeds. Some functional shifts in weed communities occurred more quickly than others. For example, it took at least 10 years of continuous application of CA for summer-germinating species to increase.

3 Seedbank dynamics

The reserve of viable weed seeds found on the soil surface and scattered throughout the soil profile is known as the weed seedbank. It contains both newly shed seeds and older seeds that have remained on/in the soil from previous years (Roberts, 1981). Additionally, tubers, bulbs, rhizomes, and other vegetative structures that allow some of the most persistent perennial weeds to perennate via propagules are buried in the soil. The soil seedbank increases through seed production and dispersal, while it is depleted through germination, predation, and decay (Baskin and Baskin, 2014). In annual crop production systems, the soil seedbank is the primary source of the yearly infestation of annual weeds and represents the majority of the weedy species in fields (Vasileiadis et al., 2007). The weed seedbank acts as a record of past

© Burleigh Dodds Science Publishing Limited, 2025. All rights reserved.

successes and failures in cropping systems, and understanding its size and species composition can help farmers anticipate and mitigate potential effects of crop-weed competition on crop yield and quality (Menalled and Schonbeck, 2009). The germinable fraction of the seedbank has the most substantial impact on weed seedbank flux, with the potential to account for more than half of seedbank losses annually (Gallandt et al., 2004). Soil and environmental characteristics, along with management practices, largely influence the density and species composition of the weed seedbank (Cardina et al., 1998). Farmers have some control over variables such as tillage system, crop rotation, and weed control practices (Ball, 1992). Tillage can bring buried seeds closer to the soil surface and promote germination, as well as bury seeds at depths that inhibit germination (Mohler and Calford, 1997). Crop rotation has been found to increase weed diversity (Stevenson et al., 1997) and reduce weed seed densities compared to monoculture systems (Liebman and Dyck, 1993).

3.1 Influence of soil disturbance

In CA, practices such as minimum soil disturbance, cover cropping, and diversified crop rotations can have a significant impact on the size and composition of the weed seedbank, and over time they can lead to changes in weed communities (Hugo et al., 2021). For example, CA has the potential to elevate the proportion of the weed seedbank located on or near the soil surface and replace annual weed species with perennial ones (Chauhan et al., 2012; Derrouch et al., 2021). Species composition and density of weed seeds in the soil can vary significantly depending on changes in tillage patterns, planting systems, and crop management practices (Swanton et al., 2000). Tillage affects the structure and properties of soil as well as the distribution of weed seeds in the soil layers; thus, tillage can potentially affect the establishment of weeds and their accessibility to soil nutrients (Hernández Plaza et al., 2015). In fact, tillage methods can lead to alterations in the distribution of seed depth, which can subsequently lead to shifts in weed species present in crops. As a result, these tillage/management practices have the potential to modify the soil environment and cause a notable shift in the soil seedbanks compared to the weed population aboveground (Armengot et al., 2016).

Many studies confirm that adopting a minimum soil disturbance system leads to an increase in the size of the seedbank, and these negative alterations in weed seedbanks and weed emergence explain why farmers often are discouraged after adopting minimum soil disturbance (Peigné et al., 2007). Without any soil disturbances in CA, i.e. NT, a high proportion of the weed seeds generally remains close to the soil surface. Due to the concentration of NT seedbanks in the upper layer of soil, there is a greater likelihood of weed seed germination in NT compared to those in conventional tillage (CT)

© Burleigh Dodds Science Publishing Limited, 2025. All rights reserved.

seedbanks (Nichols et al., 2015). Soil disturbance stimulates the germination of many species, leading to an increase in the germinable fraction. This germination response can be triggered by various factors such as exposure to light for light-sensitive species, temperature fluctuations, and movement of seeds to favorable germination sites.

Feldman et al. (1997) analyzed the soil seedbank of a wheat crop that was cultivated using four different tillage systems (moldboard plough, disk, chisel, and NT) over a period of 3 years. The moldboard ploughing system had the lowest seed density in the seedbank, whereas the NT system had the highest density seedbank, particularly in the upper part of the soil profile. Additionally, there was an increase in species diversity from the moldboard ploughing to disking to chiseling to no-tilling system, i.e. the NT system had the most diverse seedbank. Long-term studies conducted by Barberi and Lo Cascio (2001) (>12 years) and Cardina et al. (2002) (>35 years) confirmed significant tillage system effects on weed seed densities. Moonen and Barberi (2004) found that after 7 years of continuous maize cropping, seedbank density in the NT system was about five times higher than that in the CT (conventional tillage) system. Vasileiadis et al. (2007) found that in rotary tilled and chisel ploughing treatments, the upper 0–15 cm soil layer contained 61% and 59% of the total viable seeds, respectively. In contrast, the moldboard ploughing treatment had 48% of viable seeds in this soil layer. However, for the moldboard ploughing treatment, the highest concentration of viable seeds (52%) was in the 15–30 cm depth. For chisel-ploughing and rotary-tilled treatments, viable seed concentration was 41% and 39%, respectively. Moldboard ploughing significantly buried more seeds in the 15–30 cm soil layer compared to the chisel ploughing and rotary-tilled treatments, resulting in a decrease in total seed density near the surface of the moldboard ploughed plots.

In a long-term evaluation, Santín-Montanyá et al. (2016) observed that seed density and species diversity in the soil seedbank increased as a result of minimum soil disturbance techniques. This increase likely is due to specific conditions created in the soil, such as minimal soil disturbance expected with MT (minimum tillage) and NT. These conditions allowed seeds to remain nongerminated/dormant in the soil seedbank, possibly due to lack of exposure to light. Feledyn-Szewczyk et al. (2020) assessed the qualitative and quantitative changes in soil seedbank under minimum soil disturbance and NT (i.e. direct sowing) compared with traditional ploughing. Under direct sowing, the seedbank was significantly larger compared to traditional ploughing, being approximately three times smaller in the latter. In direct sowing, the majority of weed seeds were concentrated in the top (0–5 cm) soil layer, whereas in the ploughing system, most weed seeds were at depths of 5–10 cm and 10–20 cm. During the 3 years of research, the seedbank under direct sowing exhibited the highest species richness, with a total of 30 species. During the initial years

© Burleigh Dodds Science Publishing Limited, 2025. All rights reserved.

of implementation, Kumar et al. (2022) found that fields utilizing NT systems experienced increased weed infestations, which can be attributed to the fact that weed seeds were located within the upper 2 cm of the soil profile. Haidar and Siblani (2023) evaluated the size of the weed seedbank in 7-year-old NT fields. The NT fields contained almost seven times more monocot and dicot weed seeds than the conventionally tilled fields. Results of this study suggested that both the transient and persistent seedbanks at a depth of 0–5 cm in NT fields were larger than those in conventionally tilled fields.

On the other hand, some studies indicated that although NT systems might favor the initial accumulation of weed seeds at the soil surface, they led to a reduction in the seedbank over time as a result of reduced seed longevity and increased predation (Anderson, 2008). Ghosh et al. (2023) compared weed seedbank dynamics under CT and NT systems. These authors found that the dominant species in the weed seedbank were *Chenopodium album, Commelina benghalensis, Cyperus esculentus, Cyperus rotundus, Dactyloctenium aegyptium, Digera arvensis, Dinebra retroflexa, Euphorbia microphylla, Lepidium didymum, Leptochloa chinensis, Melilotus indicus,* and *Spergula arvensis;* these weed species were common in the study region. Further, a lower seedbank density was found under the NT than in the CT system. The decrease in seed numbers in NT can be linked to factors such as chemical control and an increase in predation and deterioration. Surface accumulation of seeds under minimum soil disturbance would increase predator access to seeds and therefore could increase seed removal. Lack of soil disturbance via tillage also could encourage higher predator populations. The number, diversity, and/or activity of seed-consuming fauna in NT fields can increase compared to conventionally tilled fields (Blubaugh and Kaplan, 2015), possibly due to improved habitat (Baraibar et al., 2009) or decreased mortality of weed seed predators (Shearin et al., 2007). Seed predation is typically greatest when weed seeds remain on the soil surface, and there is sufficient residue cover for predators (i.e. NT). Since NT can favor certain granivorous species over others, the associated shift in preferred seed consumption may contribute to altered seedbank composition (Nichols et al., 2015). Generalist predators such as common ground beetles or crickets can reduce weed seed emergence by 5% to 15% (White et al., 2007). Larger animals such as rodents and birds can also consume significant amounts of weed seeds. When buried and not available to predators, seed attack by pathogens may occur. Mortality of buried *Avena fatua* seeds increased as soil moisture content increased from 6% to 24% with maximum mortality reaching 55% and 88% after 2 years of the study (Mickelson and Grey, 2006).

In some long-term studies, the tillage system did not affect the size of the seedbank, but it did affect the distribution of weed seeds in the soil profile. Vasileiadis et al. (2007) found no significant differences between soil

© Burleigh Dodds Science Publishing Limited, 2025. All rights reserved.

seed density among tillage regimes (moldboard plough, chisel plough, and rotary tiller), while the distribution pattern of seeds in the soil was affected by tillage treatments. After 18 years of continuous application of moldboard ploughing or NT, Ruisi et al. (2015) concluded that the tillage system did not affect the total size of the weed seedbank, whereas it did alter both species composition and distribution of seeds within the soil profile. Guareschi et al. (2021) observed that CT resulted in a more even distribution of Italian ryegrass seeds throughout the soil profile compared to NT because CT caused seeds to be mixed up with the soil. In contrast, in the NT system, a large number of seeds remained concentrated in the 0–5 cm soil layer. Similar findings were observed for *Poa annua*, in which systems with soil inversion had a more uniform vertical distribution of seeds compared to non-inversion tillage. Soil inversion contributed to the prevention of seedling establishment because the buried seeds and unfavorable germination conditions hindered germination and seedling emergence and growth (Scherner et al., 2017).

Ghosh et al. (2023) discovered that the soil seedbank was more evenly distributed (in terms of depth) and diverse under CA than CT practices (as evident by a higher richness and evenness indices), leading them to propose that long-term implementation of CA practices could prevent seedbank enrichment and increased weed diversity in the soil seedbank. CA practices may reduce the dominance of a few specific weeds because the deposition of seeds of all weeds on the soil surface promotes the emergence of more species as compared to the CT system. A persistent seedbank with high species diversity under CA systems may reduce the negative impacts of dominant weeds (Borgy et al., 2012). Storkey and Neve (2018) hypothesized that an increase in weed diversity, regardless of the weed species, can have both agricultural and environmental benefits. Enhancing agricultural sustainability and reducing the negative impact of competitive and dominant species on productivity can be achieved through diversified weed communities. In CT, a decrease in species richness may lead to increased selection pressure on a limited number of weed species, thereby promoting the evolution of biotypes adapted to management (Neve et al., 2014). These findings strongly support the hypothesis that farming systems causing decreased soil disturbance allow a larger and more diverse soil seedbank to form.

A threatening challenge with CA systems is the introduction/proliferation of invasive and hard-to-control weed species in crops, and these may be native or introduced species. There is concern that NT cropping systems can shift the dynamics of weed species in the soil seedbank under CA (Concenco et al., 2011), and this may lead to the spread of invasive weed species. Germination of some globally distributed invasive weed species, such as *Vulpia myuros*, is hindered when their seeds are buried by tillage. Conversely, no- or minimum-till systems promote the germination of these seeds, potentially restricting

© Burleigh Dodds Science Publishing Limited, 2025. All rights reserved.

the adoption of NT-CA systems (Büchi et al., 2021). Ruisi et al. (2015) found that the continuous use of NT led to an increase in weed seeds in the upper soil layer and a significant increase in seed density of some problematic and invasive species, such as *Lactuca serriola, Papavera rhoeas*, and *Phalaris* spp. From an agronomic perspective, NT cropping systems face a significant weed management problem due to the increased abundance of species such as *L. serriola*. Feledyn-Szewczyk et al. (2020) found that despite the highest number of weed seed species being in the NT system, the weed seed pool had a low value of Shannon's diversity index and a high value of Simpson's domination index, indicating the dominance of particular weed species: *Apera spica-venti*, *Capsella bursa-pastoris*, *Veronica persica,* and *Viola arvensis*. Under the minimum soil disturbance and NT systems, there was an increased percentage of seeds of invasive species in the soil, such as *Amaranthus retroflexus, Conyza canadensis*, *Echinochloa crus-galli*, *Galinsoga parviflora,* and *Veronica persica*. Under the minimum soil disturbance and NT systems, there was a higher number of seeds of perennial species such as *Descurainia sophia Plantago major,* and *Sonchus arvensis*. Placing seeds at different depths in the soil indirectly affects their ability to germinate due to variations in temperature, oxygen, and light. For instance, *D. sophia* seeds can germinate in both light and darkness, but the presence of light promotes germination. The depth at which *D. sophia* and *Apera spica-venti* seeds are located significantly affects their ability to germinate, with the highest germination percentages occurring when seeds are on the soil surface. *Echinochloa crus-galli* seeds have the highest emergence percentage when placed at a depth of 1.5–3.0 cm.

It is worth noting that varied results regarding species diversity among studies can be largely influenced by the type of dominant weed species present in the field. Ghosh et al. (2023) found that intensive tillage in CT systems can cause an increase in the development and tuberization of the dominant sedge *Cyperus rotundus*. In fact, the extensive use of tillage could result in the fragmentation of the tubers, thereby increasing the density of the species in fields. However, by adopting minimum soil disturbance and applying herbicides under CA, it is possible to reduce the infestation of *C. rotundus* (Susha et al., 2018).

Soil microorganisms also have a significant impact on the status of soil seedbanks. Seeds interact with soil microorganisms that can promote or inhibit seed germination. Tillage enhances microbial activity by improving soil aeration and exposing labile soil organic matter fractions. An increase in microbial activity can result in increased biosynthesis of ethylene, a well-known stimulant of seed germination (Mirsky et al., 2010). The stimulating effect of soil disturbance on weed seedling recruitment has been proven in previous studies. Mulugeta and Stoltenberg (1997) found a significant increase in the emergence of *Setaria faberi* and *Chenopodium album* seedlings in tilled

© Burleigh Dodds Science Publishing Limited, 2025. All rights reserved.

soils compared to no-tilled soils, with the increase being up to six times. They also observed a significant reduction in the seedbanks of *Chenopodium album* and *Setaria faberi* within a single season, with reductions of 16 times and 6times, respectively. These results suggest that certain treatments that involve disturbance and microclimate changes may promote germination and deplete the soil seedbank more for some species than for others. It is likely that germination is the primary mechanism responsible for reducing the seedbank, as suggested by Forcella (2003). Mirsky et al. (2010) also found a significant decline in the germinable fraction of the seedbank of *Abutilon theophrasti, C. album*, and *Setaria* spp. with soil disturbance.

On the other hand, soil microorganisms play a crucial role in the decay and destruction of weed seeds. These microorganisms, including bacteria, fungi, and other microbes, interact with weed seeds in various ways, leading to their decomposition and reduction in viability. Seed longevity in the soil can, therefore, be strongly influenced by the number and variety of soil microorganisms. Soil microorganism proliferation is more likely to occur in soil with a tendency for nutrient accumulation. Additionally, factors such as vegetation cover and rhizosphere, water content, temperature, and climate can affect the soil microbial community. Previous research has shown that agricultural soils with high levels of biological activity have weed-suppressive qualities, therefore nondisturbed soils typically exhibit greater biological activity than disturbed ones due to variations in soil nutrient and micro/macro element content, organic matter, pH value, and conductibility (Parr et al., 1992; Kremer and Li, 2003; Bissett et al., 2011). Consequently, it is expected that there will be a greater difference in seed decay between disturbed and nondisturbed soils, with nondisturbed soils having stronger weed-suppressive properties (Rowell and Florence, 1993).

Nikolic et al. (2020) compared seed degradation of six common weed species (*Abutilon theopharsti*, *Alopecurus myosuroides*, *Amaranthus retroflexus*, *Digitaria sanguinalis*, *Portulaca oleracea,* and *Sorghum halepense*) in soil of a NT field and an adjacent buffer zone (nondisturbed and with higher organic matter than the NT field) in Italy. Weed seeds were placed at a soil depth of 12 cm to prevent germination. For all the species studied, the decay of buried seeds and microbial cellulolytic activity were higher in the NT field soil compared to the nondisturbed buffer strip. The level of degradation of the studied species was influenced by the physical properties of their seeds, with those having thicker seed coats (*A. theophrasti* and *S. halepense*) showing the least degradation compared to those with thinner seed coats (*D. sanguinalis* and *A. myosuroides*). The study found a positive correlation between seed coat thickness and seed longevity in the soil. Compared to the buffer strip, the NT field soil had higher microbiological activity of cellulolytic microorganisms that attacked the seed coat, which is mostly composed of lignin and cellulose. This activity was lower

© Burleigh Dodds Science Publishing Limited, 2025. All rights reserved.

in the buffer strip soil, which could explain the reduced degradation of seeds in this habitat. However, in NT soil most weed seeds are found in the surface layer where germination takes place, and damage to the seed coat by microbes could accelerate germination as well as lead to degradation of the seeds.

Accumulation of tillage effects on the composition of the weed seedbank, however, is relatively slow, as shown by previous studies (Davis et al., 2005; Légère et al., 2011a, 2011b; Lowry et al., 2021). Davis et al. (2005) found that the composition of the weed seedbank community did not differ after 12 years of continuous NT compared to CT in the Midwest region of the USA. Légère et al. (2011a, 2011b) found no variation in the composition of the weed seedbank community between zero tillage (ZT), moldboard ploughing, and chisel ploughing after 5 years of implementing different tillage systems. However, there were differences in the weed community after 18 years, so that common lambsquarters and pigweed species had large seedbanks in tilled treatments, whereas yellow foxtail and field pennycress contributed to the large seedbanks observed in NT treatments. Lowry et al. (2021) concluded that the transition to ZT wheat production did not lead to large-scale shifts in the community composition of the germinable weed seedbank after 3–4 years of ZT by farmers on the Eastern Indo-Gangetic plains of India.

3.2 Influence of cover crops and crop residues

Preserving crop residues on the soil surface is common in CA systems. Surface residues can alter the seed germination environment via changes in the soil surface temperature, soil moisture levels, and the chemical environment of the weed seeds. Surface residue lowers the maximum temperature but has little impact on the minimum temperature, resulting in cooler average soil temperatures and lower fluctuations. These changes can delay germination of agronomic crops and weeds, shortening the growing season and reducing yield. Prioritizing yield is important when considering amount of residue to leave on the soil surface, since the amount of residue can optimize yield or weed control but not both. While increased temperature fluctuations can enhance germination for certain weed species, soil temperatures buffered by surface residue can reduce germination percentages and rates and may delay time of seedling establishment of some species. Since fluctuating temperatures promote the germination of some weed seeds, the presence of residue on the soil surface might lead to reduced germination of weed species under CA (Ghosh et al., 2023).

The impact of surface residue on soil moisture conservation varies depending on the environment, as observed in studies conducted by Teasdale and Mohler (1993) and Bussière and Cellier (1994). In water-limited environments, residue may enhance weed seed germination, whereas in wetter

© Burleigh Dodds Science Publishing Limited, 2025. All rights reserved.

conditions its effect is minimal (Teasdale and Mohler, 1993; Wicks et al., 1994; Vidal and Bauman, 1996). Thus, residue has been found to be less effective in suppressing weeds in drier sites or years than in wetter sites or years (Buhler et al., 1996; Mashingaidze et al., 2012; Ngwira et al., 2014). Although the presence of plant residues on the soil surface in wetter conditions seems less important in terms of moisture conservation, residues can have inhibitory effects on the emergence and growth of weeds through changes in the quantity and quality of light, physical effects on the growth of weed seeds and also through the release of allelopathic substances (Liebman et al., 2004).

Surface residues can change the chemical environment of the weed seed via allelopathy. Allelopathic compounds can be released by living plants or when residues decompose. For example, a *Brassica* cover crop known to produce glucosinolates may suppress weed growth (Ait Kaci Ahmed et al., 2022). This discovery is important because it identifies a cover crop treatment that effectively suppresses both annual and perennial weeds commonly found in CA cropping systems (Mirsky et al., 2010). Crop residue, in particular, has a stronger negative impact on small-seeded than large-seeded weeds (Liebman et al., 2004). Greenhouse studies showed that allelopathic compounds can significantly reduce seed germination and hinder seedling growth (Khamare et al., 2022). However, it is important to note that the effects observed in the laboratory may not always translate to the field, and it can be challenging to isolate the specific allelopathic effects of residue from other changes in the environment that impact seedling emergence (Duke, 2015; Khamare et al., 2022). Nevertheless, there have been cases in which allelopathy has been found to reduce weed emergence on a large scale (Inderjit et al., 2001; Khamare et al., 2022). Various plant residues, such as rye, barley, wheat, rice, sorghum, alfalfa, sunflower, oat, and clover have been shown to exhibit allelopathic suppression of weed seed germination and emergence (Duke, 2015; Foy and Inderjit, 2001; Olofsdotter, 2001; Khamare et al., 2022). Crop residues placed near the soil surface release phytotoxins during decomposition, some of which may inhibit seed germination and early seedling growth (Khaliq et al., 2011). These allelopathic compounds affect various metabolic and physiological processes in seeds (Nelson, 1996) that are known to delay germination and reduce seedling growth (Mathiassen et al., 2006). Scherner et al. (2017) found delayed and prolonged emergence of seedlings under direct drilling and attributed this to the presence of inhibitory compounds in the soil that interfered with early weed growth.

Residues also may physically hinder contact between newly dispersed weed seeds and the soil, making conditions for seed germination less favorable (Paes and Rezende, 2001; Roller and Albrecht, 2006; Vargas et al., 2017). Additionally, stubble residues dry faster and retain heat more than soil (Liu et al., 2011), which can decrease germination of seeds retained by the residues and

© Burleigh Dodds Science Publishing Limited, 2025. All rights reserved.

may partly explain why weed emergence is prolonged under this method. It is worth noting that even with heavy crop residue, most seeds on the soil surface still receive enough light to germinate (Teasdale and Mohler, 1993). Therefore, lack of light is unlikely to significantly reduce weed seed germination when residue is retained, and, as mentioned earlier, altered soil temperatures caused by crop residues are more decisive in this regard.

Crop residues in CA also may lead to increased fungal pathogen infection and/or predation of seeds on the soil surface, resulting in lower total weed seed density compared to CT. Generally, habitat quality for potential predators is better in NT than in CT fields. Since seedbank depletion in NT systems depends on seed predation and mortality, it is very important to reduce the input of seeds in the seedbank and thus to reduce weed biomass (Moonen and Barberi, 2004). According to Kumar et al. (2022), the presence of crop residues on the surface enhances the population size of earthworms, ants, and floral/faunal diversity as residues decompose. Armengot et al. (2016) found that intensive cropping with repeated tillage (in CT) and without crop residues can decrease activities and abundance of seed predators. Thus, when crop residues are present in CA, there may be an increase in predators over time, leading to a reduction in weed seed density compared to CT. Kelton et al. (2011) found that the inclusion of high residue cover crops in a CT system can reduce the number of weed seeds in the upper 7.6 cm of the soil seedbank.

Despite suppressive effects of cover crops on the size of the seedbank, there are studies showing enhancement of seedbank size when cover crops are included in the cropping system. For example, Shrestha et al. (2015) found that the density of weed seeds increased in plots with cover crops, particularly in low disturbance tillage systems. Cover crops and their residues may enhance weed seed germination in environments with limited water availability, but they may have minimal impacts in conditions with high moisture levels. Some studies confirmed that in environments with sufficient moisture residues could even increase rates of seed decay (Nichols et al., 2015). In addition, crop residues create microenvironments that provide cover for animals that feed on them (Blanco-Canqui et al., 2015). In slow-growing perennial cover crops like red clover (*Trifolium pratense* L.), Mirsky et al. (2010) discovered that weedy species such as *Setaria* spp. were capable of growing and producing seeds within the cover crop. When this happened, the "debits" resulting from weed seed emergence were overwhelmed by the resulting seed rain "credits". In a study on the effect of diverse cover crops on weed seedbanks, Mirsky et al. (2010) found a species-specific response to cover crop treatments that resulted in restructuring of the weed community. It seems that seed traits such as size, longevity, and responsiveness to germination stimulants (disturbance and cover crop induced germination) are important elements in determining the reaction of seed germination to cover crops.

© Burleigh Dodds Science Publishing Limited, 2025. All rights reserved.

3.3 Influence of crop rotation

The influence of crop rotation on size and species composition of weed seedbanks has been confirmed by several studies. Continuous pinto bean production for 3 years resulted in the dominance of broadleaf weeds, specifically hairy nightshade (*Solanum sarrachoides*) (Ball, 1992). On the other hand, continuous corn production for 3 years decreased total weed seed count, but the seedbank primarily consisted of *Chenopodium album* and *Setaria viridis*. Another cropping sequence involving sugar beets for 2 years followed by corn production led to a weed seedbank dominated by *Amaranthus retroflexus*, *C. album*, *Eragrostis cilianensis,* and *Kochia scoparia* (Ball, 1992).

Different crop species that are included in crop rotations cause variability in environmental conditions for weed seeds in terms of factors such as moisture, light, nutrients, and chemical conditions, which can stimulate or prevent weed seed germination and ultimately affect the seedbank status. For instance, in an environment with limited water availability, a spring – irrigated crop promotes the germination of weed seeds during spring, whereas a fall-irrigated crop promotes weed seed germination in autumn. Thus, crop species have different effects on the quantity and quality of canopy-filtered light that might impact the germination of weed species, especially the small-seeded ones. Including allelopathic crops in a rotation also can decrease germination of weed seeds (Nichols et al., 2015). Furthermore, phenologically dissimilar crops involved in a rotational program have been found to increase predation pressure, and the crop cycle can influence the seasonality of seed predation (Heggenstaller et al., 2009; Westerman et al., 2011). Moreover, specific crops appear to promote predation of different weed seeds (Honek et al., 2003; Menalled et al., 2006; O'Rourke et al., 2006). O'Rourke et al. (2006) found that in maize and soybean fields, invertebrates exhibited greater predation of *Setaria faberi* seeds compared to triticale–alfalfa and alfalfa fields. Moreover, predation rates were elevated in plots with reduced input compared to conventionally managed soybean fields. In July, early August, and early October, invertebrates consumed less than 30% of seeds, but in late August and September, they consumed as much as 80–90% of seeds in maize and soybean fields. The proportion of *Setaria faberi* seeds consumed by invertebrates relative to all seed predators increased in late summer and was notably higher in maize and soybean fields compared to triticale–alfalfa and alfalfa treatments. These studies indicate that rotating crops with diverse growing seasons may enhance consumption of a wider range of seeds by predators (Nichols et al., 2015).

Diversified crop rotations practiced in CA may reduce buildup of weed seed reserves in the soil seedbank, especially for weed species with short seed longevity. Seed longevity of many weed species within the soil seedbank can serve as a reservoir for annual weed infestations (Barberi and Lo Cascio,

© Burleigh Dodds Science Publishing Limited, 2025. All rights reserved.

2001; Fracchiolla et al., 2018), thereby buffering against short-term changes to agricultural management (Lowry et al., 2021). Franke et al. (2007) found that a 1-year break from wheat production almost exhausted the available seedbank of littleseed canarygrass (*Phalaris minor*). Most seeds that fail to germinate in a given year are likely to lose viability before the start of the next growing season. Crop diversification may have a large impact on *P. minor* pressure, since winter crops other than wheat that prevent *P. minor* seed shed can strongly reduce soil seedbank density in the subsequent season. Many other annual grass weeds, such as *Bromus sterilis* (Froud-Williams, 1983) and *Avena fatua* (Miller and Nalewaja, 1990), also tend to have short seed longevities in the field.

Crop rotation has been shown in some studies to be the main factor affecting seedbank size. For instance, Guareschi et al. (2021) found that the crop rotation systems were more effective than soil tillage in reducing seedbank density. These authors found that the use of monoculture crop production during winter and summer under CT and NT systems resulted in a greater number of Italian ryegrass seeds at all soil depths sampled than with rotation systems. The inclusion of maize, wheat, and mungbean residues in succession in CA-based treatments may lead to a decrease in weed seedbank density (Susha et al., 2018; Kumar et al., 2022; Ghosh et al., 2023). Valaie et al. (2022) compared seedbank size in fields that had been under maize-wheat and fallow-wheat for 5 years and found that the seedbank size was significantly greater in maize-wheat than that of a fallow-wheat rotation. Additionally, relative density of *Portulaca oleracea* and *Amaranthus retroflexus* seeds increased in maize-wheat rotation compared with fallow-wheat. Scherner et al. (2017) found a significant build-up of grass seeds in monoculture cropping systems when combined with non-inversion tillage.

The long-term study conducted by Santín-Montanyá et al. (2016) demonstrated that cereal rotation systems implemented with minimum soil disturbance could reduce the presence of viable weed seeds in the soil. This decrease is attributed to the characteristics of the crop and the associated management practices, which create diverse environments for weed species. Menegat (2023) observed a significant decline in seedbank size of blackgrass (*Alopecurus myosuroides*) when a spring barley cropping was added to the continuous winter wheat cropping system. The reduction in seedbank size likely was due to the fact that *A. myosuroides* has two distinct germination and seedling emergence cohorts, one in autumn and another in early spring (Andersson and Åkerblom Espeby, 2009). By preparing the seedbed in spring before planting the crop, autumn-germinated plants are effectively removed, thus preventing seed production. Selective methods (either chemical or non-chemical) can control any plants emerging after crop planting, further reducing seed production. Second cohort plants likely produce fewer viable seeds due to their shorter growth phase compared to autumn-germinated plants.

© Burleigh Dodds Science Publishing Limited, 2025. All rights reserved.

Choice of crops and sequence in which they are grown can markedly influence weed community dynamics because of differences in biological cycles, end-use, competitive ability against weeds, cultural management practices (e.g. fertilisation, and seedbed preparation), and particularly weed control measures (e.g. use of herbicides and mechanical operations) (Ruisi et al., 2015). Vasileiadis et al. (2007) observed that in the sugar beet phase of the cotton–sugar beet rotation the total seedbank of *Portulaca oleracea* was significantly reduced compared with the cotton and tobacco monocultures, while the opposite occurred for *Echinochloa crus-galli*. The higher density of *E. crus-galli* in the cotton-sugar beet rotation was apparently due to the ability of this species to tolerate the herbicides (metamitron, phenmedipham, ethofumesate, and lenacil) used for weed management in sugar beet, as well as the fact that early-stage sugar beets are less competitive than cotton. On the other hand, the reduction in *P. oleracea* seed densities observed in the cotton-sugar beet rotation during the second year of the experiment can be attributed to the substantial growth reduction experienced by late-emerging *P. oleracea* plants due to the early development of the sugar beet crop canopy.

Ruisi et al. (2015) found that both the Shannon's diversity and Shannon's evenness indices were significantly higher in wheat-faba bean and wheat-berseem clover than in wheat monoculture. They found a remarkable reduction in weed seed production in the wheat-berseem clover rotation and attributed this to the fact that berseem clover is cut in spring before weeds have a chance to produce seeds. The combination of spring cutting and strong regrowth capability of berseem clover effectively restricts weed seed production. Additionally, some species, such as *Polygonum aviculare* and *Anagallis arvensis*, were markedly more abundant in monoculture wheat than in wheat-faba bean or wheat-berseem clover, whereas seedling density of *Papavera rhoeas* was lower in wheat-berseem clover than in continuous wheat or wheat-faba bean. Crop rotation markedly influences the weed seedbank by creating environmental conditions that differentially affect species emergence, development, and dissemination (Dorado et al., 1999; Blackshaw et al., 2001; Sosnoskie et al., 2006; Ruisi et al., 2015).

Murphy et al. (2006) found a positive correlation between increase in crop diversity within rotation sequences and increase in weed seedbank diversity. Likewise, Leon et al. (2015) found that overall weed seedbank diversity, species richness, and species evenness were higher in a cotton-peanut-bahia grass rotation compared to a cotton-peanut rotation that was less complex. Liebman et al. (2021) found an increase in Simpson's diversity index for weed species in the soil seedbank when considering whole rotation systems, i.e. diversifying a 2-year maize–soyabean rotation with two additional crop phases, oat, and lucerne to form a 4-year rotation. Diversifying to a 4-year rotation increased maize and soyabean productivity,

© Burleigh Dodds Science Publishing Limited, 2025. All rights reserved.

while maintaining total crop energy production and profitability and reducing damage to the environment and human health. The reason for the greater rotation-wide diversity of weed seeds in the 4-year rotation compared to the 2-year rotation was due to a higher number of weed species found in oat and lucerne compared to maize and soybean. This could be attributed to oat and lucerne not being treated with herbicides, unlike maize and soybean, or it could be influenced by other management practices, as well as by differences in crop growth patterns and resource utilization. In this experiment, oats were planted earlier and at higher density than maize and soybeans, and they were harvested in mid-July instead of late September or October like the latter two crops. Unlike maize, soybean, and oat, lucerne is a perennial forage species with deep roots, and it competes effectively with annual weeds even when harvested regularly.

4 Implications for ecological weed management

One of the challenges that most farmers face when transitioning from traditional agriculture to CA is the presence of high numbers of weeds, especially in the first years of the conversion period. Several preventive, cultural, biological, and chemical methods have been proposed to solve this problem (Bajwa et al., 2015; Eslami, 2014; Nichols et al., 2015; Singh et al., 2015). In many cases, an effective strategy for sustainable weed management in CA requires an integration of non-chemical and chemical weed management tools. However, here we aim to examine weed management methods based on ecological principles in response to the changes that occur in weed population dynamics in the soil seedbank and aboveground after CA has been adopted. Ecological weed management does not entirely rule out the use of herbicides. Instead, it strives to reduce their usage by establishing agricultural systems that suppress weed growth. Similar to conventional approaches, ecological weed management cannot completely eradicate weeds. Nevertheless, it holds the promise of significantly decreasing weed density, limiting the competitive ability of weeds on crop species, and preventing undesirable shifts in composition of the weed community. An ecological approach also decreases consumption of nonrenewable resources, decreases risks to human health and the environment and offers a net advantage to local and national economies (Liebman et al., 2004). Storkey and Westbury (2007) suggest that an effective approach to weed management should aim to minimize the negative effects of weeds on crop production while also maintaining the positive contributions of weeds to the trophic network and local environmental conditions. To elucidate the ecological principles for weed management in CA, we first reassess the effectiveness of the three pillars (principles) of CA in reducing weeds pressure.

© Burleigh Dodds Science Publishing Limited, 2025. All rights reserved.

5 Conservation Agriculture techniques and weed management

The fundamental principles of CA can provide a suitable means to reduce weed pressure on farms, but they need to be implemented by appropriate methods with adjustments depending on the practices of the farms in each region. However, there is still a lack of well-established knowledge regarding the effective implementation of the CA pillars for weed control to maximize long-term regulation (Cordeau, 2022). Despite many benefits of minimum soil disturbance and NT operations, problems can arise after long-term usages of these strategies in CA systems, such as increased incidence of weeds, inadequate weed control, buildup of herbicide-resistant weeds, soil compaction, soil acidification, runoff of dissolved nutrients and herbicides, increased incidence of soil and stubble-borne diseases, and stratification of nutrients and organic carbon in the topsoil (Dang et al., 2015; Blanco-Canqui and Wortmann, 2020; Cordeau et al., 2020; Server, 2021; Mia et al., 2023). The accumulation of fertilizers, lime, and crop residues near the soil surface (nutrient stratification), is often a problem with CA systems (McPheeters et al., 2022). This stratification can have negative effects on both crop production and the environment (Baker et al., 2017; Norton, 2020). Herbicides are commonly used to eliminate cover- and perennial crops and manage weeds in both annual and perennial crop rotations under NT management. However, relying heavily on herbicides can lead to development of herbicide-resistant weeds and have negative environmental consequences, which contradict the objectives of CA. Another issue worth noting is the potential glyphosate ban in Europe, which necessitates finding an alternative method for weed control, especially in CA systems. Antier et al. (2020) reported that various non-chemical alternatives to glyphosate are available (e.g. use of bio-herbicides, use of cover crops and of a roller-crimper for their destruction, mulching, crop rotation diversification, delaying crop sowing dates, higher seed rates, increasing crop competitiveness, use of early-ripening varieties and weed seed removal during harvest), but their effectiveness, cost, and implications for adoption vary greatly and/or are difficult to quantify.

5.1 Soil disturbance

After long-term reduction of soil disturbance in CA systems, farmers often encounter soil compaction that affects the growth of weeds and crops. According to Yu et al. (2020), tillage resulted in a 5.19% decrease in soil bulk density (BD) and a 5.69% improvement in soil porosity. Previous studies by Rai et al. (2018) and Peixoto et al. (2020) have shown that low disturbance tillage leads to lower BD, soil total carbon, and soil penetration resistance, while increased tillage intensity results in higher water infiltration rates and

© Burleigh Dodds Science Publishing Limited, 2025. All rights reserved.

soil porosity. Minhas et al. (2023) also found that conventional tillage resulted in less dense and more porous soil, whereas NT exerted opposite effects on soil porosity. Thus, NT may lead to the development of bare soil and increased infestation of invasive weeds.

Soil compaction also is likely to have an impact on occurrence, density, and competition between weed species and crops. However, some weed species may be less affected by compaction than crop species (Mia et al., 2023). Everaarts (1992) found that soil compaction by the planter press wheels did not influence weed growth. A study on barley crops showed that increasing compaction led to a decrease in weed dry biomass and density. *Eleusine indica*, a widely distributed weed of many agricultural systems, tolerates low soil oxygen (Shekoofa et al., 2020). Perennial species, such as *Spergula arvensis* and *Tripleurospermum perforatum*, were found to be less affected than annual species. *Fumaria officinalis* experienced the greatest decrease in density due to compaction, while *Chenopodium album* increased in dry biomass as a percentage of total weeds with compaction (Reintam and Kuht, 2012).

Azam et al. (2014) found that in compacted soil, garden pea crops primarily developed roots near the soil surface, whereas perennial species like *Acacia salicina* had roots distributed more evenly throughout the soil column. Place et al. (2008) observed that roots of two weed species *Amaranthus palmeri* and *Senna obtusifolia* were able to better penetrate a compacted soil layer and efficiently acquire nitrogen from below it compared to soybean crops. Additionally, overall shoot growth of weeds was sustained even when root growth beyond the compacted layer was restricted, while soybean shoot growth declined in such conditions. McCurdy et al. (2022) also concluded that *Juncus tenuis*, a warm-season, perennial weed of turfgrass, could persist in compacted soils.

Moreover, soil compaction might affect the efficacy of soil applied herbicides. Souza et al. (2019) concluded that soil compaction, regardless of texture, might favor the ready availability of herbicide in the soil solution, and thus plants were exposed to larger amounts of herbicide causing damage to soybean plants. To deal with this issue, some researchers have suggested using strategic disturbance events like occasional inversion tillage and shallow cultivation to terminate perennials or cover crops. Strategic tillage (ST), also called one-time tillage, occasional tillage (OT), targeted tillage, single-inversion, means tilling on an occasional basis, for example once every 4–and 10 years. While strategic tillage has been proposed by some researchers, implementing this method is not recommended due to its departure from the fundamental principles of CA and the significant changes it imposes on a CA system. Therefore, in cases where soil compaction increases, alternative methods such as using cover crops with deep roots, incorporating cash crops in rotations that reduce soil compaction should be employed to overcome this issue.

© Burleigh Dodds Science Publishing Limited, 2025. All rights reserved.

5.2 Diversified crop rotations

Using multi-species rotations to diversify cropping systems in CA can increase the variety of disturbances and stress factors that regulate weed population dynamics. This approach can effectively reduce weed population densities and alleviate reliance on herbicides and physical weed control methods for crop protection (Teasdale, 2018; Weisberger et al., 2019; Liebman and Nichols, 2020). Additionally, diversified crop rotation systems can help decrease the development of herbicide resistance in weed populations (Beckie et al., 2014; Beckie and Harker, 2017; Liebman et al., 2021). In a study on the impact of different tillage implements on weed seedbank composition in three crop rotational sequences, Ball (1992) concluded that cropping sequence was the most dominant factor influencing species composition in the seedbank. This was partly due to herbicide use in each cropping sequence producing a shift in the weed seedbank in favor of species less susceptible to applied herbicides. In a meta-analysis conducted by Weisberger et al. (2019), 298 paired observations from 54 studies across six continents were examined to compare the effects of simple and diverse crop rotations on weed responses. The study found that increasing the variance in crop planting dates through diversification was more successful in suppressing weeds compared to simply increasing the number of crop species. Additionally, increasing rotational diversity had a greater impact on reducing weed density under zero-tillage conditions (65%) compared to tilled conditions (41%), regardless of environmental factors and herbicide use.

Almost any weed can cause problems under certain conditions, especially with continuous cropping. To prevent such problems, it is crucial to consistently monitor the presence of weed species and take action to control them when their populations exceed certain levels. Unger et al. (1999) demonstrated that implementing crop rotations could effectively reduce weed problems, especially when the rotation included both winter and summer crops. By having different growing seasons, these crops allowed for weed control through tillage and/or herbicide application during non-growing periods. Another approach would be to alternate between dicot and monocot crops, which would enable the use of a wider range of herbicides for effective weed control. Alternatively, selective herbicides can be utilized specifically for in-crop weed management.

The findings of Menegat's (2023) studies indicate that the use of rotational direct seeding can prevent the accumulation of seeds in the soil. Ecological factors such as seed predation and exposure to environmental stressors significantly diminish the number of viable seeds that enter the seedbank. Combining this approach with integrated weed management techniques, such as increased seeding rates of crops, competitive crop cultivars, and cover crops, can further minimize the impact of weed competition on crops. Additionally, the study highlights the effectiveness of spring cropping for controlling

© Burleigh Dodds Science Publishing Limited, 2025. All rights reserved.

Alopecurus myosuroides, underscoring the importance of diversifying cereal-based cropping systems. Although direct seeding may lead to reduced yields, the decreased weed pressure in the system can compensate for this, resulting in a lower need for direct weed control measures and associated labor and energy costs.

As mentioned earlier, soil compaction is a challenge in fields under long-term CA. To ameliorate soil conditions and mitigate this problem, a potential strategy in addition to occasional tillage would be to include legumes as a cover or cash crop in the rotational system because they can produce roots deep in the soil profile and thus increase soil porosity. Minhas et al. (2023) found that a fallow-wheat cropping system resulted in higher soil bulk density (BD) and lower soil porosity, while a mungbean-wheat cropping system had the opposite result. This difference is due to the tap root system of mungbeans, which creates channels in the soil and improves soil structure and porosity. Additionally, mungbean roots emit exudates that aid in breaking down soil aggregates, thereby mitigating soil compaction. Several studies, including those by Shahzad et al. (2016a) and Naeem et al. (2021a), also have shown that including mungbean in wheat- and barley-based cropping systems can improve soil porosity and reduce bulk density. Better root growth in soils with higher porosity allows for greater nutrient and water absorption, leading to improved plant growth and yield.

When choosing crop plants for a rotation program, whether they serve as a cover or cash crop, it is important to consider not only their impact on weeds but also their influence on the following crops. In a study on cover crop-corn sequence, Liu et al. (2022) found that the inclusion of *Lolium multiflorum* as a cover crop in the rotation provided weed suppression but lowered the corn yield, which was attributed to the phytotoxic potential of this cover crop.

In certain agricultural regions practicing CA, there is a challenge of increasing pest populations. According to Kumar et al. (2022), implementing CA practices has increased the number of pests (insects), thereby posing threats to CA-based systems in the Indo-Gangetic Plain of India. Multiple studies in the Indo-Gangetic Plain have reported shifts in pest populations and crop losses due to an increase in the number of insects in both full and partial CA systems (Singh et al., 2014; Jasrotia et al., 2021). To ensure CA success, farmers need to consider management strategies such as crop rotation, periodic tillage-based residue management, and weed control within and in the vicinity of crops (Mishra et al., 2019). Introducing leguminous crops into rice-wheat rotations can effectively disrupt the continuous cycle of preferred host plants by pests. Reports indicate that diversifying with leguminous plants not only reduces pests but also enhances soil health and water-use efficiency (Arora et al., 2020).

© Burleigh Dodds Science Publishing Limited, 2025. All rights reserved.

5.3 Cover cropping

Cover cropping, a fundamental component of CA, plays a significant role in weed suppression within fields as long as appropriate planting dates, seeding rates, and termination dates are applied in each region, which are guided by previous research findings. Saini et al. (2008) found that there was an increase in weed biomass in cotton with earlier termination and late planting of the rye cover crop. Kelton et al. (2011) found that using cover crops in reduced tillage systems could potentially allow for the adoption of a conservation tillage system without needing to increase herbicide usage to control overall weed density.

Although most studies confirm suppressive effects or no effects of cover crops on weeds, cover crops also may cause increases in weed density. While cover crops can create shaded areas that inhibit weed seed germination, they also can increase soil moisture, which is conducive to weed seed germination. Shrestha et al. (2015) found that the types of weed species present varied depending on whether or not cover crops were used. Thus, cover crops are always successful in suppressing weeds, and they cannot be the only tool for weed management due to various factors that influence their effectiveness. Factors influencing the effectiveness of cover crops include the type of cover crop used, when it is planted, its density and amount of biomass produced, when it is terminated, the type of cash crop in rotation, and various climatic conditions. Moreover, the success of weed suppression may differ between annual and perennial cropping systems. Some studies have shown that using a mix of cover crop species is more effective in suppressing weeds than using a single species (Baraibar et al., 2018; Inderjit and Asakawa, 2001).

Previous studies primarily focused on introduced species as cover crops, but it is possible that native species could provide greater benefits due to their adaptation to the environment and natural ability to compete with invasive plants in the area. Although some studies have shown partial success in weed suppression, complete success cannot be guaranteed. Therefore, it is important to supplement cover crops with other weed management tools, since they may not be sufficient on their own. Nonetheless, cover crops remain an essential component of integrated weed management strategies (Fernando and Shrestha, 2023).

To effectively thwart weeds by acting as a physical barrier, mulch should be thick. Growing enough cover crop biomass and generating an adequate amount of mulch for weed control becomes a challenge due to dry summers and increasing climate variability (Colbach and Cordeau, 2022). This challenge is particularly pronounced in regions where cover crop biomass production is limited by a short growing season, and the resulting mulch is insufficient as the sole means of weed suppression. The inclusion of inorganic mulch can assist in suppressing weeds in certain types of farming systems, albeit deviating from

© Burleigh Dodds Science Publishing Limited, 2025. All rights reserved.

one of the CA pillars (maintaining permanent soil organic cover) (Cordeau, 2022).

Another challenge observed in the context of CA is delayed emergence of certain weeds, which reduces effectiveness of soil-applied herbicides. In CA systems, several weed species, including *Amaranthus tuberculatus* (Govindasamy et al., 2020), *Apera spica-venti* (Scherner et al., 2017), *Sorghum halopense* (Govindasamy et al., 2020), and *Vulpia myuros* (Scherner et al., 2017), delayed seedling emergence due to reduced temperatures at the soil surface caused by the presence of cover crop residues, although soil moisture increased. To manage such weeds, especially in directly-drilled winter cereals using soil-applied herbicides, it is recommended to consider sequential herbicide applications to ensure adequate coverage throughout the autumn seedling emergence period (Scherner et al., 2017).

6 Ecological mitigation of weed crop competition

Weeds and crops compete for resources such as light, water, and nutrients. Various strategies have been proposed to enhance resource acquisition by cultivated crops and weaken weeds in this competition, such as the use of crop varieties with high competitive ability, earlier emergence of crops compared to weeds, increased crop density, placement of resources close to crops for easy accessibility but away from weeds, and providing nutritional resources (notably nitrogen) in a form usable by crops while rendering them nonusable for weeds. Nonetheless, paying attention to the differences in resource acquisition patterns between crops and weeds in each region is of utmost importance. In situations where resources are scarce, and both crops and weeds share the same root depth and have concurrent resource requirements, weeds lead to significantly increased yield losses (Adeux et al., 2019). However, if resources are diversified in terms of availability and timing it is possible to mitigate yield loss (Smith et al., 2010).

Smith et al. (2010) introduced the Resource Pool Diversity Hypothesis (RPDH), a concept that elucidates how the diversity of soil resource pools can impact the competition between crops and weeds for these resources. According to the RPDH, the intensity of competition between species in plant communities hinges on the ability to distinguish their ecological niches and allocate resources effectively. Agricultural systems are unique because management practices like crop rotation, fertilization methods, and weed control contribute to the soil's resource pool, which can be divided between crops and weeds. When weed species differ from crops in traits related to resource acquisition, they should be able to take advantage of alternative resource pools as pool diversity increases (Smith et al., 2010). Conversely, when weeds and crops have similar resource acquisition traits, opportunities

© Burleigh Dodds Science Publishing Limited, 2025. All rights reserved.

for resource niche differentiation between crops and weeds will decrease and reductions in weed–crop competitive intensity resulting from increasing resource pool diversity likely will be lessened.

This hypothesis suggests that both the variety of crops and the way farms are managed (such as the type of fertilizers used, tillage practices, and machinery) have an impact on the competition between weeds and crops (Menalled et al., 2020). The hypothesis proposes that diversified systems lead to reduced weed-crop competition. Ryan et al. (2010) provided evidence supporting this hypothesis by showing that farm management practices can affect soil resources. They observed that organic cropping systems using legume- or manure-based fertilizer, despite having significantly more weed biomass, exhibited higher soil carbon levels and faster crop growth compared to a less diverse crop rotation in a conventional system. Remarkably, maize yields were similar in both organic and conventional systems, indicating that the diverse crop rotation and organic fertilizers in organic cropping systems effectively mitigated weed-crop competition. Similarly, Smith et al. (2010) compared nine organic and conventional cropping systems with varying levels of crop diversity and fertility sources. They found that as crop diversity increased intensity of weed-crop competition tended to decrease.

The microbial status of the soil also can influence competition between crops and weeds. The relationship between soil microbes, plant diversity, and weed-crop competition, however, is not well understood. Soil microbes can play a role in how specific species acquire resources, and they can influence how different plants occupy different niches belowground. The microbial-mediated resource hypothesis proposes that diverse associations between microbes and plants allow for greater utilization of soil resources (Reynolds et al., 2003). Miller and Menalled (2015) found support for this theory and suggested that interactions between plants and microbes affect competition dynamics.

The legacies of previous crops could potentially impact the competition between weeds and crops. Menalled et al. (2020) investigated how crop diversity legacies influence the intensity of weed-crop competition by altering soil nutrient availability and plant–soil microbe interactions in a greenhouse experiment. They used field soils (with nutrient and microbial legacies), a sterile greenhouse potting mix inoculated with microorganisms from the field soils (microbial legacies), and a sterile greenhouse potting mix in their research and found that weed-crop competition increased with crop diversity legacy in both field- and inoculated soil treatments in the annual system. In the perennial system, differences in weed-crop competition intensity were driven by crop yield potential. In the perennial field soil treatment, soils with the highest crop diversity legacy had the greatest crop yield potential, while in the perennial inoculated soil treatment soils with the lowest crop diversity legacy had the highest crop yield potential. Overall, these results suggest that there may be

© Burleigh Dodds Science Publishing Limited, 2025. All rights reserved.

negative effects of crop diversity on weed-crop competition when a crop is taxonomically related to its predecessors in a rotation. Thus, microbial and nutrient legacies from crop rotations can have contrasting effects on yield.

Therefore, it seems that in addition to adopting strategies that shift the competition dynamics in favor of cultivated crops paying attention to niche differentiation between crops and dominant weeds in the field as well as considering that previous crop legacies could contribute significantly to the management of weed-crop competition.

7 Management of seedbank and soil nutrients

Approximately 34% of global yields of major crops could potentially be reduced by weeds (Oerke, 2006). To ensure sustainable crop production, it is crucial to develop a long-term weed management strategy (Das et al., 2020; Nath et al., 2021). A key aspect of weed biology and ecology is understanding the dynamics of weed seedbanks, which is essential for ecological weed management (Storkey and Neve, 2018). Changes in tillage patterns, planting systems, and crop management practices can significantly affect the species composition and density of weed seeds in the soil (Swanton et al., 2000), potentially altering the soil environment and resulting in changes in the soil seedbanks and aboveground weed flora (Armengot et al., 2016).

Analyzing the seedbank of weeds in long-term trials is extremely valuable for understanding how agricultural practices impact the dynamics of weed communities. This knowledge is not easily obtained from short-term studies that rely on the actual weed flora, which can vary significantly over time and space. By looking at the data from long-term weed seedbank studies, we can make predictions to enhance soil, crop, and weed management (Barberi and Lo Cascio, 2001).

A number of weed species are adapted to specific crop sequences and levels of disturbance. Any alteration in the ecosystem, such as changes in crop rotations, soil disturbance, and fertilization levels, will impact populations of weed seeds in the seedbank. Weed species richness, seedbank abundance, and diversity of weed species in different soil depths are greatly influenced by the levels of disturbance (Hosseini et al., 2014). Farming practices affect both the seedbank and the plants that emerge from it. However, the seedbank also is affected by the presence of weed seeds in the soil and is more influenced by soil properties, such as the percentage of organic carbon and total nitrogen, than by management practices (Hawes et al., 2010). Additionally, changes in environmental factors from year to year also can greatly impact the weed seedbank (Harbuck et al., 2009). Reducing the weed seedbank is a vital aspect of an integrated weed management system. The most straightforward approach to decreasing the weed seedbank is to minimize the number of seeds

© Burleigh Dodds Science Publishing Limited, 2025. All rights reserved.

added to it. Any technique that diminishes the size and quantity of weeds that produce seeds will result in a decrease in the number of seeds being added to the seedbank. It is worth noting that alternative methods can be employed to manage the weed seedbank, such as increasing seed mortality or promoting germination when the weeds can be easily controlled (Hossain and Begum, 2015).

Ecological weed management involves reducing the size of the soil seedbank through stimulating seed death, encouraging germination, limiting seed production of emerged plants, or removing above-ground biomass (Mirsky et al., 2010). Most seedbank losses are attributed to seed germination, which accounts for more than half of annual reductions. Hence, focusing on ecologically based weed management (EBWM) targeting the germinable seedbank and seed production is key to shrinking the size of the seedbank. Tillage-induced germination and control are suggested by Gallandt (2006) as an effective way to achieve these reductions. Using synchronized management tactics and controlling the periodicity of germination can effectively manage emerging plants. Growers practicing EBWM have typically utilized methods to promote the germination of weed seeds such as preparing a seedbed that induces weed germination before crop planting (Lonsbary et al., 2003). However, since germination timing varies among weed species, implementing EBWM can alter the weed community. Identifying weed management practices that consistently reduce seedbanks across various weed life histories (e.g. broadleaved vs. grasses, annuals vs. perennials, early vs. late-emerging weeds) will further encourage EBWM adoption.

CA technology, particularly no-tillage, has a known drawback of promoting the growth of weed species. The use of non-selective herbicides solely for weed control can be problematic in the long run since it may facilitate the invasion of dominant weed species. To address this issue and maintain the sustainability of the agricultural system, Santín-Montanyá et al. (2016) proposed utilizing the weed seedbank as an early warning tool. By effectively managing it, the dominance of weed species can be prevented. Surveys have shown significant variations in the frequency of species within seedbanks depending on soil texture and organic matter content (Pareja et al., 1985; Andreasen et al., 1991; Cardina et al., 1991).

In CA systems, a large number of weed seeds accumulates on the soil surface. Farmers must develop strategies for managing weeds to decrease the number of weed seeds present in the field, which will ultimately result in fewer weeds during the crop-growing season. One way to achieve this is by effectively managing the weed seedbank in the top layer of the soil. It is crucial to reduce seed rain as part of weed seed management, and using PRE herbicides can help prevent weed seed germination or emergence, thereby increasing weed mortality. In particular, more attention could be directed

© Burleigh Dodds Science Publishing Limited, 2025. All rights reserved.

toward ways to optimize weed seed mortality in NT systems. Additionally, it is necessary to apply foliar herbicides during the growing season to minimize seed rain from weeds (Haidar and Siblani, 2023).

To avoid the problem of a single species dominating in CA, Santín-Montanyá et al. (2016) suggested breaking the dormancy of weed seeds present in the soil seedbank. This management plan involves manipulating soil conditions such as porosity, water-stable aggregates, and organic matter content, as well as other selected physical properties that influence resistance to freezing/thawing and rates of drying. These factors can affect soil surface cracking, which in turn promotes the emergence of diverse weed species and increases competition among them. Nevertheless, due to the complex nature of seed dormancy and its interaction with environmental factors, it is not possible to completely cleanse the soil of all weed seeds. However, it is possible to significantly reduce their numbers through agrotechnical treatments and a decrease in the supply of new seeds (Benech-Arnold et al., 2000). Weeds can exhibit primary and secondary dormancy, and seeds with nondeep physiological dormancy undergo annual dormancy/nondormancy cycles in the soil (Baskin and Baskin, 1985, 2014), which can result in sporadic germination and make it difficult to control weeds effectively (Wei et al., 2010; Srivastava, 2012). This ability allows weeds to evade control measures implemented on specific dates (Grundy, 2003; Conn et al., 2006; Karlsson et al., 2006; Ali et al., 2011).

Using a single herbicide application is insufficient for managing a diverse weed community that varies in terms of their physical characteristics and biological processes, as well as their germination and emergence timing during different stages of crop growth (Stewart et al., 2010; Tuti and Das, 2011; Vijayalaxmi et al., 2012). Relying solely on one active ingredient repeatedly can lead to changes in weed species composition, including the development of resistant biotypes or an increase in perennial weeds that are naturally less susceptible to many herbicides. Researchers are currently exploring ways to manipulate seed dormancy dynamics, such as using chemical germination stimulants in the soil to effectively eradicate all weeds simultaneously (Feledyn-Szewczyk et al., 2020). Das and Das (2018) discovered that a combination treatment involving the use of KNO3 (6% w v^{-1}) as a dormancy breaker, along with pre-emergence pendimethalin at a rate of 0.75 kg a.i. ha^{-1} + imazethapyr at a rate of 0.10 kg a.i. ha^{-1}, achieved 99% control of annual weeds and reduced the growth of *C. rotundus* by 83%. This treatment also resulted in significantly higher soybean yield and net returns. In the same way, when KNO3 (6%) was applied, followed by a tank mixture of clodinafop (0.06 kg a.i. ha^{-1}) + metsulfuron (0.006 kg a.i. ha^{-1}) at 30 days after sowing, weeds were effectively controlled in wheat crops.

An alternative approach to reducing the seedbank size involves enhancing soil biological activity. Sustaining a high level of soil biological activity through

© Burleigh Dodds Science Publishing Limited, 2025. All rights reserved.

effective organic soil management has the potential to shorten the lifespan of weed seedbanks, as soil microorganisms are known to aid in the decay of weed seeds. Additionally, introducing a legume or another cover crop may yield two potential outcomes. It could either trigger weed seed germination by elevating soil nitrate levels or facilitate the decomposition of weed seeds or seedlings due to soil microorganisms actively consuming green manure residues (Menalled and Schonbeck, 2009).

It is crucial to remember that none of these tactics can completely eradicate the weed seedbank. Moreover, the approach to managing the seedbank may need adjustments as the seedbank itself undergoes changes. The challenge in managing the weed seedbank lies in its diversity, typically comprising 20 to 50 different weed species within a single field. This means farmers may have to address the survival strategies of numerous weeds. Consequently, some weeds are likely to persist or even thrive despite the combination of seedbank management techniques employed by the farmer. For instance, light promotes seed germination of some weed species, so using dark cultivation methods reduces seedling emergence of these weeds. Similarly, careful nitrogen (N) management can mitigate issues with weeds responsive to nitrate, but it may not affect non-responsive weeds and could even benefit weed species well adapted to low soluble N levels (Liebman et al., 2004).

The most effective approach to managing the weed seedbank is to tailor a strategy that focuses on the four or five most troublesome weed species present. It also is essential to monitor changes in the weed population over time and take note of any new weed species emerging as the original target weeds decline. Then, adjustments can be made to the seedbank management strategy. It is likely that adjustments in the management plan will be needed every few years, and it is good to maintain a sense of curiosity and humor when dealing with weeds (Menalled and Schonbeck, 2009).

In an ecological weed management system, in addition to paying attention to the dynamics of the weed population on the farm and in the seedbank, special attention should be paid to the properties of the soil as a substrate for the growth and nutrition of plants. Manipulation of soil nutrients might be an effective strategy in weed management. Uddin et al. (2020) concluded that manipulating soil nutrients (nitrogen) might be an effective strategy to control invasive species and that this management tool is essential to restore degraded ecosystems. Their results suggest that activated carbon (AC) addition to the soil followed by frequent cutting of the shoots of the invasive species *Phragmites*, gave *Melaleuca* a competitive advantage over *Phragmites*. It is assumed that soil nitrogen immobilization due to AC addition may be responsible for the reduction in *Phragmites* growth, thus leading to increased *Melaleuca* growth. This strategy also can be generalized to agroecosystems, and by adding

© Burleigh Dodds Science Publishing Limited, 2025. All rights reserved.

high-residue cover crops we possibly can promote weed suppression through nitrogen (N) immobilization.

Research indicates that N limitation caused by high residue cover crops can play a significant role in suppressing weeds (Kumar et al., 2008, 2011). In organic soybean systems that used roller-crimped cover crops, the summer annual broadleaf weed *Amaranthus retroflexus* experienced reduced availability of N and subsequent N stress (Wells et al., 2013). This reduction in plant-available N in systems with high cover crop residue may contribute to overall weed suppression. While this residue-induced N immobilization can limit the uptake of N by both cash crops and weeds, leguminous cash crops like soybeans are not solely dependent on plant-available soil N for growth and therefore may not be negatively affected by this immobilization. A study conducted by Williams et al. (2018) on mulched NT systems found evidence supporting residue-induced nitrogen immobilization as a mechanism for weed suppression, with pigweed biomass and nitrogen uptake being reduced in the presence of cereal rye roots and shoots. It is likely that the immobilization occurred through fungal transportation of N to the applied cereal rye shoots, while soybean acted as a stronger sink for soil inorganic N. The availability of soil N did not impact the growth and development of soybean.

8 Conclusion

The shift to CA presents both advantages and challenges in managing weeds. While CA reduces soil disturbance and provides sustainability benefits, it introduces intricacies in weed dynamics. Weed density and species composition undergo notable changes, demanding a nuanced approach to ecological weed management. The reliance on herbicides in the absence of tillage creates challenges, resulting in increased weed densities and herbicide resistance. Furthermore, the introduction of invasive species and shifts in the weed flora require vigilant management. Despite these hurdles, the success of CA lies in understanding and efficiently managing the weed seedbank, ecologically mitigating weed-crop competition, and optimizing soil nutrient management. The chapter underscores the significance of strategies tailored for each aspect of weed ecology under CA to ensure its long-term sustainability and effectiveness in contemporary agricultural practices.

Thus, addressing the challenges of CA necessitates a strategic departure from the fundamental CA pillars. To enhance weed control effectiveness, there is an urgent need to reconsider strict adherence to CA principles such as no soil disturbance, permanent soil cover, and diversified crop rotations. Strategic interventions, including strictly limited soil disturbance, use of synthetic mulches, and targeted cover cropping, could be explored to disrupt weed growth cycles without compromising the overall benefits of CA.

© Burleigh Dodds Science Publishing Limited, 2025. All rights reserved.

The findings of this chapter highlight the necessity for research in the following areas:

- Explore strategies and ecological interventions to mitigate the introduction and proliferation of invasive and challenging-to-control species under CA.
- Investigate and devise alternative approaches to herbicides for weed management in CA, emphasizing ecological sustainability and the reduction of herbicide resistance.
- Evaluate the lasting impact of CA on weed dynamics, taking into account changing ecological conditions, climate variations, and their consequences for sustainable weed management.
- Conduct additional research to comprehend the intricate dynamics of crop-weed competition within the CA framework, enabling the formulation of precisely targeted ecological solutions.

9 Where to look for further information

9.1 Books

- Radosevich, S. R., Holt, J. S., Ghersa, C. M. (2007), *Ecology of Weeds and Invasive Plants: Relationship to Agriculture and Natural Resource Management* (3rd Edition)
- A comprehensive reference on weed ecology, evolution, and management strategies across agricultural and natural ecosystems
- Liebman, M., Mohler, C. L., Staver, C. P. (2001), *Ecological Management of Agricultural Weeds*. Cambridge University Press.
- A seminal work that applies ecological principles to weed management, offering insights particularly relevant to reduced tillage and diversified cropping systems.
- Zimdahl, R. L. (2018), *Fundamentals of Weed Science* (5th Edition.). Academic Press.
- A comprehensive textbook that delves into weed biology, population dynamics, and management, with applications in both conventional and conservation agriculture.
- Chauhan, B. S., Mahajan, G. (2014) *Recent Advances in Weed Management. Springer.*
- Contains recent research findings and methodologies that complement ecological weed management approaches in conservation systems.

9.2 Key Journals

The following journals regularly publish high-impact articles on weed ecology, soil-weed-crop interactions, and weed management in low-disturbance and conservation systems:

© Burleigh Dodds Science Publishing Limited, 2025. All rights reserved.

- *Weed Science*
- *Weed Research*
- *Agriculture, Ecosystems and Environment*
- *Field Crops Research*
- *Soil & Tillage Research*
- *Biological Agriculture & Horticulture*
- Crop Protection

9.3 Research networks

- CIMMYT – Conservation Agriculture Program (https://www.cimmyt.org): Offers research-based insights and long-term trial results focused on CA in global cropping systems, including weed management dynamics.
- FAO Conservation Agriculture Knowledge Hub (http://www.fao.org/conservation-agriculture): Provides global data, case studies, and technical publications relevant to CA systems, including weed-related components.
- CGIAR Initiatives on Agroecology (https://www.cgiar.org): Delivers interdisciplinary research addressing weed management within resilient and regenerative farming systems.

10 References

Adeux, G., et al. (2019), 'Mitigating crop yield losses through weed diversity'. *Nat. Sustain.*, 2, 1018–1026.

Ait Kaci Ahmed, N., et al. (2022), 'Ecosystem services provided by cover crops and biofumigation in sunflower cultivation', *Agron.*, 12(1), 120.

Ali, H. H., et al. (2011), Methods to break seed dormancy of *Rhynchosia capitata*, a summer annual weed', *Chil. J. Agric. Res.*, 71(3), 483– 487.

Anderson, R. L. (2008), 'Diversity and no-till: keys for pest management in the U.S. Great Plains', *Weed Sci.*, 56(1), 141–145.

Anderson, R. L., et al. (1998), 'Weed community and species response to crop rotation, tillage, and nitrogen fertility', *Weed Technol.*, 12(3), 531–536.

Andersson, L. and Åkerblom Espeby, L. (2009), 'Variation in seed dormancy and light sensitivity in *Alopecurus myosuroides* and *Apera spica-venti*', *Weed Res.*, 49(3), 261–270.

Andreasen, C., et al. (1991), 'Soil properties affecting the distribution of 37 weed species in Danish fields', *Weed Res.*, 31(4), 181–187.

Antier, C., et al. (2020), A survey on the uses of glyphosate in European countries. INRAe, Paris Cedex, France.

Armengot, L., et al. (2016), 'Tillage as a driver of change in weed communities: a functional perspective', *Agric. Ecosyst. Environ.*, 222, 276–285.

Arora, S., et al. (2020), *Handbook of Soil Health and Water Management,* Vol. 313, 1–550. Agrotech Publishing Academy, Udaipur, India.

Azam, G., et al. (2014), 'Comparison of the penetration of primary and lateral roots of pea and different tree seedlings growing in hard soils', *Soil Res.*, 52(1), 87–96.

© Burleigh Dodds Science Publishing Limited, 2025. All rights reserved.

Bajwa, A., et al. (2015), 'Nonconventional weed management strategies for modern agriculture', *Weed Sci.*, 63(4), 723–747.

Baker, C., et al. (2018), 'Weed species composition and density under conservation agriculture with varying fertilizer rate', *S. Afr. J. Plant. Soil.*, 35(5), 329–336.

Baker, C. J., et al. (2007), *No-.Tillage Seeding in Conservation Agriculture* 2nd ed. Cambridge: FAO and CAB International.

Baker, D. B., et al. (2017), 'Vertical stratification of soil phosphorus as a concern for dissolved phosphorus runoff in the Lake Erie Basin', *J. Environ. Qual.*, 46(6), 1287–1295.

Ball, D. A. (1992), 'Weed seedbank response to tillage, herbicides and crop rotation sequence', *Weed Sci.*, 40(4), 654–659.

Balota, E. L., et al. (2004), 'Soil enzyme activities under long-term tillage and crop rotation systems in subtropical agroecosystems', *Braz. J. Microbiol,* 35(4), 300–306.

Baraibar, B., et al. (2018), 'Weed suppression in cover crop monocultures and mixtures', *Weed Sci.*, 66, 121–133.

Baraibar, B., et al. (2009), 'Effects of tillage and irrigation in cereal fields on weed seed removal by seed predators', *J. Appl. Ecol.*, 46(2), 380–387.

Barberi, P., et al. (2001), 'Weed density and composition in winter wheat as influenced by tillage systems', *First World Congress on Conservation Agriculture*, Madrid, Spain, October1st to 5th., pp. 451–455.

Barberi, P. and Lo Cascio, B. (2001), 'Long-term tillage and crop rotation effects on weed seedbank size and composition', *Weed Res.*, 41(4), 325–340.

Baskin, C. C. and Baskin, J. M. (2014), *Seeds Ecology, Biogeography, and Evolution of Dormancy and Germination*. 2nd ed. San Diego, CA: Academic Press/Elsevier.

Baskin, J. M. and Baskin, C. C. (1985), 'The annual dormancy cycle in buried weed seeds: a continuum', *BioScience*, 35(8), 492–498.

Beckie, H. J. and Harker, K. N. (2017), 'Our top 10 herbicide-resistant weed management practices', *Pest Manage. Sci.*, 73(6), 1045–1052.

Beckie, H. J., et al. (2014), 'Selection and evolution of acetyl-CoA carboxylase (ACC)-inhibitor resistance in wild oat (*Avena fatua* L.) in a long-term alternative cropping systems study', *Can. J. Plant Sci.*, 94(4), 727–731.

Benech-Arnold, R. L., et al. (2000), 'Environmental control of dormancy in weed seed banks in soil', *Field Crop. Res.*, 67(2), 105–122.

Bissett, A., et al. (2011), 'Long-term land use effects on soil microbial community structure and function', *Appl. Soil Ecol.*, 51(1), 66–78.

Blackshaw, R., et al. (2001), 'Tillage intensity and crop rotation affect weed community dynamics in a winter wheat cropping system', *Can. J. Plant Sci.*, 81(4), 805–813.

Blanco-Canqui, H., et al. (2015), 'Cover crops and ecosystem services: insights from studies in temperate soils', *Agron. J.*, 107, 2449–2474.

Blanco-Canqui, H. and Wortmann, C.S. (2020), 'Does occasional tillage undo the ecosystem services gained with no-till? A review', *Soil Tillage Res.*, 198, 104534.

Blubaugh, C. K. and Kaplan, I. (2015), 'Tillage compromises weed seed predator activity across developmental stages', *Biol. Control.*, 81, 76–82.

Borgy, B., et al. (2012), 'Non-random distribution of weed species abundance in arable fields', *Weed Res.*, 52(4), 383–389.

Büchi, L., et al. (2021), '*Vulpia, myuros*, an increasing threat for agriculture', *Weed Res.*, 61(1), 13–24.

© Burleigh Dodds Science Publishing Limited, 2025. All rights reserved.

Buhler, D. D., et al. (1996), 'The effect of maize residues and tillage on emergence of *Setaria faberi*, *Abutilon theophrasti*, *Amaranthus retroflexus* and *Chenopodium album*. *Weed Res.*, 36(2), 153–165.

Bussière, F. and Cellier, P. (1994), 'Modification of the soil temperature and water content regimes by a crop residue mulch: experiment and modelling', *Agric. For. Meteorol.*,68(1–2), 1–28.

Calzarano, F., et al. (2018), 'Durum wheat quality, yield and sanitary status under conservation agriculture', *Agriculture,* 8(9), 140.

Campiglia, E., et al. (2018), 'Floristic composition and species diversity of weed community after 10 years of different cropping systems and soil tillage in a Mediterranean environment', *Weed Res.*, 58(4), 273–283.

Cardina, J., et al. (2002), 'Crop rotation and tillage system effects on weed seedbanks', *Weed Sci.*, 50(4), 448–460.

Cardina, J., et al. (1991), 'Long-term tillage effects on seed banks in three Ohio soils', *Weed Sci.*, 39(2), 186–194.

Cardina, J., et al. (1998), 'Long-term tillage and rotation effects on soil seedbank characteristics', *Asp. Appl. Biol.*, 51, 213–220.

Chancellor, R. (1985), 'Changes in the weed flora of an arable field cultivated for 20years', *J. Appl. Ecol.*, 22(2), 491–501.

Chauhan, B. S., et al. (2012), 'Ecology and management of weeds under conservation agriculture: a review', *Crop Prot.* 38, 57–65.

Chetan, F., et al. (2022) 'Influence of soil tillage systems on the yield and weeds infestation in the soybean crop', *Land*, 11(10), 1–13.

Colbach, N. and Cordeau, S. (2022), 'Are no-till herbicide-free systems possible? A simulation study', *Front. Agron.*, 4, 823069.

Concenco, G., et al. (2011), 'Effect of long-term agricultural management systems on occurrence and composition of weed species', *Planta Daninha*, 29(3), 515–522.

Conn, J. S., et al. (2006), 'Seed viability and dormancy of 17 weed species after 19.7 years of burial in Alaska', *Weed Sci.*, 54(3), 464–470.

Cordeau, S. (2022), 'Conservation agriculture and agroecological weed management', *Agronomy*, 12(4), 867.

Cordeau, S., et al. (2020),'Is tillage a suitable option for weed management in conservation agriculture?', *Agron.*, 10(11), 1746.

Cox, W. J., et al. (1990), 'Tillage effects on some soil physical and corn physiological characteristics', *Agron. J.*, 82, 806.

Cussans, J. (2022), 'The threat of weeds for regenerative agriculture', *Groundswell Agriculture*, 22–23 June, https://www.youtube.com/watch?v=Zb9kEUsVhYA (Accessed 8 December 2022).

Dang, Y. P., et al. (2015), 'Strategic tillage in no-till farming systems in Australia's northern grains-growing regions: I. Drivers and implementation', *Soil Tillage Res.*, 152, 104–114.

Das, T. K. and Das, D. K. (2018), 'Using chemical seed dormancy breakers with herbicides for weed management in soyabean and wheat', *Weed Res.*, 58(3), 188–199.

Das, T. K., et al. (2020), 'Conservation agriculture in rice-mustard cropping system for five years: impacts on crop productivity, profitability, water-use efficiency, and soil properties', *Field Crops Res.*, 250, 107781.

Davis, A. S., et al. (2005) 'Weed seedbank and community shifts in a long-term cropping systems experiment', *Weed Sci.,* 53(3), 296–306.

© Burleigh Dodds Science Publishing Limited, 2025. All rights reserved.

Délye, C., et al. (2008), 'Cross-resistance patterns to ACCase-inhibiting herbicides conferred by mutant ACCase isoforms in *Alopecurus myosuroides* Huds. (black-grass), re-examined at the recommended herbicide field rate', *Pest Manage. Sci.*, 64, 1179–1186.
Derksen, D. A., et al. (1993), 'Impact of agronomic practices on weed communities: tillage systems', *Weed Sci.* 41(3),409–417.
Derksen, D. A., et al. (1994), 'Impact of agronomic practices on weed communities: fallow within tillage systems', *Weed Sci.*, 42(2), 184–94.
Derrouch, D., et al. (2021), 'Functional shifts in weed community composition following adoption of conservation agriculture', *Weed Res.*, 62(2), 103–112.
Derrouch, D., et al. (2020), 'Weed management in the transition to conservation agriculture: farmers' response', *Agronomy*, 10(6), 843.
Dorado, J., et al. (1999), 'Weed seedbank response to crop rotation and tillage in semiarid agroecosystems', *Weed Sci.*, 47(1), 67–73.
Duke, S. O. (2015), 'Providing allelopathy in crop – weed interactions', *Weed Sci.*, 63, 121–132.
Eslami, S. V. (2014), 'Weed management in conservation agriculture systems'. In Chauhan, B. S., and Mahajan, G., (Eds.), *Recent Advances in Weed Management.* New York: Springer Publishing, pp. 87–124.
Everaarts, A. P. (1992), 'Response of weeds to the method of fertilizer application on low-fertility acid soils in Suriname', *Weed Res.*, 32(5), 391–397.
Feldman, S. R., et al. (1997), 'The effect of different tillage systems on the composition of the seedbank', *Weed Res.*, 37(2), 71–76.
Feldman, S. R., et al. (1998), 'Gap colonization by weeds in a wheat crop grown under different cultivation regimes', *Weed Res.*, 38(1), 35–45.
Feledyn-Szewczyk, B., et al. (2020), 'Weed flora and soil seed bank composition as affected by tillage system in three-year crop rotation', *Agriculture*, 10(5), 186.
Fernando, M. and Shrestha, A. (2023), 'The potential of cover crops for weed management: a sole tool or component of an integrated weed management system?', *Plants*, 12(4), 752.
Forcella, F. (2003), 'Debiting the seedbank: priorities and predictions', *Asp. Appl. Biol*, 69, 151–162.
Foy, C. L. and Inderjit (2001), 'Understanding the role of allelopathy in weed interference and declining plant diversity', *Weed Technol.*, 15, 873–878.
Fracchiolla, M., et al. (2018), 'Effects of conservative tillage and nitrogen management on weed seed bank after a seven-year durum wheat-faba bean rotation', *Plants*, 7(4), 82.
Franke, A. C., et al. (2007), '*Phalaris minor* seedbank studies: longevity, seedling emergence and seed production as affected by tillage regime', *Weed Res.*, 47(1), 73–83.
Fried, G., et al. (2012), 'Trajectories of weed communities explained by traits associated with species' response to management practices', *Agric. Ecosyst. Environ.*, 158, 147–155.
Friedrich, T., et al. (2009), Conservation agriculture in developing countries: the role of mechanization; innovation for sustainable agricultural mechanisation. Club of Bologna, Hannover, Germany, November; FAO.
Froud-Williams RJ (1983), 'The influence of straw disposal and cultivation regime on the population dynamics of *Bromus sterilis*', *Ann. Appl. Biol.*, 103(1), 139–148.

© Burleigh Dodds Science Publishing Limited, 2025. All rights reserved.

Gaba, S., et al. (2014), 'Agroecological weed control using a functional approach: a review of cropping systems diversity', *Agron. Sustain. Dev.*, 34(1), 103–119.
Gallandt, E. R. (2006), 'How can we target the weed seedbank?', *Weed Sci.*, 54(3), 588–596.
Gallandt, E. R., et al. (2004), 'Effect of tillage, fungicide seed treatment, and soil fumigation on seed bank dynamics of wild oat (*Avena fatua*)', *Weed Sci.*, 52(4), 597–604.
Ghosh, S., et al. (2023), 'Weed seedbank, above-ground weed community and crop yields under conventional and conservation agriculture practices in maize–wheat–mungbean rotation', *Weed Res.*, 63(4), 270–281.
Gill, K. S. and Arshad, M. A. (1995), 'Weed flora in the early growth period of spring crops under conventional, reduced, and zero tillage systems on a clay soil in northern Alberta, Canada', *Soil Tillage Res.*, 33(1), 65–79.
Govindasamy, P., et al. (2020), 'No-tillage altered weed species dynamics in a long-term (36-year) grain sorghum experiment in southeast Texas', *Weed Sci.*, 68(5), 476–484.
Grey, T. L., et al. (2017), 'Evaluation of control of napiergrass (*Pennisetum purpureum*) with tillage and herbicides', *Invasive Plant Sci. Manage.*, 8(4), 393–400.
Grundy, A. C. (2003), 'Predicting weed emergence: a review of approaches and future challenges', *Weed Res.*, 43(1), 1–11.
Guareschi, A., et al. (2021), 'Crop rotation and tillage system on the Italian ryegrass seedbank after long-term management, *Sci. Agric.*, 78(5), e20190266.
Haidar, M. and Siblani, W. (2023), 'Weed seed bank density in seven-year-old conservation in no-till system', *J. Agric. Stud.*, 11(1), 41–48.
Harbuck, K. S. B., et al. (2009), 'Impact of cropping systems on weed seed banks in the northern Great Plains, USA', *Weed Biol. Manage.*, 9(2), 160–168.
Hawes, C., et al. (2010), 'Arable plant communities as indicator of farming practice', *Agric Ecosyst. Environ.*, 138(1), 17–26.
Heggenstaller, A. H., et al. (2009), 'Growth analysis of biomass production in sole-crop and double-crop corn systems', *Crop Sci.*, 49(6), 2215–2224.
Hernandez-Restrepo, M., et al. (2016), 'Take-all or nothing', *Stud. Mycol.*, 83, 19–48.
Holland, J. (2004), 'The environmental consequences of adopting conservation tillage in Europe: reviewing the evidence', *Agric. Ecosyst. Environ.*, 103, 1–25.
Honek, A., et al. (2003), 'Ground beetles (*Carabidae*) as seed predators', *Eur. J. Entomol.*, 100(4), 531–544.
Hossain, M. M. and Begum, M. (2015), 'Soil weed seed bank: Importance and management for sustainable crop production – a review', *J. Bangladesh Agric. Univ.*, 13(2), 221–228.
Hosseini, P., et al. (2014), 'Weed seed bank as affected by crop rotation and disturbance', *Crop Prot.*, 64(1), 1–6.
Hugo, E., et al. (2021), 'Weed species diversity and shifts in conservation agriculture-based crop rotation systems on the Highveld area of South Africa', *S. Afr. J. Plant Soil*, 38(3), 264–275.
Hume, L., et al. (1991), 'Tillage and rotation influences on weed community composition in wheat (*Triticum aestivum* L.) in southwestern Saskatchewan', *Can. J. Plant Sci.*, 71(3), 783–789.
Inderjit and Asakawa, C. (2001), 'Nature of interference potential of hairy vetch (*Vicia villosa* Roth) to radish (*Raphanus sativus* L.): does allelopathy play any role?', *Crop Prot.*, 20, 261–265.
Inderjit, et al. (2001), 'On the significance of filed studies in allelopathy', *Weed Technol.*, 15, 792–797.

© Burleigh Dodds Science Publishing Limited, 2025. All rights reserved.

Jasrotia, P., et al. (2021), 'Tillage intensity influences insect-pest and predator dynamics of wheat crop grown under different conservation agriculture practices in rice-wheat cropping system of indo-Gangetic plain', *Agronomy*, 11(6), 1087.

Johansen, C., et al. (2012), 'Conservation agriculture for small holder rainfed farming: opportunities and constraints of new mechanized seeding systems', *Field Crops Res.*, 132, 18–32.

Kadziene, G., et al. (2020), 'Tillage and cover crop influence on weed pressure and Fusarium infection in spring cereals', *Crop Prot.*, 127, 104966.

Karlsson, L. M., et al. (2006), 'Seed dormancy and germination in the summer annual *Galeopsis speciosa*', *Weed Res.*, 46(5), 353–361.

Kelton, J. A., et al. (2011), 'Weed seed bank density and composition in a tillage and landscape variability study', *Commun. Biomet. Crop Sci.*, 6(1), 21–30.

Khalil, Y. (2018), 'Effect of crop residues on interception and activity of prosulfocarb, pyroxasulfone, and trifluralin', *PLoS ONE*, 13(12), e0208274.

Khaliq, A., et al. (2011), 'Influence of wheat straw and rhizosphere on seed germination, early seedling growth and bio-chemical attributes of *Trianthema portulacastrum*', *Planta Daninha*, 29(3), 523–533.

Khamare, Y., et al. (2022), 'Allelopathy and its application as a weed management tool: a review', *Front. Plant Sci.*, 13, 1034649. DOI: 10.3389/fpls.2022.1034649.

Kremer, R. J. and Li, J. (2003), 'Developing weed-suppressive soils through improved soil quality management', *Soil Tillage Res.*, 72(2), 193–202.

Kumar, N., et al. (2022), 'Long-term impact of zero-till residue management in post-rainy seasons after puddled rice and cropping intensification on weed seedbank, above-ground weed flora and crop productivity', *Ecol. Eng.*, 176, 106540.

Kumar, V., et al. (2008), 'Suppression of Powell Amaranth (*Amaranthus powellii*), shepherd's purse (*Capsella bursa- Pastoris*), and corn chamomile (*Anthemis arvensis*) by buckwheat residues: role of nitrogen and fungal pathogens', *Weed Sci.*, 56(2), 271–280.

Kumar, V., et al. (2011), 'Buckwheat residue effects on emergence and growth of weeds in winter-wheat (*Triticum aestivum*) cropping systems. *Weed Sci.*, 59(4), 567–573.

Latif, S., et al. (2017), 'Allelopathy and the role of allelochemicals in plant defence', *Adv. Bot. Res.*, 82, 19–54.

Lee, N. and Thierfelder, C. (2017), 'Weed control under conservation agriculture in dryland smallholder farming systems of southern Africa. A review', *Agron. Sustain. Dev.*, 37(5), 48.

Légère, A., et al. (2005), 'Diversity and assembly of weed communities: contrasting responses across cropping systems', *Weed Res.*, 45(4), 303–315.

Légère, A., et al. (2011a) 'The selective memory of weed seedbanks after 18 years of conservation tillage', *Weed Sci.*, 59(1), 98–106.

Légère, A., et al. (2011b), 'A corn test crop confirms beneficial effects of crop rotation in three tillage systems', *Can. J. Plant Sci.*, 91(5), 943–946.

Lemieux, C., et al. (1993), 'Distribution and survival of quackgrass (*Elytrigia repens*) rhizome buds', *Weed Sci.*, 41(4),600–606.

Leon, R. G., et al. (2015), 'Weed seed banks are more dynamic in a sod-based rotation, than in a conventional, peanut-cotton rotation', *Weed Sci.*, 63(4), 877–887.

Liebman, M. and Dyck, E. (1993), 'Crop rotation and intercropping strategies for weed management', *Ecol. Appl.*, 3(1), 92–122.

© Burleigh Dodds Science Publishing Limited, 2025. All rights reserved.

Liebman, M. and Nichols, V. A. (2020), 'Cropping system redesign for improved weed management: a modeling approach illustrated with giant ragweed (*Ambrosia trifida*)', *Agronomy*, 10(2), 262.

Liebman, M., et al. (2004), *Ecological Management of Agricultural Weeds*. 2nd ed. Cambridge: Cambridge University Press.

Liebman, M., et al. (2021), 'Weed seedbank diversity and sustainability indicators for simple and more diverse cropping systems', *Weed Res.*, 61(3), 164–177.

Liu, C., et al. (2011), 'Effects of irrigation, fertilization and crop straw management on nitrous oxide and nitric oxide emissions from a wheat-maize rotation field in northern China', *Agric. Ecosyst. Environ.*, 140(1), 226–233.

Liu, S., et al. (2022), 'The impact of different winter cover crops on weed suppression and corn yield under different tillage systems', *Agron.*, 12(5), 999.

Lonsbary, S. K., et al. (2003), 'Stale-seedbed as a weed management alternative for machine-harvested cucumbers (*Cucumis sativus*)', *Weed Technol.*, 17(4), 724–730.

Lowry, C. J., et al. (2021), 'Weed germinable seedbanks of rice–wheat systems in the Eastern Indo-Gangetic Plains: Do tillage and edaphic factors explain community variation?', *Weed Res.*, 61(6), 475–485.

Maillard, E., et al. (2016), 'Greater accumulation of soil organic carbon after liquid dairy manure application under cereal-forage rotation than cereal monoculture', *Agric. Ecosyst. Environ.*, 233(3), 171–178.

Martin, A.L., et al. (2002), 'Weed management in conservation crop productionsystems'. *Weed Biol. Manag.*, 2:123–132.

Mashingaidze, N., et al. (2012), 'Crop yield and weed growth under conservation agriculture in semi-arid Zimbabwe', *Soil Tillage Res.*, 124, 102–110.

Mathiassen, S. K., et al. (2006), 'Herbicidal effects of soil-incorporated wheat', *J. Agric. Food Chem.*, 54(4), 1058–1063.

Mayor, J.-P. and Maillard, A. (1995), Resultats d'un essai de culture sans labour depuis plus de 20 ans a changins. IV. Stock semencier et maitrise de la flore adventice. *Revue Suisse d'agric.*, 27, 229–236.

McCurdy, J. D., et al. (2022), 'Effects of soil compaction and moisture on the growth of *Juncus tenuis*', *Int. Turfgrass Soc. Res. J.*, 14(1), 776–782.

McPheeters, D., et al. (2022), 'Integrated weed management with strategic tillage can maintain soil quality in continuous living cover systems', *Front. Sustain. Food Syst.*, 6, 907590.

Menalled, F. D., et al. (2001), Weed aboveground and seedbank community responses to agricultural management systems. *Ecol. Appl.*, 11(6), 1586–1601.

Menalled, F. D., et al. (2006), 'The ecology of weed seed predation in herbaceous crop systems'. In Singh, H.P., et al. (Eds.), *Handbook of Sustainable Weed Management*, Food Products Press, an imprint of The Haworth Press Inc., New York, USA, pp. 297–327.

Menalled, F. D. and Schonbeck, M. (2009), 'Manage the weed seed bank-minimize "deposits" and maximize "withdrawals"', https://eorganic.org/node/2806 (Accessed 3 October 2022).

Menalled, U. D., et al. (2020), 'Soil-mediated effects on weed-crop competition: Elucidating the role of annual and perennial intercrop diversity legacies', *Agronomy*, 10(9), 1373.

Menegat, A. (2023), 'Minimal soil disturbance combined with spring cropping can halt soil seedbank accumulation of *Alopecurus myosuroides*', *Weed Res.*, 63(2), 115–122.

© Burleigh Dodds Science Publishing Limited, 2025. All rights reserved.

Mhlanga, B., et al. (2015), 'Weed community responses to rotations with cover crops in maize-based conservation agriculture systems of Zimbabwe. *Crop Prot.*, 69(1), 1–8.
Mia, M. S., et al. (2023), 'Strategic tillage in Australian conservation agricultural systems to address soil constraints: how does it impact weed management?', *Weed Res.*, 63(1), 12–26.
Mickelson, J. A. and Grey, W. E. (2006), 'Effect of soil water content on wild oat (*Avenafatua* L.) seed mortality and seedling emergence', *Weed Sci.* 54(2), 255–262.
Miller, S. D. and Nalewaja, J. D. (1990), 'Influence of burial depth on wild oats (*Avena fatua*) seed longevity', *Weed Technol*, 4(3), 514–517.
Miller, Z. J. and Menalled, F. D. (2015), 'Impact of species identity and phylogenetic relatedness on biologically-mediated plant-soil feedbacks in a low and a high intensity agroecosystem', *Plant Soil*, 389(1–2), 171–183.
Minhas, W. A., et al. (2023), 'Weed infestation and productivity of wheat crop sown in various cropping systems under conventional and conservation tillage', *Front. Plant Sci.*, 14, 1176738.
Mirsky, S. B., et al. (2010), 'Reducing the germinable weed seedbank with soil disturbance and cover crops', *Weed Res.*, 50(4), 341–352.
Mishra, J. S., and Singh, V. P. (2012), 'Tillage and weed control effects on productivity of a dry seeded rice–wheat system on a Vertisol in Central India', *Soil Tillage Res.,* 123, 11-20.
Mishra, J. S., et al. (2019), 'Rice mealybug (*Brevennia rehi*): a potential threat to rice in a long-term rice-based conservation agriculture system in middle Indo-Gangetic plains', *Curr. Sci.*, 117(4), 566–568
Mohler, C. L. and Calford, A. E. (1997), 'Weed seedling emergence and seed survival: separating the effects of seed position and soil modification by tillage', *Weed Res.*, 37(3), 147–155.
Moonen, A. and Barberi, P. (2004), 'Size and composition of the weed seedbank after 7 years of different cover-crop-maize management systems', *Weed Res.*, 44(3), 163–177.
Moyer, J., et al. (1994), 'Weed management in conservation tillage systems for wheat production in North and South America', *Crop Prot.*, 13(4), 243–259.
Mulugeta, D. and Stoltenberg, D. E. (1997), 'Weed and seedbank management with integrated methods as influenced by tillage', Weed Sci., 45(5), 706–715.
Murphy, S., et al. (2006), 'Promotion of weed species diversity and reduction of weed seedbanks with conservation tillage and crop rotation', *Weed Sci.*, 54(1), 69–77.
Naeem, M., et al. (2021a), 'Weed flora composition of different barley-based cropping systems under conventional and conservation tillage practices', *Phytoparasitica*, 49(1), 1–19.
Naeem, M., et al. (2021b), 'Impact of different barley-based cropping systems on soil physicochemical properties and barley growth under conventional and conservation tillage systems', *Agronomy*, 11(1), 8.
Nath, C. P., et al. (2021), 'Long-term impact of legume-based cropping with chemical and integrated fertilisation on viable weed seed density, diversity and community structure', *Weed Res.*, 61(5), 360–374.
Nelson, C. J. (1996), 'Allelopathy in cropping systems', *Agron. J.*, 88(6), 991–996.
Neve, P., et al. (2014) 'Expanding the ecoevolutionary context of herbicide resistance research', *Pest Manage. Sci.*, 70(9), 1385–1393.
Ngwira, A., et al. (2014), 'Adoption and extent of conservation agriculture practices among smallholder farmers in Malawi', *J. Soil Water Conserv.*, 69(4), 107–119.

© Burleigh Dodds Science Publishing Limited, 2025. All rights reserved.

Nichols, V., et al. (2015), 'Weed dynamics and conservation agriculture principles: a review, *Field Crops Res.*, 183(1), 56–68.

Nikolic, L., et al. (2018), 'Weed infestation and biodiversity of winter wheat under the effect of long-term crop rotation', *Appl. Ecol. Environ. Res.*, 16(2), 1413–1426.

Nikolić, N., et al. (2020), 'Weed seed decay in no-till field and planted riparian buffer zone', *Plants*, 9(3), 293.

Norton, R. M. (2020), 'Challenges and opportunities in fertilizer placement in no-till farming systems'. In Dang, Y., Dalal, R. and Menzies, N., (Eds.), *No-Till Farming Systems for Sustainable Agriculture*. Cham: Springer, pp. 65–81.

Oerke, E. C. (2006), 'Crop losses to pests', *J. Agric. Sci.*, 144(1), 31–43.

Olofsdotter, M. (2001), 'Getting closer to breeding for competitive ability and the role of allelopathy – an example from rice (*Oryza sativa*)', *Weed Technol.*, 15, 798–806.

O'Rourke, M. E., et al. (2006), 'Post-dispersal weed seed predation by invertebrates in conventional and low-external-input crop rotation systems', *Agric. Ecosyst. Environ.* 116(3), 280–288.

Owen, M. (2008), 'Weed species shifts in glyphosate-resistant crops', *Pest Manag. Sci.*, 64(4), 377–387.

Paes, J. M. V. and Rezende, A. M. (2001), 'Manejo de plantas daninhas no sistema plantio direto na palha', *Informativo Agropecuario*, 22, 37–42.

Pareja, N. M. R., et al. (1985), 'Distribution of weed seed among soil structural units', *Weed Sci.*, 33(2), 182–189.

Parr, F. J., et al. (1992), 'Soil quality: attributes and relationship to alternative and sustainable agriculture', *Am. J. Altern. Agric.*, 7(1–2), 5–11.

Peigné, J., et al. (2007), 'Is conservation tillage suitable for organic farming? A review', *Soil Use Manage.*, 23(2), 129–144.

Peixoto, D. S., et al. (2020), 'Occasional tillage in no-tillage systems: a global meta-analysis', *Sci. Total Environ.*, 745, 140887.

Place, G., et al. (2008), 'Root penetration through a high bulk density soil layer: differential response of a crop and weed species', *Plant Soil*, 307(1), 179–190.

Plaza, E. H., et al. (2010), 'Tillage system did not affect weed diversity in a 23-year experiment in Mediterranean dryland', *Agric. Ecosyst. Environ.*, 140(1–2), 102–105.

Plaza, E. H., et al. (2015), 'Intensity of soil disturbance shapes response trait diversity of weed communities: the long-term effects of different tillage systems', *Agric. Ecosyst. Environ.*, 207(1), 101–108.

Rai, V., et al. (2018), 'Effect of conservation agriculture on soil physical health', *Int. J. Curr. Microbiol. Appl. Sci.* 7(2), 373–389.

Recasens, J., et al. (2023), 'Dynamics of *Cynodon dactylon* and weed community composition in different cover crops in a vineyard', *Weed Res.*, 63(4), 261–269.

Refsell, D. E. and Hartzler, R. G. (2009), 'Effect of tillage on common waterhemp (*Amaranthus rudis*) emergence and vertical distribution of seed in the soil', *Weed Technol.*, 23(1), 129–133.

Reintam, E. and Kuht, J. (2012), 'Weed responses to soil compaction and crop management'. In Price, A.J., (Ed.), *Weed Control*. Rijeka, Croatia: InTech, pp. 243–264.

Reynolds, H. L., et al. (2003), 'Grassroots ecology: plant-microbe-soil interactions as drivers of plant community structure and dynamics', *Ecol.*, 84(9), 2281–2291.

Riaz Marral, M. W., et al. (2020), 'The influence of transgenic (Bt) and non-transgenic (non-Bt) cotton mulches on weed dynamics, soil properties and productivity of different winter crops', *PLoS ONE*, 15(9), e0238716.

© Burleigh Dodds Science Publishing Limited, 2025. All rights reserved.

Roberts, H. A. (1981), 'Seed banks in soils', *Adv. Appl. Biol.* 6, 1–55.

Roller, A. and Albrecht, H. (2006), 'Effects of tillage systems on the seed bank persistence and seedling emergence of ten arable weeds', *J. Plant Diseases Prot.,* 11, 257–265.

Rowell, M. J. and Florence, L. Z. (1993), 'Characteristics associated with differences between undisturbed and industrially-disturbed soils', *Soil Biol. Biochem.*, 25(11), 1499–1511.

Ruisi, P., et al. (2015), 'Weed seedbank size and composition in a long-term tillage and crop sequence experiment', *Weed Res.*, 55(3), 320–328.

Ryan, M. R., et al. (2010), 'Elucidating the apparent maize tolerance to weed competition in long-term organically managed systems', *Weed Res.*, 50(1), 25–36.

Saini, M., et al. (2008), 'Planting and termination dates affect winter cover crop biomass in a conservation-tillage corn-cotton rotation: Implications for weed control and yield', In Endale, D. M., (Ed.), *Southern Conservation Agricultural Systems.* published by Sustainable Agriculture Research and Education (SARE), Tifton, GA, pp. 137–141.

Santín-Montanyá, M. I., et al. (2016), 'Management of weed emergence and weed seed bank in response to different tillage, cropping systems and selected soil properties', *Soil Tillage Res.*, 161, 38–46.

Scherner, A., et al. (2017), 'Reducing tillage intensity affects the cumulative emergence dynamics of annual grass weeds in winter cereals', *Weed Res.*, 57(5), 314–322.

Schwartz-Lazaro, L. M. and Cope, J. T. (2019), 'A Review of the soil seedbank from a weed scientist perspective', *Agronomy*, 9(7), 369.

Server, M. (2021), 'Strategic tillage has its place in no-till agriculture', *Crops and Soils Magazine*, November-December issue. DOI: 10.1002/crso.20150.

Shahzad, M., et al. (2016a), 'Weed spectrum in different wheat-based cropping systems under conservation and conventional tillage practices in Punjab, Pakistan', *Soil Tillage Res.*, 163, 71–79.

Shahzad, M., et al. (2016b), 'Impact of different crop rotations and tillage systems on weed infestation and productivity of bread wheat', *Crop Prot.*, 89, 161–169.

Sharma, T., et al. (2023), 'Long-term conservation agriculture influences weed diversity, water productivity, grain yield, and energy budgeting of wheat in North-Western Indo-Gangetic Plains', *Sustainability*, 15(9), 7290.

Shearin, A., et al. (2007), 'Direct effects of tillage on the activity and density of ground beetle (Coleoptera: Carabidae) weed seed predators', *Environ. Entomol.*, 36(5), 1140–1146.

Shekoofa, A., et al. (2020), 'Environmental effects on efficacy of herbicides for postemergence goosegrass (*Eleusine indica*) control', *Sci. Rep.*, 10(1), 20579.

Shrestha, A., et al. (2002), 'Effect of tillage, cover crop and crop rotation on the composition of weed flora in a sandy soil. *Weed Res.*, 42(1), 76–87.

Shrestha, A., et al. (2015), 'Weed seedbank characterization in long-term cotton–tomato rotations in California', *Agron. J.*, 107(2), 597–604.

Şimon, A., et al. (2019) 'Influence of tillage systems on the degree of weeding on agricultural crops at ARDS Turda, *Agriculture,* 76(2), 99–106.

Singh, B., et al. (2014), 'Relative abundance and damage of some insect pests of wheat under different tillage practices in rice–wheat cropping in India', *Crop Prot.*, 61, 16–22.

Singh, V. P., et al. (2015), *Weed Management in Conservation Agriculture System.* Jabalpur, India: ICAR - Directorate of Weed Research.

© Burleigh Dodds Science Publishing Limited, 2025. All rights reserved.

Smith, R. G., et al. (2010), 'A new hypothesis for the functional role of diversity in mediating resource pools and weed–crop competition in agroecosystems', *Weed Res.*, 50(1), 37–48.
Sorte, C. J., et al. (2013), 'Poised to prosper? A cross-system comparison of climate change effects on native and non-native species performance'. *Ecol. Lett.*, 16, 261–270.
Sosnoskie, L. M., et al. (2006), 'Weed seedbank community composition in a 35-yr-old tillage and rotation experiment', *Weed Sci.* 54(2), 263–273.
Souza, A. S., et al. (2019), 'Interference of soil compaction on diclosulam residual', *Rev. Bras. Herbic.*, 18(4), 687.
Srivastava, R. (2012), 'Diversity of weed soil seed bank in Indian dryland and irrigated agroecosystems. *Indian J. Fundam. Appl. Life Sci.*, 2, 30–37.
Steponaviciene, V., et al. (2021), 'The effect of different soil tillage systems and crop residues on the composition of weed communities', *Agronomy*, 11, 1276.
Stevenson, F. C., et al. (1997), 'Weed species diversity in spring barley varies with crop rotation and tillage, but not with nutrient source'. *Weed Sci.*, 45, 798–806.
Stewart, C. L., et al. (2010), 'Environment and soil conditions influence pre- and postemergence herbicide efficacy in soyabean', *Weed Technol.*, 24(3), 234–243.
Storkey, J. and Neve, P. (2018), 'What good is weed diversity?', *Weed Res.*, 58(4), 239–243.
Storkey, J. and Westbury, D. B. (2007), 'Managing arable weeds for biodiversity', *Pest. Manage. Sci.*, 63(6), 517-523.
Susha, V. S., et al. (2018), 'Impacts of tillage and herbicide mixture on weed interference, agronomic productivity and profitability of a maize–wheat system in the North-western Indo-Gangetic Plains', *Field Crop. Res.*, 219(10), 180–191.
Swanton, C., et al. (1993), 'Weed succession under conservation tillage: A hierarchical framework for research and management', *Weed Technol.*, 7(2), 286-297.
Swanton, C. J., et al. (1999), 'Effect of tillage systems, N, and cover crop on the composition of weed flora', *Weed Sci.*, 47(4), 454–461.
Swanton, C. J., et al. (2000), 'Influence of tillage type on vertical weed seedbank distribution in a sandy soil', *Can. J. Plant Sci.*, 80(2), 455–457.
Taa, A., et al. (2004), 'Effects of stubble management, tillage andcropping sequence on wheat production in the south-eastern highlands of Ethiopia', *Soil Tillage Res.*, 76(1), 69–82.
Teasdale, J. R. (2018), 'The use of rotations and cover crops to manage weeds'. In Zimdahl, R., (Ed.), *Integrated Weed Management for Sustainable Agriculture*. Cambridge: Burleigh Dodds Science, pp. 227–260.
Teasdale, J. R. and Mohler C. L. (1993), 'Light transmittance, soil temperature and soil moisture under residue of hairy vetch and rye', *Agron J.*, 85(3), 673–680.
Thomas, A. G., et al. (2017), 'A multistudy approach to understanding weed population shifts in medium- to long-term tillage systems', *Weed Sci.*, 52(5), 874–880.
Tillmann, M., et al. (2017), 'Crop rotation effects on incidence and diversity of Fusarium species colonizing stem bases and grains of winter wheat', *J. Plant Diseases Prot.*, 124(2), 121–130.
Tørresen, K. S., et al. (2003), 'Long-term experiments with reduced tillage in spring cereals. I. Effects on weed flora, weed seedbank and grain yield', *Crop Prot.*, 22(1), 185–200.
Tourdonnet, S. and Nozières, A. (2007), 'Conservation agriculture, organic farming and GM crops in France. Comprehensive inventory and assessment of existing knowledge on sustainable agriculture in the European platform of KASSA', Deliverable1.1–Appendix A1. Published by CIRAD, Paris.

© Burleigh Dodds Science Publishing Limited, 2025. All rights reserved.

Travlos, I. S., et al. (2018), 'Weed-species abundance and diversity indices in relation to tillage systems and fertilization', *Front. Environ. Sci.*, 6,11.

Trichard, A., et al. (2013), 'Identification of weed community traits response to conservation agriculture', *Agric. Ecosyst. Environ.*, 179, 179–186.

Tuesca, D., et al. (2001), 'A long-term study of weed flora shifts in different tillage systems', *Weed Res.*, 41(4), 369–382.

Tuti, M. D. and Das, T. K. (2011), 'Sequential application of metribuzin on weed control, growth and yield of soyabean (*Glycine max*),' *Indian J. Agron.*, 56, 57–61.

Uddin, M. N., et al. (2020), 'Nitrogen immobilization may reduce invasibility of nutrient enriched plant community invaded by *Phragmites australis*', *Sci. Rep.*, 10, 1601.

Unger, P. W., et al. (1999), 'Weed seeds in long-term dryland tillage and cropping system plots', *Weed Res.*, 39(3), 213–223.

Valaie, N., et al. (2022), 'Weed seed bank and seedling dynamics in response to crop rotation in conventional agroecosystems', *Iran. J. Plant Prot. Res.*, 36(2), 239–258.

Vargas, L. A., et al. (2017), 'Soil seed bank phytosociology in no-tillage systems in the southwestern Amazon region', *Am. J. Plant Sci.*, 8(13), 3399–3413.

Vasileiadis, V. P., et al. (2007), 'Vertical distribution, size and composition of the weed seedbank under various tillage and herbicide treatments in a sequence of industrial crops', *Weed Res.*, 47(3), 222 –230.

Vidal, R. A. and Bauman, T. T. (1996), 'Surface wheat (*Triticum aestivum*) residues, giant foxtail (*Setaria faberi*), and soybean (*Glycine max*) yield', *Weed Sci.*, 44(4), 939–943.

Vijayalaxmi, G. S., et al. (2012), 'Sequential application of pre- and post-emergence herbicides in soyabean (*Glycine max* L.)', *Karnataka J. Agric. Sci.*, 25, 262–263.

Wei, S., et al. (2010), 'Rapid and effective methods for breaking seed dormancy in buffalobur (*Solanum rostratum*)', *Weed Sci.*, 58(2), 141–146.

Weisberger, D., et al. (2023), 'Do cover crops suppress weeds in the U.S. Southeast? A meta-analysis', *Weed Sci.*, 71(3), 244–254.

Weisberger, D., et al. (2019), 'Does diversifying crop rotations suppress weeds? A meta-analysis', *PLoS ONE*, 14(7), e0219847.

Wells, M. S., et al. (2013), 'The reduction of plant-available nitrogen by cover crop mulches and subsequent effects on soybean performance and weed interference', *Agron. J.*, 105(2), 539–545

Westerman, P., et al. (2011), 'Weed seed predation in a phenologically late crop', *Weed Res.*, 51(2), 157–164.

White, S. S., et al. (2007), 'Feeding preferences of weed seed predators and effect on weed emergence', *Weed Sci.* 55(6), 606–612.

Wicks, G. A., et al. (1994), 'Influence of wheat (*Triticum aestivum*) straw mulch and metolachlor on corn (*Zea mays*) growth and yield', *Weed Sci.*, 42(1), 141–147.

Williams, A., et al. (2018), 'Establishing the relationship of soil nitrogen immobilization to cereal rye residues in a mulched system', *Plant Soil*, 426(1–2), 95–107.

Woźniak, A. (2018), 'Effect of tillage system on the structure of weed infestation of winter wheat', *Span. J. Agric. Res.*, 16(4), e1009.

Woźniak, A. (2020), 'Effect of cereal monoculture and tillage systems on grain yield and weed infestation of winter durum wheat', *Int. J. Plant Prod.*, 14(1), 1–8.

Woźniak, A. and Soroka, M. (2015) 'Biodiversity of weeds in pea cultivated in various tillage system', *Roman. Agric. Res.*, 32, 231–237.

© Burleigh Dodds Science Publishing Limited, 2025. All rights reserved.

Woźniak, A. and Soroka, M. (2018) 'Effect of crop rotation and tillage system on the weed infestation and yield of spring wheat and on soil properties' *Appl. Ecol. Env. Res.,*16(3), 3087–3096.

Woźniak, A. and Stępniowska, A. (2017), 'Yield and quality of durum wheat grain in different tillage systems', *J. Elem.,* 22(3), 817–829.

Yasin, M., et al. (2017), 'The effect of reduced light intensity on grass weeds', *Weed Sci.*, 65(5), 603–613.

Yenish, J. P., et al. (1992), 'Effects of tillage on vertical distribution and viability of weed seed in soil', *Weed Sci.*, 40(3), 429–433.

Yu, Q., et al. (2020), 'A suitable rotational conservation tillage system ameliorates soil physical properties and wheat yield: an 11-year in-situ study in a semi-arid agroecosystem', *Soil Tillage Res.*, 199(1), 104600.

Zanin, G., et al. (1997), 'Ecological interpretation of weed flora dynamics under different tillage systems', *Agric. Ecosyst. Environ.,* 66, 177–188.

Zhang, J. and Wu, L. F. (2021), 'Impact of tillage and crop residue management on the weed community and wheat yield in a wheat–maize double cropping system', *Agriculture*, 11(3), 265.

Zhang, J., et al. (2021), 'Weed responses to crop residues management in a summer maize cropland in the North China Plain', *Agriculture*, 11(8), 746.

© Burleigh Dodds Science Publishing Limited, 2025. All rights reserved.

Chapter 2

Modeling weed population dynamics in conservation agriculture systems

Jose L. Gonzalez-Andujar, Instituto de Agricultura Sostenible (CSIC), Spain; and Cesar Fernandez-Quintanilla, Instituto de Ciencia Agrarias (CSIC), Spain

1 Introduction

Conservation agriculture (CA) includes three pillars: minimal soil disturbance, permanent soil cover, and crop rotation (Hobbs, 2007; FAO, 2015). Although benefits are context specific, CA has been identified as an effective tool for sustainability, increasing yields, and reducing environmental impacts in many parts of the world (Hobbs, 2007; Giller at al., 2015; Pittelkow et al., 2015). Minimal soil disturbance has repercussions on the provision of ecosystem services, including climate regulation through carbon storage and greenhouse gas emissions, as well as regulation and supply of water through changes in soil's physical, chemical, and biological attributes. Furthermore, it can impact the underlying biodiversity that is crucial for supporting various ecosystem services (Palm et al., 2014). Choudary et al. (2020) reported that CA was a useful strategy for climate change mitigation and adaptation. Nyanga et al. (2020) have shown that CA reinforces ecosystem services through a number of pathways. The practice of minimum tillage reinforces various regulatory services such as carbon sequestration, climate regulation, and control of soil erosion. Crop rotation, minimum soil disturbance, and plant residue retention support services through nutrient recycling and soil formation habitat provision (Nichols et al., 2015).

http://dx.doi.org/10.19103/AS.2024.0145.02
© Burleigh Dodds Science Publishing Limited, 2025. All rights reserved.

The systemic approach of simultaneously implementing all three pillars in CA generates ecological habitats that are different from those generated by conventional systems. Although many studies have investigated tillage intensity, cover cropping, and crop rotation diversification individually, very few studies have investigated the effect of the joint implementation of all three pillars (Cordeau, 2022). In spite of this limitation, available literature shows that habitat changes generated by implementing CA practices have a direct and indirect influence on the population dynamics of weeds.

Numerous studies have proved that the suppression of tillage associated to CA systems results in substantial changes in the weed communities present in different crops (Froud-Williams et al., 1983; Koskinen and McWhorter, 1986; Thomas and Frick, 1993; de la Fuente et al., 2006; Nichols et al., 2015). Implementing no-till without crop rotation frequently results in severe weed problems, particularly with perennials, grasses, wind-borne species, and volunteer crops. However, greater rotational crop diversity results in easier weed management (Nichols et al., 2015).

The implementation of CA results in significant changes in weed population dynamics by influencing seed distribution, persistence, germination, and mortality rates. No-till systems accumulate seeds near the soil surface where they are more likely to germinate. However, they are also exposed to greater mortality risks through weather variability and predation. Assuming no seed input in the system, germinable seed banks under no-till decrease more rapidly than under conventional tillage.

Understanding these dynamics is essential for devising effective weed management strategies tailored to CA systems, thereby ensuring sustainable and productive agricultural practices. Weed population dynamic modeling helps inform decision-making in conservation agriculture by evaluating management scenarios, predicting long-term outcomes, and cost-benefit analysis.

2 Weed population dynamics

A population is a group of individuals of the same species that interchange genetic material and live in a specific site and time. The population's individuals exhibit ecological equivalence, which is characterized by identical life cycles and equivalent ecological processes. The rates of these processes or the likelihood of ecological events are largely uniform when individuals inhabit the same environment, although some individual variability may occur.

Although a population has a common genetic background, its composition is continually changing as a result of gene interchanges among individuals. This genetic background and these interchanges are the raw materials used by natural selection to drive population evolution (Solbrig, 1980).

© Burleigh Dodds Science Publishing Limited, 2025. All rights reserved.

A population interacts with the environment present in a given site. The site can be defined according to different spatial scales: a field plot, a landscape, or a geographic region. Although a particular farmer may be focused on the populations present in a given plot or in his whole farm, dispersal processes that take place at landscape and regional scales will affect his local populations.

All the previous facts indicate that weed populations are changing continuously. Population dynamics in a farm setting encompasses the study of various changes, including alterations in genetic structure, abundance, and distribution of weeds. Understanding and forecasting these changes have clear practical implications.

2.1 Population processes

Populations are formed by individuals in different functional stages (seeds, seedlings, adult plants, etc.), phenological stages, ages, sizes, and genotypes. Consequently, the starting point to study the population dynamics of any species is to establish the population structure. In other words, we need to know the number of individuals present in different states at a given time and site (Cousens and Mortimer, 1995).

The abundance and persistence of an annual weed in a given site is determined by six processes that act as valves that filter the number of individuals that move from one stage to the next (Fig. 1): (1) seedling recruitment; (2) seedling mortality; (3) plant growth and reproduction; (4) seed predation (pre- and post-dispersal); (5) seed dispersal (immigration and emigration); (6) seed mortality in the soil.

For annual weeds, the only functional stage present during all of the year is the seed bank. This component represents, at the same time, the historic memory of weed infestations in the past and the cause of future infestations. Starting from this seed bank, different cohorts (individuals recruited at the same time) of seedlings are produced. The survivorship, growth, and reproduction of each of these cohorts is different (Fernandez-Quintanilla et al., 1986). Although most of the new seeds produced usually fall close to the mother plant (the seed rain), a proportion may be dispersed by natural agents (wind, animals) or by human factors (tillage, harvesters). The actual relevance of the different dispersal factors is variable depending on the species. Barroso et al. (2006), working with *Avena fatua* and *Avena sterilis* (wild oats) in winter cereals, found that natural dispersal under no-tillage was very limited, without mechanical aid seed dispersal is concentrated around the mother plant. However, ploughing resulted in an annual patch displacement of 2–3 m in the tillage direction. Since wild oat seed rain took place mainly before harvest, combine harvesters dispersed a few seeds. In contrast, Ballare et al. (1987) showed that dispersal by combine harvesters was the major demographic factor explaining the

© Burleigh Dodds Science Publishing Limited, 2025. All rights reserved.

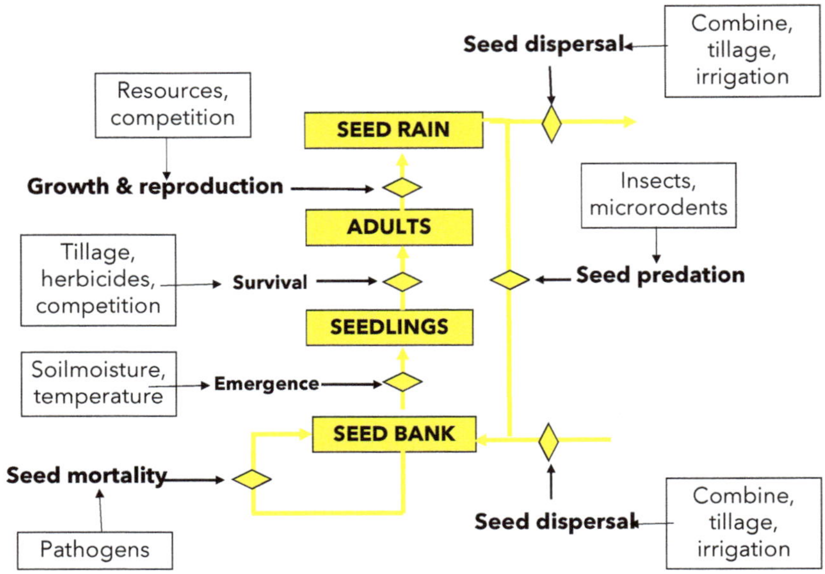

Figure 1 Flowchart of the life cycle of an annual weed, considering the main functional stages, demographic processes, and external factors.

population growth of *Datura ferox* (jimsonweed) in soybean crops. The seeds remaining on the soil surface are the target of numerous predators (insects, birds, micro-rodents). Post-dispersal weed seed predation reduces the number of weed seeds entering the seed bank after being dispersed from the mother plant (Baraibar et al., 2011; Westerman et al., 2012).

In the case of perennial weeds, the situation is more complex. Due to their multiple forms of reproduction and dispersal, it is possible to have, simultaneously, individuals in different stages and with different ages (Zambrano-Navea et al., 2016). Age-dependence of the demographic rates, survival, fecundity, and individual growth is a fundamental aspect of the population biology of perennial plants (Baden et al., 2021).

2.2 Factors regulating population growth

An important debate in ecological theory has been about the effect of endogenous processes and exogenous variables on the temporal changes in populations. Currently, there is a consensus that both endogenous (density-dependent) and exogenous (density-independent) factors contribute to population dynamics (Ziebarth et al., 2010). Endogenous factors include processes such as intra-specific competition or delayed competition (Gonzalez-Andujar et al., 2006; Hernandez-Plaza et al., 2012), whereas exogenous processes include changes such as weather or nutrient flows (Molofsky et al., 2014).

© Burleigh Dodds Science Publishing Limited, 2025. All rights reserved.

2.2.1 Endogenous factors

Endogenous regulation of population refers to mechanisms and processes within a population that naturally control its growth, maintaining a balance between births and deaths. A key concept to understand endogenous regulation is the carrying capacity of a site. Carrying capacity refers to the maximum population size that an environment can sustain over the long term without causing significant detrimental effects. As population size increases, resources become scarcer, leading to increased intraspecific competition for food, water, and other essential resources. This competition can result in decreased birth rates, increased mortality rates, or both, providing negative feedback and restoring balance to the population.

2.2.2 Exogenous factors

Exogenous factors refer to external influences that regulate populations, impacting their growth, decline, and overall dynamics. There are three major types of exogenous factors regulating population growth: climatic, biotic, and management. Since these three factors act together, it is difficult to isolate the individual effects.

Climatic factors. Soil temperature and moisture regulate the dormancy state of the seeds present in the soil as well as their germination. The requirements for breakage of dormancy and germination are very variable depending on the species (Benech-Arnold et al., 2000). Heavy frosts may have a clear effect on the survivorship of young seedlings that emerged in late autumn and winter. Severe droughts may have the same effect on young seedlings, reducing drastically growth and reproduction (Debaeke, 1988).

Biotic factors. Within the agroecosystems, weeds interact with numerous organisms. Some of these interactions are positive for the weeds (e.g. pests that weaken the competing crops) but some are negative. There are many examples of insects and microrodents that feed on weed seeds (Baraibar et al., 2011) and bacteria, fungi, virus, and nematodes that attack weed plants (Gomez et al., 2013).

Management factors. Crop rotations are probably the most determining factor. Rotation design, crop sequencing, tillage practices, crop residue management, and competitive crop canopies are important aspects to consider in this regard (Anderson, 2005). Since each crop has a specific planting and harvesting period, the synchronicity of these periods with the life cycle period of a weed species determines its probability of success. This explains the convenience of rotating winter and summer crops. For suppression of creeping perennial weed species, it may be convenient including a perennial crop such as alfalfa (Gramig, 2024)

© Burleigh Dodds Science Publishing Limited, 2025. All rights reserved.

Tillage practices also have a major influence on weed populations. Weed population composition and dynamics in CA environments are likely to be different compared with traditional cultivated environments. Minimum tillage may have contrasting effects on weed populations. Usually, tillage places seeds in a "safer" environment, the soil matrix, compared with the soil surface (Mohler et al., 2012). In this sense, tillage can contribute to the buildup of the weed seed bank (Buhler et al., 1997). However, tillage can also be used as a weed management strategy if seeds are buried to a depth from which they cannot successfully emerge (fatal germination) (Chauhan et al., 2006) or if they are attacked by soil microorganisms (seed decay) (Gomez et al., 2013). In semi-arid systems, populations of *Messor barbarus* (harvester ant), an important seed predator, are particularly large in no-till and minimally tilled fields (Baraibar et al., 2011).

Herbicides have different impacts on demographic processes depending on the weed growth stage and application time. Pre-emergence herbicides, generally applied at planting time, inhibit seedling emergence. Early treatments with post-emergence herbicides increase drastically seedling mortality whereas late treatments stop growth and reduce or inhibit reproduction of survivors (Fernandez-Qunitanilla et al. 1987).

3 Modeling weed population dynamics

A model is a simplified representation of reality. Models in ecology aim to project and understand the functioning of a population. How the number of individuals evolves over time and space and how it is affected by different types of inferences. The construction of weed population models has emerged as a valuable tool for integrating the available information on weed growth, evaluating weed control scenarios, and identifying critical research gaps (Holst et al., 2007; Somerville et al., 2020; Bagavathiannan et al., 2020).

Ecological models are specifically geared toward unraveling the intricacies of population dynamics. They delve into how populations evolve over time and space, elucidating the impacts of various factors. Weed population modeling has proven to be a valuable tool for synthesizing existing knowledge on weed growth, evaluating weed control strategies, and identifying research gaps. (Holst et al., 2007; Somerville et al., 2020; Bagavathiannan et al., 2020).

The roots of modeling weed population dynamics can be traced back to the 1970s (Sagar and Mortimer, 1976; Mortimer et al., 1980). These pioneering efforts entailed relatively basic demographic models delineating the different phases of weed life cycles. Over time, subsequent studies enriched these models with added intricacies, encompassing elements such as density-dependent mortality, seed production (Watkinson, 1980), vertical seed bank

© Burleigh Dodds Science Publishing Limited, 2025. All rights reserved.

distribution (Cussans and Moss, 1982; Cousens and Moss, 1990), buried seed age (Wilson et al., 1984), seed dispersal (Ballare et al., 1987), and seedling emergence influenced by environmental variables (Alm et al., 1993), among others. The latter part of the 20th century witnessed notable strides in weed population modeling, fueled by advancements in computer technology, statistics, and data accessibility. Researchers began infusing evolutionary biology, genetics, and spatial ecology principles into their models, striving for a more comprehensive understanding of weed population dynamics.

Based on their treatment of spatial aspects population dynamics models can be categorized into two groups: those that consider spatial locations of populations or individuals and those that do not.

3.1 Non-spatial models

These models solely consider the temporal evolution of populations. Their simplest description is through the following equation:

$$N_{t+1} = \lambda N_t \tag{1}$$

where the future population N_{t+1} is determined by a previous population N_t and λ is the geometric growth factor or the per capita change in population size over a discrete interval. This form of growth, characterized by a constant growth rate, is typical in the early stages of development for many populations and can be represented by exponential curves. The value of λ dictates the population's trajectory. If $\lambda > 1$, the population will increase, if $\lambda < 1$, the population will decrease, and if $\lambda = 1$, the population will remain stable (Fig. 1).

The growth factor λ encompasses the net result of seed production and various mortality factors acting at different stages throughout the plant cycle. Equation (1) has been expanded to acknowledge the variability within populations. Rather than assuming all individuals are identical, as described in Equation (1), it should be considered that individuals may differ in survival, growth, competition, and fecundity. To accommodate these differences and their impact on population growth, the life cycle of organisms has been partitioned into discrete intervals, each representing a cohort or functional state (Fig. 1). These cohorts are approximately the same age and exhibit similar life expectancies and fecundity rates.

With this in mind, λ can be decomposed into different factors, as shown in Equation (2):

$$N_{t+1} = (\lambda f \lambda g \lambda s ....) N_t \tag{2}$$

where N represents adult plants per square meter, λ_f refers to fecundity (seeds/plant), λ_g is the percentage of seeds that germinate, and λ_s denotes the

© Burleigh Dodds Science Publishing Limited, 2025. All rights reserved.

proportion of seedlings that survive to become adult plants. Additional factors, such as losses due to predation or emigration, can also be incorporated.

A seminal work by Cussans and Moss (1982) illustrates the traditional approach to integrating these factors into demographic models (Fig. 1). These two researchers modified the basic demographic model (Figs. 1 and 2) to include the effects of tillage practices, straw management, and herbicide application. In Fig. 3, we can see how, in the specific case of *Alopecurus myosuroides* (blackgrass) in a winter wheat crop, the proportion of seeds lost in the stubble is dependent, initially, on the tillage system used. With direct drilling, where seeds have maximum exposure on bare stubble, only 30% of the seeds were assumed to survive. In contrast, with ploughing a seed survivorship value of 75% was considered. Since straw burning may modify these figures, a second factor was included in the model. In the case straw is burned, a survivorship factor of 67% (33% of the seeds are assumed to be destroyed by fire) must be multiplied by the previous factor. They also introduced a mortality factor to represent weed control by the herbicide. By iterative simulations with this model, they estimated the level of herbicide performance required to just prevent populations from increasing ($\lambda = 1$). Their results showed that this level varies widely (from 50% to 92%) depending on the type of tillage system used and the management of the straw of the previous crop. Obviously, this behavior is species specific. In the case of *Avena fatua* (wild oat), the same model predicted little differences between the percentage control by herbicide required for population stability in ploughed and minimum tilled fields.

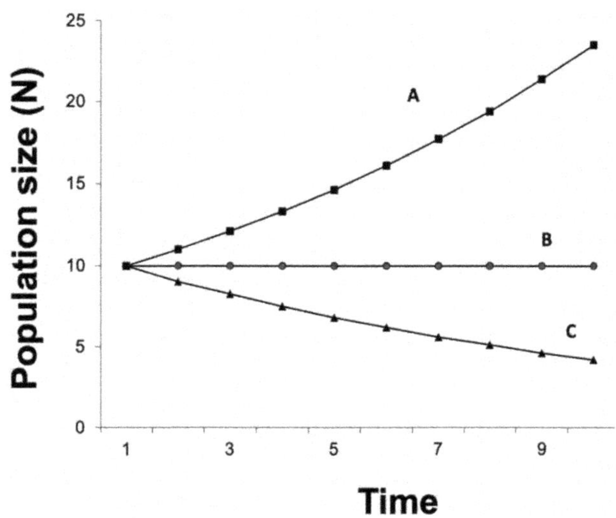

Figure 2 Population growth (N) as a function of the value of λ. A: $\lambda > 1$, B: $\lambda = 1$; C: $\lambda < 1$.

© Burleigh Dodds Science Publishing Limited, 2025. All rights reserved.

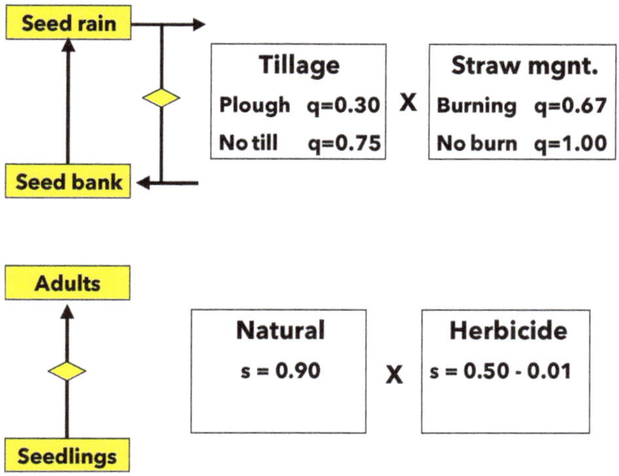

Figure 3 Integration of the effects of tillage, straw management, and herbicide application in a population dynamics model for *Alopecurus myosuroides,* s and q are seedling and seebank survivorship rates, respectively.

Applying a similar methodology, Wilson et al. (1984) simulated the effect of different cultivation methods on *Avena fatua* populations with the aim of reducing herbicide application. Murdoch and Roberts (1982) compared two strategies: reducing *A. fatua to* a specific density that does not compete with the crop, versus complete eradication. Additionally, Gonzalez-Andujar et al. (2005) determined the growth rate of *Phalaris brachystachys* (canary grass) in winter wheat without any form of control. Their findings revealed a remarkably high growth rate (λ = 75.96), indicating the species' significant potential for population expansion in the absence of control measures.

3.1.1 Density-dependent models

Unlimited population growth is a theoretical construct. In reality, populations grow until one or more limiting resources become scarce enough to inhibit reproduction, leading to a plateau in population growth. These limiting resources may include light, water, prey, nutrients, or other factors affecting intraspecific competition. Population growth is thus regulated by its density, with every population eventually reaching its carrying capacity–the maximum number of individuals an environment can sustain. To consider the effect of the density-dependent regulation, Equation (1) can be transformed as follows:

$$N_{t+1} = \lambda * N_t * f(N_t) \tag{3}$$

© Burleigh Dodds Science Publishing Limited, 2025. All rights reserved.

where $f(N_t)$ represents the density-dependent relationship. The shape of this function is determined by plotting N_t against $N_{t+1.}$ Various forms have been proposed for $f(N_t)$, with one of the most common being the hyperbolic model:

$$f(N_t) = 1/(1 + a * N_t) \tag{4}$$

where *a* model the per capita effects of intraspecific. Combining Equations (6) and (7), the general density-dependence model becomes:

$$N_{t+1} = (\lambda * N_t)/(1 + a * N_t) \tag{5}$$

This model has been extended to consider the seed bank.

$$N_{t+1} = (\lambda * N_t)/(1 + a * N_t) + sN_t \tag{6}$$

where *s* represents survival in the seed bank. Competition with other species has also been modeled (Firbank and Watkinson, 1986).

$$N_{i't+1} = \lambda_i N_{i,t/}(1 + \Sigma\alpha_{ij} N_j) \tag{7}$$

Here, the parameters include the finite rate of increase, λ, defined as the maximal mean rate of population increase from low densities, α, which models the per capita effects of intraspecific (α_{ii}) and interspecific (α_{ij}) competitors. This formulation can readily be extended to include a seed bank (Freckleton et al., 2000)

Population models of this type have been applied to various weed species, including *Alopecurus myosuroides* (Doyle et al., 1986), *Anisantha sterilis* (Smith et al., 1999), *Avena fatua* (Jones and Medd, 1997), *Chenopodium album* (Freckleton and Watkinson, 1998), *Vulpia bromoides* (Freckleton et al., 2000), as well as others summarized by Gonzalez-Andujar et al. (2023).

While the previous models subsume the demographic processes that occur within generations and that may interact to determine population dynamics as a whole, life-cycle models (or multi-state models) offer more detailed insights by dividing the life cycle into different functional states within a generation (Fig. 1). Life-cycle models have been widely used to evaluate alternative weed management strategies (Fig. 4). Mokhtassi-Bidgoli et al. (2013) developed a model to evaluate seven integrated weed management (IWM) programs targeting *Bromus diandrus* (great brome) within a winter barley no-tillage system. Their findings underscored the efficacy of an IWM approach blending cultural and biological controls, showcasing a remarkable 56% reduction in *B. diandrus* seed bank over a span of seven years. Similarly, Torra et al. (2008) conducted a modeling process focusing on the population dynamics of *Papaver rhoeas* (poppy), employing the model to appraise various IWM programs. Among the ten evaluated programs, four demonstrated notable success in curtailing *P. rhoeas* seed bank densities. Simulation outcomes indicated that delayed seeding, fallow periods, and pre-emergence herbicides

© Burleigh Dodds Science Publishing Limited, 2025. All rights reserved.

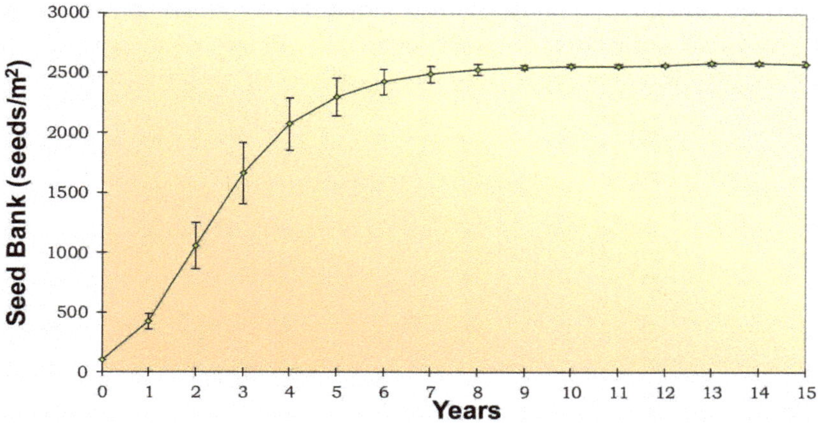

Figure 4 Temporal evolution of *Avena sterilis* in winter wheat without control measures (modified from Gonzalez-Andujar and Bastida, 2018).

stood out as optimal techniques within IWM programs for *P. rhoeas* control. The FLORSYS model predicts changes in weed community composition influenced by organic management and different crop diversification or minimum tillage methods. It provides multiple outputs, including crop yield and soil compaction, making it a valuable tool for designing multi-objective cropping systems, especially in contexts involving minimum tillage (Colbach et al., 2017).

3.1.2 Matrix population models

Matrix population models provide a concise, analytically tractable, and elegant approach for simulating the growth of age- or stage-structured populations over discrete time steps (Caswell, 2001). Despite their widespread use in basic population ecology and conservation biology, the application of matrix models to weed management questions has been relatively limited (Mortimer et al., 1980; Gonzalez-Andujar et al., 1986; Jordan et al., 1995; Mertens et al., 2002; Davis et al., 2004). One of the key advantages of employing matrix simulation models in ecological weed management is their capacity to integrate the effects of multiple management tactics on various life stage transitions. This integration allows for the observation of weed population growth as an emergent property of management practices. By assessing how different management strategies influence the survival, growth, and reproduction of weeds at different life stages, matrix models offer a powerful tool for predicting and optimizing weed control efforts.

A matrix model is represented by a column vector ($\mathbf{N}_t$), showing the number in each category. The various survival probabilities and fecundities

© Burleigh Dodds Science Publishing Limited, 2025. All rights reserved.

for each state can be summarized in a "projection" or "transition matrix" (**L**), such as Leslie and Lefkovitch matrix models, where the columns denote the current states and the rows represent the probabilities of survival within those states.

$$\begin{bmatrix} N_1 \\ N_2 \\ N_3 \\ N_4 \end{bmatrix}(t+1) = \begin{bmatrix} F_1 & F_2 & F_3 & F_4 \\ S_1 & 0 & 0 & 0 \\ 0 & S_2 & 0 & 0 \\ 0 & 0 & S_3 & 0 \end{bmatrix} \begin{bmatrix} N_1 \\ N_2 \\ N_3 \\ N_4 \end{bmatrix}(t)$$

$\mathbf{N_{t+1}}$ **L** $\mathbf{N_t}$

L= Leslie Matrix
N= Population

Where $\mathbf{N_t}$ is a population vector at time *t*, S_i is the proportion of seeds surviving in cohort or state *i* ($i = 1,...,n$), F_i is the number of seeds produced by an adult plant in cohort *i*.

In shorthand terms, the above expression can be written as:

$$\mathbf{N_{t+1} = L\,N_t} \tag{8}$$

The population growth rate (λ) is determined by the dominant eigenvalue of the stage matrix **L**.

Several researchers have employed this model type. McMahon and Mortimer (1980) divide the life cycle of *Elymus repens* (couch grass) into five states: seeds, rhizome, buds, inmature adults and mature adults. Shea and Kelly (1998) use a matrix model to assess the impact of *Rhinocyllus conicus* (nodding thistle receptacle weevil) on *Carduus nutans* (nodding thistle) in New Zealand. Maxwell et al. (1988) use this type of model to study the influence of different control measures on *Euphorbia esula* (leafy spurge).

Equation (8) does not consider the length of the time step between successive multiplications or through different environments. To address this, periodic matrix models offer a robust framework for explicitly capturing variations in life-stage transition probabilities across a sequence of environments that repeat themselves over time (Caswell, 2001). By tracking populations through complete life cycles in different crop environments, researchers can identify critical junctures where management interventions might be most effective.

$$\mathbf{N_{t+1} = (L_1 L_2 ... L_n)\,N} \tag{9}$$

Gonzalez-Andujar et al. (1986) adapted this methodology to analyze the life cycle of *Avena sterilis* by transforming Equation (8) into a periodic matrix, where

© Burleigh Dodds Science Publishing Limited, 2025. All rights reserved.

L_i (i = 1..3) corresponds to three periods of the year: L_1 corresponds to the period autumn-spring, L_2 to spring-summer, and L_3 to summer-autumn period. This model predicted an annual rate of increase of the seed bank of $\lambda = 6.93$. Davis et al. (2004) applied periodic-matrix population models to simulate the impacts of red clover and tillage timing on the population growth rate of *Setaria faberi* (giant foxtail), explicitly accounting for variations in life stage transition probabilities over time.

Matrix model can incorporate density-dependence; the single values elements of the matrix can be replaced by equations. Holmes et al. (2022) utilized a density-dependent matrix model to assess the effect of planting certain crops on the evolution of herbicide resistance across various management scenarios.

A recent approach to matrix modeling is the density-structured population model. Unlike traditional population models that treat all individuals as homogeneous entities, density-structured models recognize variations in population density and how these affect demographic processes such as birth, death, dispersal, and interactions among individuals. These models typically partition the population into discrete density classes or categories, with individuals grouped based on their spatial distribution, age, size, or other relevant attributes. By accounting for density-dependent processes, such as competition for resources or intraspecific interactions, density-structured models provide a more realistic representation of population dynamics, particularly in environments where population density varies across space or time (Freckleton et al., 2011). Goodsell et al. (2021) develop a density-structured model for *Alopecurus myosuroides,* through transient analysis, two-step ahead projections, and stochastic simulations, they demonstrate that using break crops in rotations dominated by wheat can reduce weed densities and inter-year variability, with some break crops providing greater reductions than others.

3.2 Spatio-temporal models

To gain a better understanding of population dynamics models, it is necessary to integrate spatial dynamics within them. This entails incorporating components that account for seeds or pollen dispersion processes, as well as delineating a spatial framework through which the population interacts and moves. Although spatiotemporal models have not been extensively utilized in weed science compared to temporal models, pioneering efforts in the late 1980s and early 1990s laid the groundwork for their expansion (Ballaré et al., 1987; Perry and Gonzalez-Andujar, 1993). These advancements, bolstered by technological capabilities, particularly in computing, have facilitated their adoption in various domains. These models have found application

© Burleigh Dodds Science Publishing Limited, 2025. All rights reserved.

across various domains, notably in field-level management scenario studies (Gonzalez-Andujar and Perry, 1995; Paice et al., 1998; Blanco-Moreno, 2004; Jurado-Exposito et al., 2005; van Mourik et al., 2008; Izquierdo et al., 2009) and within the context of herbicide resistance expansion (Roux et al., 2008; Dauer et al., 2009). Moreover, they have been instrumental in deepening our understanding of spatial processes and their impacts on weed populations (Ballare et al., 1987; Perry and Gonzalez-Andujar, 1993).

These models, developed across different studies, exhibit methodological diversity. Cousens et al. (2008) proposed a classification based on the treatment of time and space, distinguishing between models that incorporate continuous time and space, such as reaction-diffusion models, those considering continuous space and discrete time, like integro-differential models, and models treating discrete time and space, such as cellular models. While models handling continuous time and space offer analytical tractability, reaction-diffusion models often fall short in capturing discrete growth and dispersion processes typical in plant populations. Integro-differential models, on the other hand, prove more suitable, as evidenced by their application in several studies (Allen et al., 1996; Woolcock and Cousens, 2000). For instance, Gonzalez-Andujar and Saavedra (1999) employed an integro-differential model to examine the critical population size and expansion rate of various weed species, shedding light on their potential as ground covers in olive groves. These models typically comprise a growth function and a dispersal kernel, enabling a comprehensive representation of population dynamics (Lutscher, 2019). Integro-differential model is composed of a dispersal kernel (k) and a growth model (g).

$$N_{t+1}(j,k) = \sum_{j} \sum_{k} \underbrace{k(|x-j|,|y-k|)}_{\textbf{Dispersal model}} \underbrace{g(N_{t}(j,k))}_{\textbf{Growth model}}$$

Where j and k are coordinates.

While the allure of obtaining analytical solutions is undeniable, these models have inherent limitations when complexity is introduced into the system. For studies demanding higher levels of complexity, realism, and detail such as the consideration of spatial heterogeneity or space fragmentation– cellular models emerge as the more suitable option (Cousens et al., 2008). In cellular models, space is divided into discrete units of equal size known as cells (Fig. 5), each adopting a particular state (mathematically speaking) at every moment in time.

This state changes according to a rule of local transition, meaning that each cell is associated with others that form its neighborhood, so the state of one of them at time $t+1$ depends on the states in which its neighbors were at

© Burleigh Dodds Science Publishing Limited, 2025. All rights reserved.

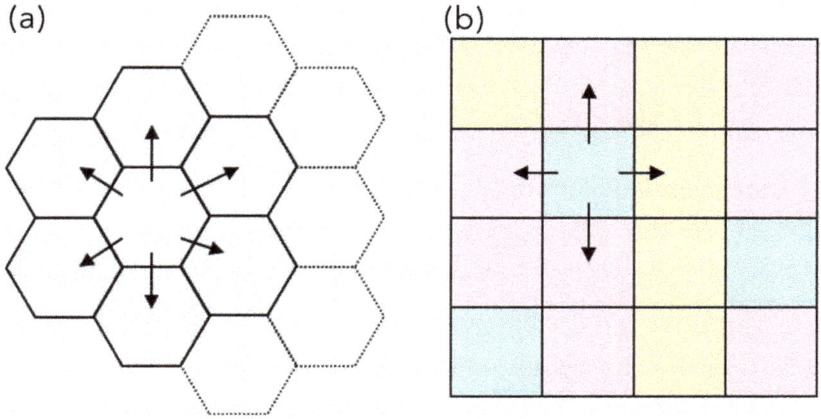

Figure 5 Spatial framework consisting of (a) hexagonal and (b) square cells showing the equivalent distance between neighboring cells.

the previous time, *t*. Cells typically adopt a square shape (Paice et al., 1998; Roux et al., 2008), given the greater ease of programming and visualization of these structures, although hexagonal cells have also been used in some works (González-Andujar and Perry, 1995; Gonzalez-Andujar et al., 1999). Hexagonal cells offer the advantage of equidistant spacing from neighboring cells' centers, unlike square cells where diagonal neighbors are more distantly spaced than adjacent ones (Fig. 5). Conversely, square cells may better model anisotropic dispersion processes (e.g. influenced by machinery or tillage), whereas hexagonal cells facilitate natural dispersion modeling with greater ease (Paice et al., 1998).

Cellular models can be further categorized into cellular automata and coupled map lattice models. In cellular automata, each cell represents a single individual, while coupled map lattice models utilize the local density of plants or seeds as the state variable. Within each cell, the life cycle of the weed unfolds, and interconnectivity between cells occurs through seed and pollen dispersal processes.

Seed dispersal between cells hinges on various factors such as seed production, characteristics, and dispersal vectors (Bastida et al., 2021), governing spatial dispersal at the relevant scale (Ghersa and Roush, 1993). Typically, seed dispersal between cells is delineated through probability density functions by the weed species and its dispersal mechanisms for that scale (Cousens et al., 2008).

Individual-based models (IBMs) are computational tools that simulate the behavior and interactions of individual weed plants within a population. These models concentrate on depicting the life cycle, growth, reproduction, and interactions of individual weeds, thereby offering a detailed understanding of

© Burleigh Dodds Science Publishing Limited, 2025. All rights reserved.

how various factors influence weed populations over time. Schut et al. (2017) discuss the long-term effects of herbicide applications on weed populations using IBM models, providing insights into how repeated herbicide use can impact the dynamics of weed populations.

3.3 Landscape-level models

Farmers often plan agricultural management at the field scale in a highly individualistic way. However, many studies have shown that the ecological processes that sustain agriculture operate at higher landscape levels (Dainese et al., 2019), and thus management practices could be optimized by implementation at the landscape level (González-Díaz et al., 2015).

The majority of spatio-temporal models have been developed at the field level, with many geared toward precision agriculture applications. However, there has been a notable scarcity of studies focused on examining population dynamics at the landscape scale. By coupling spatially explicit population dynamics with landscape structure and environmental heterogeneity, these models can help identify landscape features and management strategies that are most influential in shaping weed population dynamics (Dauer et al., 2009; Ricci et al., 2018) (Fig. 6).

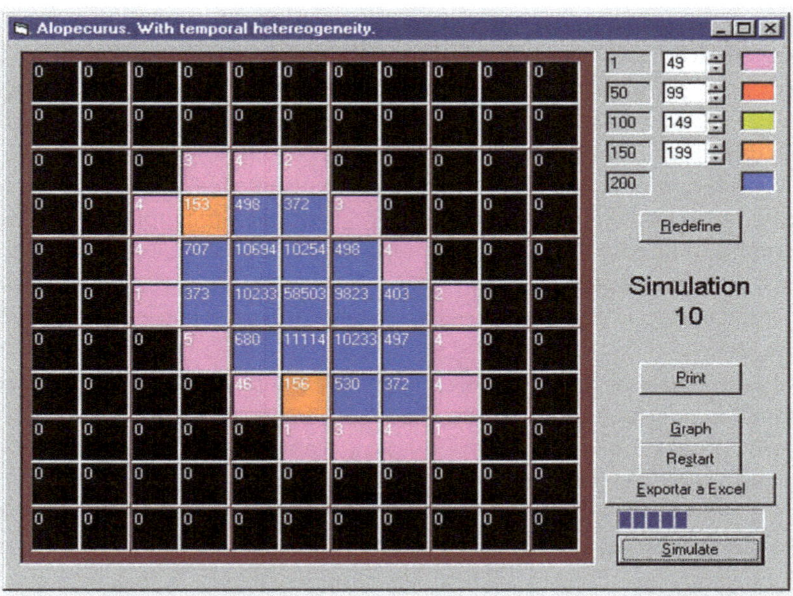

Figure 6 Landscape simulation of the spreading of *Aloperucus myosuroides* population from a point source in the middle of the grid space after 10 years. Each cell represents a field of 1 ha.

© Burleigh Dodds Science Publishing Limited, 2025. All rights reserved.

In a seminal study, Auld and Coote (1990) utilized a simulation model to investigate the impact of weed control at the landscape scale. Their findings revealed that to achieve effective weed control in a fully infested agricultural landscape, a minimum of 20% of the fields would need treatment. Similarly, Gonzalez-Andujar et al. (2006) developed a simple agricultural landscape model comprising two fields of similar size and quality interconnected through weed seed dispersal. Their objective was to analyze the influence of field-level weed management decisions on landscape-level weed population dynamics. Their findings suggested a potential 16% rise in *A. sterilis* population density in untreated fields compared to non-spatial models. These results highlight the importance of inter-field dispersal in shaping weed population dynamics at the landscape level.

More advanced and complex models have been developed. Jayasuriya et al. (2008) develop a spatially explicit framework for regional-level modeling of alien plant spread in arable paddocks. The model considers the speed of infestation in an area of 10 000 km^2 of five species, namely *Raphanus raphanistrum, Solanum elaeagnifolium, Orobanche ramosa, Aegilops cylindrica,* and *Echinocloa colona*. González-Díaz et al. (2015) conclude that an effective crop rotation deployment might limit, and in some cropping years even eliminate, the need for herbicide applications to control the dominant weed species.

4 Integration of population dynamic models in decision support systems

The integration of weed population models into decision support systems (DSS) represents a significant advancement in weed management strategies, offering stakeholders enhanced decision-making capabilities and improved outcomes. By combining predictive modeling with real-time data and expert knowledge, DSS provides valuable insights into weed population dynamics and supports the development of effective management strategies tailored to specific agricultural contexts (Chantre & Gonzalez-Andujar, 2020). One example of the integration of weed population models into a DSS is AVENANET (Gonzalez-Andujar, 2020) (Fig. 7). AVENANET is based on a bioeconomic model that integrates three submodels: population dynamics, weed-crop competition, and economics. The model includes a database of 45 weed-management strategies, encompassing chemical, mechanical, and cultural methods. By integrating these predictions into a user-friendly online platform, AVENANET assists farmers and technicians in making timely decisions regarding *A. sterilis* management practices.

Another notable example is the DSS developed by Molinari et al. (2022). This model uses bioecological and agronomic information as inputs,

© Burleigh Dodds Science Publishing Limited, 2025. All rights reserved.

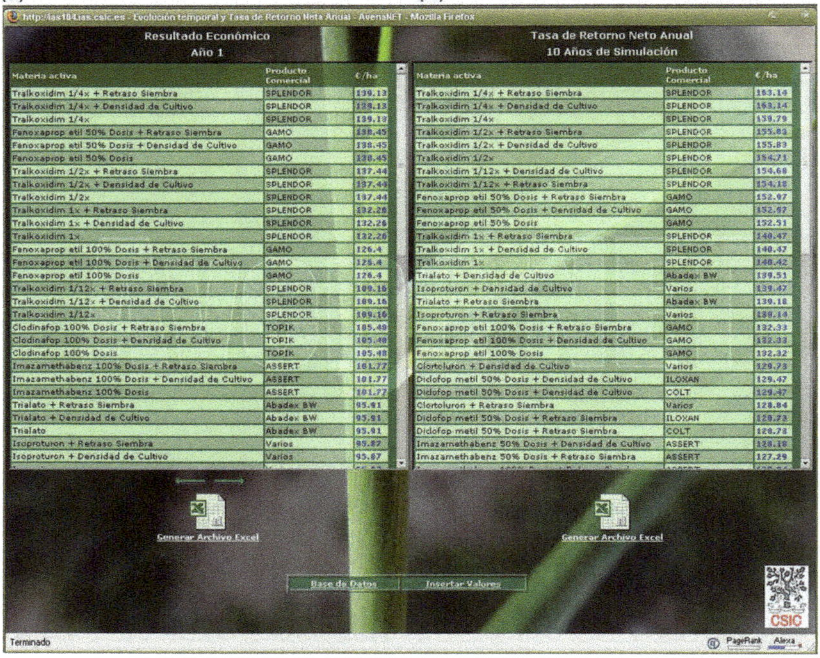

Figure 7 The economic output (euros/ha) of AvenaNET considering (A) short-term (1 year) and (B) long-term (10 years) horizons.

such as daily weather records, weed population dynamics data, weed-management tactics (chemical, mechanical, and cultural methods), and crops' ecophysiological requirements to control *Euphorbia davidii* infesting soybean crops. Furthermore, the AgroEcoModel DSS developed by Barberi et al. (2006) integrates weed population models with agroecological principles to support sustainable weed management practices in organic farming systems. AgroEcoModel simulates weed population dynamics under different crop rotations, cover crop strategies, and soil management practices, allowing farmers to evaluate the long-term impacts of their management decisions on weed populations, soil health, and crop productivity.

Overall, the integration of weed population models into DSS offers valuable tools for improving weed management practices across various agricultural systems. By leveraging predictive modeling, real-time data, and expert knowledge, DSS empowers stakeholders to make informed decisions, optimize resource allocation, and develop sustainable weed management strategies that enhance agricultural productivity and environmental stewardship.

© Burleigh Dodds Science Publishing Limited, 2025. All rights reserved.

In general, the systems-based approach inherent to CA can be enhanced by the utilization of DSS, thereby conferring a number of benefits:

- Facilitates enhanced decision-making processes. The use of a DSS enables the implementation of data-driven decisions that enhance the effectiveness of CA practices, thereby optimizing the utilization of resources while maintaining the ecological benefits associated with such practices.
- Increased adoption of CA practices: By demonstrating the potential benefits of CA practices through scenario analysis and economic forecasting, DSS can encourage more farmers to adopt CA practices.
- Reduction in environmental impact: DSS helps farmers minimize soil disturbance, reduce chemical inputs, reduce herbicide resistance, and optimize nutrient management, leading to a lower environmental footprint and improved biodiversity.

5 Challenges in modeling population dynamics and summary

The utilization of models in weed science has proven invaluable for predicting the long-term effects of weed management practices and understanding population behaviors. However, the field of weed modeling faces several challenges, with model validation being a significant concern. Despite its importance, model validation remains underexplored (Holst et al., 2007). Consequently, the accuracy of model predictions remains uncertain. To address this, researchers must dedicate more effort to both the development and validation of models (Freckleton and Stephens, 2009; Gonzalez-Andujar et al., 2010; Egea-Cobrero et al., 2020).

Weed populations possess a spatial dimension that extends beyond the field scale and exhibit spatial and temporal variability due to factors such as soil heterogeneity, climate fluctuations, and management practices. Modeling this variability requires spatially explicit and dynamic models capable of accounting for spatial and temporal scales relevant to weed dynamics. Therefore, it is imperative to develop models that account for various spatial scales and how management practices influence them. A multi-scale approach, as advocated by Gonzalez-Diaz et al. (2015) and Evans et al. (2016), would enhance precision weed management and landscape ecology.

Another factor to consider is human behavior. Human activities, such as weed management practices, land-use changes, and agricultural policies, play a significant role in shaping weed population dynamics (Doohan et al., 2010). However, modeling human behavior and its effects on weed populations

© Burleigh Dodds Science Publishing Limited, 2025. All rights reserved.

is complex and requires interdisciplinary approaches that integrate social, economic, and ecological factors.

Addressing these challenges requires interdisciplinary collaboration among scientists, agronomists, mathematicians, and stakeholders, as well as the development of innovative modeling techniques and data-driven approaches. By overcoming these challenges, researchers can develop population dynamics models that provide valuable insights into weed management strategies and support sustainable agricultural practices.

In summary, modeling weed population dynamics in the context of conservation agriculture is a powerful tool for optimizing sustainable weed management strategies, providing insights into how weed populations change and adapt under CA. These models help identify practices that minimize weed pressure, promote crop productivity, and sustain long-term soil health. Despite the challenges of modeling complex ecological systems, advances in weed population modeling are making it increasingly feasible to develop predictive tools that support effective decision-making in CA.

6 Conclusions

This chapter examines how conservation agriculture practices alter weed dynamics by affecting seed distribution, persistence, germination, and mortality rates, thereby influencing the overall effectiveness of management strategies. Understanding these changes is essential for devising sustainable weed management techniques that are tailored to conservation agriculture systems. The text reviews various modelling approaches–including non-spatial, density-dependent, matrix, spatio-temporal, and individual-based models–which provide valuable insights into weed population behavior and help simulate growth patterns under different conditions. These models not only predict responses to diverse management tactics but also support decision-making by forecasting long-term outcomes. Integrating population models into Decision Support Systems further enhances farmers' and agronomists' ability to implement targeted and effective control measures. Despite significant advances, challenges remain in terms of model validation, addressing spatial heterogeneity, and incorporating the effects of human behavior. Overcoming these hurdles will require interdisciplinary collaboration and the development of multi-scale approaches that combine environmental, biological, and socio-economic factors. Overall, refined modelling serves as a crucial tool for achieving sustainable agriculture by balancing productivity with environmental stewardship.

© Burleigh Dodds Science Publishing Limited, 2025. All rights reserved.

7 Where to look for further information

Further readings

- *Dynamics of weed Populations* (Cousens & Mortimer, 1995). This is a reference book for anyone interested in this topic.
- Good sources of information on the subject are: Holst et al., 2007 and Bagavathiannan et al., 2020 (see references).
- A good introduction to ecological modeling is An Introduction to Ecological Modelling: Putting Practice into Theory (Gillman and Hails, 1997).

Key journals/conferences

The two main journals in weed science where it is possible to find articles related to weed population dynamics are *Weed Research* and *Weed Science*. Additionally, relevant articles can be found in agronomy-focused publications such as *Agronomy*, as well as other specialized weed science journals like *Advances in Weed Management*.

Conferences hosted by esteemed organizations like the Weed Science Society of America (WSSA), the European Weed Research Society (EWRS), and the International Weed Science Society (IWSS) routinely feature communications on weed population dynamics.

8 References

Allen, L. J. S., et al. (1996), A mathematical model for weed dispersal and control, *Bull. Math. Biol.* 58: 815–834.

Alm, D. M., et al. (1993), An index model for predicting seed germination and emergence rates. *Weed Technology*, 7, 560–569.

Anderson, R. L. (2005), A multi-tactic approach to manage weed population dynamics in crop rotations, *Agronomy J.* 97: 1579–1583.

Auld, B. A. and Coote, B. G. (1990), INVADE: towards the simulation of plant spread. *Agriculture, Ecosystems and Environment* 30, 121–128.

Baden, H. M., et al. (2021), The effects of age on the demography of a perennial plant depend on interactions with size and environment, *J. Ecol.* 109: 1068–1077.

Bagavathiannan, M. V., et al. (2020), Simulation models on the ecology and management of arable weeds: structure, quantitative insights, and applications, *Agronomy 10*: 1611.

Ballaré, C. L., et al. (1987), The population ecology of *Datura ferox* in soybean crops: a simulation approach incorporating seed dispersal, *Agric. Ecosyst. Environ.* 19: 177–188.

Baraibar, B., et al. (2011), Unravelling the process of weed seed predation: developing options for better weed control, *Biol. Control* 56: 85–90.

Barberi, P., et al. (2006), Design and testing of a multifunctional Decision Support System for organic farming systems, *Comput. Electron. Agr.* 50(1): 70–93.

© Burleigh Dodds Science Publishing Limited, 2025. All rights reserved.

Barroso, J., et al. (2006). Dispersal of Avena fatua and Avena sterilis patches by natural dissemination, soil tillage and combine harvesters. *Weed Research* 46:118–128.

Bastida, F., et al. (2021), Season-long seed dispersal patterns of the invasive weed *Erigeron bonariensis* in South-Western Spain, *Crop Prot.* 148: 105720.

Benech-Arnold, R. L., et al. (2000), Environmental control of dormancy in weed seed banks in soil, *Field Crops Res.* 67: 105–122.

Blanco-Moreno, J. M., et al. (2004), Spatial distribution of *Lolium rigidum* seedlings following seed dispersal by combine harvesters, *Weed Res.* 44: 375–387.

Buhler, D. D., et al. (1997), Implications of weed seedbank dynamics to weed management, *Weed Sci.* 45: 329–336.

Caswell, H. (2001), *Matrix Population Models: Construction, Analysis and Interpretation*, 2nd edition, Sunderland, Massachusetts, Sinauer Associates.

Chauhan, B. S., et al. (2006), Influence of environmental factors on seed germination and seedling emergence of rigid ryegrass (*Lolium rigidum*), *Weed Sci.* 54: 1004–1012.

Choudary, M., et al. (2020), Conservation Agriculture and Climate Change: An Overview, In Bisht, J., et al. (eds.), *Conservation Agriculture*, Singapore, Springer.

Colbach, N., et al. (2017), A methodology for multi-objective cropping system design based on simulations: application to weed management, *Eur. J. Agron.* 87: 59–73.

Cordeau, S. (2022), Conservation agriculture and agroecological weed management, *Agronomy* 12: 867.

Cousens, R. D. and Moss, S. R. (1990), A model of the effects of cultivation on the vertical distribution of weed seeds within the soil. *Weed Research* 30, 61–70.

Cousens, R. D. and Mortimer, M. (1995), *Dynamics of Weed Populations*, Cambridge, Cambridge University Press.

Cousens, R. D., et al. (2008), *Dispersal in Plants: A Population Perspective*, New York, Oxford University Press.

Cussans, G. W. and Moss, S. R. (1982), Population dynamics of annual grass weeds, In *Proceedings of the Symposium Decision Making in the Practice of Crop Protection*, pp. 91–98.

Dainese, M., et al. (2019), A global synthesis reveals biodiversity-mediated benefits for crop production, *Sci Adv.* 5: eaax012.

Dauer, J. T., et al. (2009), Effects of landscape composition on spread of an herbicide-resistant weed, *Landsc. Ecol.* 24: 735–747.

Davis, A. S., et al. (2004), Using matrix models to determine cropping system effects on annual weed demography, *Ecol. Appl.* 14: 655–668.

Debaeke, P. (1988), Dynamique de quelques dicotylédones adventices en culture de céréale. II. Survie, floraison et fructification, *Weed Res.* 28: 265–279.

De la Fuente, E. B., et al. (2006), Soybean weed community composition and richness between 1955 and 2003 in the Rolling Pampas (Argentina), *Agric. Ecosyst. Environ.* 115: 229–236.

Doohan, D., et al. (2010), Investigating the human dimension of weed management: new tools of the trade, *Weed Sci.* 58(4): 503–510.

Doyle, C. J., et al. (1986), A model of the economics of controlling *Alopecurus myosuroides* huds in winter wheat, *Crop Prot.* 5: 143–150.

Egea-Cobrero, V., et al. (2020) Validation of predictive empirical weed emergence models of Abutilon theophrasti Medik. based on intercontinental data. *Weed Res.* 60: 297–302.

© Burleigh Dodds Science Publishing Limited, 2025. All rights reserved.

Evans, J. A., et al. (2016), Managing the evolution of herbicide resistance, *Pest Manag. Sci.* 72: 74–80.

FAO (2015), Conservation Agriculture, Food and Agriculture Organization of the United Nations (FAO). Available at: https://www.fao.org/conservation-agriculture/en/; accessed February 8, 2024.

Fernandez-Quintanilla, C., et al. (1986), Seedling recruitment and age-specific survivorship and reproduction in populations of *Avena sterilis ssp. Ludoviciana, J. Appl. Ecol.* 23: 945–955.

Fernandez-Qunitanilla, C., et al. (1987). The influence of herbicide treatments on the population of Avena sterilis ssp. ludoviciana in winter wheat. *Weed Research*, 27: 375–383.

Firbank, L. G. and Watkinson, A. R. (1986), Modelling the population dynamics of an arable weed and its effects upon crop yield, *J. Appl. Ecol.* 23: 147–159.

Freckleton, R. P. and Watkinson, A. R. (1998), Predicting the determinants of weed abundance: a model for the population dynamics of *Chenopodium album* in sugar beet, *J. Appl. Ecol.* 35: 904–920.

Freckleton, R. P. and Watkinson, A. R. (2000), Designs for greenhouse studies of interactions between plants: an analytical perspective, *J. Appl. Ecol.* 88: 386–391.

Freckleton, R. P. and Stephens, P. A. (2009), Predictive models of weed population dynamics, *Weed Res.* 49: 225–232.

Freckleton, R. P., et al. (2011), Density-structured models for plant population dynamics, *Am. Nat.* 177(1): 1–17.

Froud-Williams, R. J., et al. (1983), Influence of cultivation regime on weed floras of arable cropping systems, *J. Appl. Ecol.* 20: 187–197.

Ghersa, C. M. and Roush, M. L. (1993), Searching for solutions to weed problems, *BioScience* 43: 104–109.

Giller, K., et al. (2015), Beyond conservation agriculture, *Front. Plant Sci.* 6: 870.

Gillman, M., and R. Hails. (1997) Introduction to Ecological Modelling. Putting practice into Theory. Wiley-Blackwell. 216 pages. ISBN: 978-0-632-03634-9

Gomes, R. R., et al. (2013). Diaporthe: a genus of endophytic, saprobic and plant pathogenic fungi. *Persoonia* 31, 1–41. doi: 10.3767/003158513x666844

Gonzalez-Andujar, J. L. (2020), AvenaNET and VallicoNET: DSS for *Avena sterilis* and *Lolium rigidum* Control in Spanish Dryland Cereal Crops, In *Decision Support Systems for Weed Management*, Springer, pp. 299–308.

Gonzalez-Andujar, J. L. and Bastida, F. (2018), Modeling the population dynamics of a community of two grass weeds of winter wheat in a Mediterranean area, *J. Plant Prod.* 12: 219–223.

Gonzalez-Andujar, J. L. and Perry, J. N. (1995), Models for the herbicidal control of the seed bank of *Avena sterilis*: the effects of spatial and temporal heterogeneity and of dispersal, *J. Appl. Ecol.* 32: 578–587.

Gonzalez-Andujar, J. L. and Saavedra, M. (1999), Estimación del tamaño crítico y velocidad de expansión de los rodales de malas hierbas, In *Actas del Congreso de la Sociedad Española de Malherbología,* pp. 123–126.

Gonzalez-Andujar, J. L., et al. (1986), Modelización y simulación de la dinámica de poblaciones de plantas mediante el modelo matricial de Leslie, *Investigación Agraria, Producción y Protección Vegetales* 1(2): 209–218.

© Burleigh Dodds Science Publishing Limited, 2025. All rights reserved.

Gonzalez-Andujar, J. L., et al. (2005), Demography and population dynamic of the arable weed *Phalaris brachystachys L.* (short-spiked canary grass) in winter wheat, *Crop Prot.* 24: 581–584.
Gonzalez-Andujar, J. L,, et al. (2006), Population cycles in an annual plant produced by delay-density dependence. *Am. Nat.* 168: 318–322.
Gonzalez-Andujar, J. L., et al. (2010), Field Evaluation of a Decision Support System for herbicidal control of Avena sterilis ssp. ludoviciana in winter wheat. *Weed Research* 50:83–88.
Gonzalez-Andujar, J. L., et al. (2023), Quantifying and disentangling the competition effect of a weed community in a long-term biennial cereal-legume rotation, *Agronomy* 13: 1432.
Gonzalez-Diaz, L., et al. (2015), Spatially-explicit bioeconomic model for weed management in cereals: validation and evaluation of management strategies, *J. Appl. Ecol.* 52: 240–249.
Goodsell, R. M., et al. (2021), Developing hierarchical density-structured models to study the national-scale dynamics of an arable weed, *Ecol. Monogr.* 91: e01449.
Gramig, G., et al. (2024), Diversified crop sequencing for suppression of creeping perennial weeds in organic production systems, *Farm. Sys.* 2: 100070.
Hernandez Plaza, E., et al. (2012), Fluctuations in plant populations: role of exogenous and endogenous factors. *J. Veg. Sci.* 23: 640–646.
Hobbs, P. R. (2007), Conservation agriculture: what is it and why is it important for future sustainable food production?, *J. Agric. Sci.* 145(2): 127.
Holmes, K. H., et al. (2022), Modeling the evolution of herbicide resistance in weed species with a complex life cycle, *Ecol. Appl.* 32: e02473.
Holst, N., et al. (2007), Field weed population dynamics: a review of model approaches and applications, *Weed Res.* 47: 1–14.
Izquierdo, J., et al. (2009). Spatial distribution of weed diversity within a cereal field. *Agron. Sustain. Dev.* 29: 491–496.
Jayasuriya, R. T., et al. (2008), *An Economic Decision Tool for Responding to New Weed Incursion Risks in the Australian Grains Industry*, Adelaide, Technical Series no. 14, CRC for Australian Weed Management.
Jones, R. and Medd, R. (1997), Economic analysis of integrated management of wild oats involving fallow, herbicide and crop rotational options, *Aust. J. Exp. Agric.* 37: 683–691.
Jordan, N., et al. (1995), Simulation analysis of crop rotation effects on weed seedbanks, *Am. J. Bot.* 82: 390–398.
Jurado-Exposito, M., et al. (2005). Characterizing population growth rate of *Convolvulus arvensis* in wheat-sunflower no-tillage systems, *Crop Sci.* 45: 2106–2112.
Koskinen, W. C. and McWhorter, C. G. (1986), Weed control in conservation tillage, *J. Soil Water Conserv.* 38: 253–254.
Lutscher, F. (2019), Modeling with integrodifference equations, *Interdiscip. Appl. Math.* 49: 9–22.
Maxwell, B. D., et al. (1988), Population modeling approach for evaluating leafy spurge (Euphorbia esula) development and growth. *Weed Technology* 2, 132–138.
Mertens, S. K., et al. (2002), Weed populations and crop rotations: exploring dynamics of a structured system, *Ecol. Appl.* 12: 1125–1141.
Mohler, C. L., et al. (2012). Reduction in weed seedling emergence by pathogens following the incorporation of green crop residue. *Weed Research*. 52, 467–477.

© Burleigh Dodds Science Publishing Limited, 2025. All rights reserved.

Mokhtassi-Bidgoli, A., et al. (2013), Modelling the population dynamic and management of *Bromus diandrus* in a non-tillage system, *Crop. Prot.* 43: 128–133.

Molinari, F. A., et al. (2022), A weed population dynamics model for integrated weed-management decision-making support: *Euphorbia davidii* subils in Soybean crops as a simulation study, *Agronomy* 12: 2369.

Molofsky, J., et al. (2014), Nutrient enrichment alters dynamics in experimental plant populations. *Popul. Ecol.* 56: 97–107.

Mortimer, A. R., et al. (1980), The prediction of weed infestations and cost of differing control strategies, In *Proceedings of British Crop Conference - Weeds*, pp. 415–422.

Murdoch, A. J. and Roberts, E. H. (1982), Biological and financial criterium of long-term control strategies for annual weeds, In *Proceedings of British Crop Conference - Weeds*, pp. 741–748.

Nichols, V., et al. (2015), Weed dynamics and conservation agriculture principles: a review, *Field Crops Res.* 183: 56–68.

Nyanga, P. H., et al. (2020), Reinforcing Ecosystem Services Through Conservation Agriculture in Sustainable Food Systems, In Rusinamhodzi, L. (ed.), *The Role of Ecosystem Services in Sustainable Food Systems*, Kathmandu, Academic Press, pp. 119–133.

Paice, M. E. R., et al. (l998), A stochastic simulation model for evaluating the concept of patch spraying, *Weed Res.* 38: 373–388.

Palm, C., et al. (2014), Conservation agriculture and ecosystem services: an overview, *Agric. Ecosyst. Environ.* 187: 87–105.

Perry, J. N. and Gonzalez-Andujar, J. L. (l993), Dispersal in a metapopulation neighbourhood model of an annual plant with a seedbank, *J. Ecol.* 81: 453–463.

Pittelkow, C. M., et al. (2015), Productivity limits and potentials of the principles of conservation agriculture, *Nature*, 517: 365–368.

Ricci, B., et al. (2018), How effective is large landscape-scale planning for reducing local weed infestations? A landscape-scale modelling approach, *Ecol. Modell.* 384: 221–232.

Roux, F., et al. (2008), Delaying weed adaptation to herbicide by environmental heterogeneity: a simulation approach, *Pest Manag. Sci.* 64: 16–29.

Schut, M., et al. (2017), Modeling the long-term effects of repeated clopyralid applications on the performance of *Cirsium arvense* populations, *Ecol. Modell.* 347: 40–48.

Sagar, G. R., and Mortimer A. M. (1976), An approach to the study of the population dynamics of plants with special reference to weeds. *Applied Biology* 1: 1–47.

Shea, K. and Kelly, D. (1998), Estimating biocontrol agent impact with matrix models: *Carduus nutans* in New Zealand, *Ecol. Appl.* 8: 824–832.

Smith, G. L., et al. (1999), The population dynamics of *Anisantha sterilis* in winter wheat: comparative demography and the role of management, *J. Appl. Ecol.* 36: 455–471.

Solbrig, O. T. (1980), Genetic Structure of Plant Populations, In Solbrig, O. T. (ed.), *Demography and Evolution in Plant Populations*, Oxford, Blackwell Scientific Publications, pp. 49–65.

Somerville, G. J., et al. (2020), Spatial modelling of within-field weed populations: a review, *Agronomy* 10: 1044.

Thomas, A. G. and Frick, B. L. (1993), Influence of tillage systems on weed abundance in southwestern Ontario, *Weed Technol.* 7: 699–705.

Torra, J., et al. (2008), Modelling the population dynamics of *Papaver rhoeas* under various weed management systems in a Mediterranean climate, *Weed Res.* 48: 136–146.

© Burleigh Dodds Science Publishing Limited, 2025. All rights reserved.

Van Mourik, T. A., et al. (2008), Long-term management of *Striga hermonthica*: strategy evaluation with a spatio-temporal population model, *Weed Res.* 48: 329–339.
Watkinson, A. R. (1980). Density-dependence in single species populations of plants. *Journal of Theoretical Biology* 83:345–357.
Westerman, P. R., et al. (2012), Differential weed seed removal in dryland cereals, *Arthropod-Plant Inte.* 6: 591–599.
Wilson, B. J., et al. (1984), Exercises in modelling population of *Avena fatua* to aid strategic planning for the long-term control of this weed in cereals, In *Proceedings of the 7th International Symposium on Weed Biology, Ecology and Systematics*, pp. 287–294.
Woolcock, J. L. and Cousens, R. (2000), A mathematic analysis of factors affecting the rate of spreads of patches of annual weeds in an arable field, *Weed Sci.* 48: 27–34.
Zambrano-Navea, C., et al. (2016), A cohort-based stochastic model of the population dynamic and long-term management of *Conyza bonariensis* in fruiting tree crops, *Crop Prot.* 80: 15–20.
Ziebarth, N. L., et al. (2010), Weak population regulation in ecological time series, *Ecol. Lett.* 13: 21–31.

© Burleigh Dodds Science Publishing Limited, 2025. All rights reserved.

Part 1

Cultural, physical and chemical weed management techniques

Chapter 3

Cultural techniques to control weeds in Conservation Agriculture systems

Francisco Skora Neto, No-till Brazilian Federation (FEBRAPDP)/ formerly Agronomic Institute of Parana (IAPAR), Brazil

1 Introduction

During their evolutionary path, plants have improved their ability to adapt to different environments. Identifying limiting factors (light, temperature, water, and nutrients) triggers responses that ensure maximum biological productivity with what is available. The intense competition for environmental resources has motivated plant species to develop different mechanisms to capture these resources for their preservation. The efficiency of these mechanisms can determine the survival and dominance of the species. In this way, species specialize in habitats favorable to their survival and maintenance of descendants for future generations. A group of species called weeds have specialized in occupying particular niches in agroecosystems.

In agricultural systems, weeds are primary plants in an ecological succession of vegetation, which have specialized in occupying these altered environments. Through efficient mechanisms of dispersion, establishment, and multiplication, they have evolved to be able to colonize agricultural areas, competing with cultivated plants. In contrast, cultivated plants are human-selected species with characteristics intended to provide products for human consumption. Many wild features were eliminated during domestication, requiring human

http://dx.doi.org/10.19103/AS.2024.0145.04
© Burleigh Dodds Science Publishing Limited, 2025. All rights reserved.

intervention to establish crops and express their yield potential. Other plants that are not of interest are typically excluded to optimize capture of resources from the environment.

In the early days of agriculture, manual and mechanical weed control predominated to allow crops to grow without competition – a task that usually requires a huge amount of energy. After a few years of cultivation, it was customary to use years of fallow to recover fertility and reduce the presence of competing plants. Later, when land use pressure was intensified, farmers adopted combined agriculture/livestock systems, and ley pastures replaced fallow periods that, when well-managed, kept the weed population at reasonable levels (Hakanson, 2003; Meiss et al., 2010; Schuster et al., 2020). More recently, with the advent of herbicides along with other pesticides, and widespread use of synthetic fertilizers, more simplified production systems have become predominant.

These systems, while very efficient in using plants to obtain energy for the ever-increasing human population, have nevertheless become questionable from an environmental and social point of view. The need for more sustainable production systems has become more evident. In addition to various other agronomic, environmental, and social consequences of high herbicide use, herbicide-resistant weeds have become widespread and a severe problem.

Recommendations for integrated weed management (IWM) and strategies to prevent the emergence and spread of resistant weeds have not had the expected effect, with rates of resistance globally continuing to increase (HRAC, 2023; Heap, 2023). It was expected that the development of herbicide-resistant crops would mean less input of chemicals into the environment and that it would also help solve the problem of resistant weeds. However, after initial success, new cases of resistance have continued to multiply, requiring greater use of chemicals that these new crops were intended to eliminate. Currently, genetically modified crops are typically combined with a mixture of various herbicides, undermining the expectation that chemical applications would be reduced (Holkem et al., 2022) and creating even greater dependence of farmers on herbicides provided by large companies. As chemical control is now becoming increasingly problematic, there is growing recognition of the need to consider more sustainable practices which incorporate ecological concepts in agriculture.

An alternative approach is to understand weeds as a natural component of agroecosystems and to explore ways weeds and crops can co-exist (Kudsk, 2022). In managing the coexistence of cultivated plants with competing weed species, incorporating ecological concepts must be part of the weed management program. Two ways of incorporating ecological concepts are:

© Burleigh Dodds Science Publishing Limited, 2025. All rights reserved.

- Managing space occupation; and
- Reduction of the presence of competing species.

These are discussed in more detail below.

2 Cultural methods: occupation of space

One way of managing the relationship between crops and weeds is controlling the space occupied by either at any one time in a way that means weeds are unable to outcompete crops for resources that significantly affects crop growth, yield, and quality. A key issue in weed–crop competition studies is the timing of weed control measures (Knezevic et al., 2002). Current research suggests there are no fixed periods for managing coexistence between crops and weeds within individual agricultural systems. Managing crop–weed relationships depends on a large number of variables, such as type and arrangement of crops, predominant weed species, density and arrangement of weeds, time of emergence, fertility, climate, etc. Simple rules include the following (Charles and Taylor, 2021):

- Crops that take longer to cover the soil must remain without weed competition longer than fast-growing crops that cover the ground earlier; and
- In the presence of less competitive weeds and/or at low density and/or with late emergence, initial coexistence is more viable and makes it possible to delay weed control.

Low weed density still provides better control not only with manual hoeing (Van Der Schans et al., 2006) or mechanical practices but also with pre-emergence (Martin and Namuth, 2017) and post-emergence herbicides (Fipke and Vidal, 2019).

2.1 Crop spacing and density

The spatial arrangement of the plants of a crop must consider the optimum exploitation of space required by plants to express their greatest production potential. In some crops, what matters is the needs of individual plants whilst, in others, it may mean what a group of plants requires. However, crops must maximize the occupation of space.

When a crop has increased density, with reduced spacing between plants, there is a faster and more complete occupation of space and, as a result, the ability of a crop to compete with weeds increases (Olsen et al., 2005; Marin and Weiner, 2014). A practical effect is a reduction of the Critical Weed-Free Period

© Burleigh Dodds Science Publishing Limited, 2025. All rights reserved.

(CWFP) – that is, the period of the crop cycle after which weeds no longer interfere with yield (Mulugeta and Boerboom, 2000; Rosset and Gulden, 2020). It is usually not enough to dispense with all control measures after the CWFP, but it means greater competitive efficiency of the crop after any initial control measures (Yelverton and Coble, 1991; Teasdale, 1995; Harder et al., 2007).

A space occupation strategy in reducing spacing and increasing density must consider the response of different crops (Bradley, 2006; Ethridge et al., 2022; McCollough and Melander, 2022). Phytotechnical parameters may also make it difficult to make changes without affecting costs or yields (e.g. cost of seed or specialist equipment required to achieve a given density) or affecting pest/disease management, for example (which might be more difficult at higher densities). However, it does highlight the importance of obtaining and maintaining an adequate and evenly distributed plant population.

A space occupation strategy fits very well when looking for biomass production rather than grain yield. In the case of cover crops, it is desirable to aim for a population density that provides rapid occupation of the biological space and maximum biomass production. The population density must be adjusted to the 'Constant Final Yield' (Weiner and Freckleton, 2010), which establishes that the total biomass initially increases proportionally to plant density, reaches a maximum plateau, and remains constant with increasing density. Densities beyond this threshold create practical and economic problems, e.g. they require more seeds and, consequently, a higher cost than is needed to achieve the required density. However, aiming to reduce costs by reducing the seed rate can lead to plant density below the level necessary to obtain the benefit of weed suppression.

2.2 Varieties

Cultivated plants are differentiated by varieties that generally have genetic differences in their ability to occupy space and, therefore, differences in their ability to compete with weeds. Plants can demonstrate different adaptive forms of competition. Researchers investigating plant competition distinguish two aspects:

- the competitive effect; and
- the competitive response.

The competitive effect of a plant on a neighbor is the reduction of resources available to the neighbor because of the plant in question. A plant's competitive response to its neighbors is its ability to grow, develop, and reproduce despite resource depletion from its neighbors. In theory, a plant can be a good competitor by significantly reducing the resources available to neighbors

© Burleigh Dodds Science Publishing Limited, 2025. All rights reserved.

or by thriving despite low resource levels caused by its neighbors (Weiner et al., 2010; Cahill et al., 2005), as the competitive response leads to species occupying different 'niches' (Miller and Werner, 1987).

Light is a resource that can become scarce very quickly, hence the importance of the ability of plants to capture space in the initial coexistence between crops and weeds. There are different plant strategies to deal with shade: shade avoidance, shade tolerance (Ruberti et al., 2012), or even an offensive strategy of 'shading' to deprive competing plants of light (Weiner et al., 2010).

When exposed to shade, plants must adjust to the reduction in the quality and quantity of light. In shade avoidance, plants allocate their energies to stem growth to reach the light source, decrease chlorophyll production, and reduce the leaf area. In the case of shade tolerance, they react by increasing leaf area, the number of chloroplasts, and chlorophyll to boost photosynthesis, and decreasing leaf thickness. Offensive shading is a strategy for plants with rapid initial growth resulting from an ecological adaptation often associated with some biological cost. The 'decision-making' about which strategy (avoid or stand up to) to use depends on genetic characteristics and the complex integration of receptor, signaling, and messenger molecules.

In early growth, most of our understanding of weed competition is focused on plant shoots and capturing of light. This highlights characteristics of good competitors such as greater ground cover in the initial stages, high accumulation of dry matter in the shoot, and greater plant height. Place et al. (2011), e.g., demonstrated that the most competitive soybean genotype had the highest seed weight, biomass accumulation, soil cover in 3 weeks, and plant height at the beginning of the season. High initial plant height, dry matter of leaves at 20 days after emergence, and dry matter of stems and branches were also the characteristics found by Trezzi et al. (2013) in the evaluation of different soybean varieties. However, they also found that the cultivar least affected by competition was not the one with the highest productivity, emphasizing that it is necessary to reconcile high competitiveness with high grain productivity.

In corn, Marin and Weiner (2014) found that the cultivar that best-suppressed weeds at high density had the slightest variation in the insertion angle of the oldest living leaf at harvest, supporting the hypothesis that reduced phenotypic plasticity may be advantageous for weed suppression at high density and spatial uniformity. Other research also suggests that reduced phenotypic plasticity is a good trait for weed suppression under high density and spatial uniformity, possibly because weed suppression is a group activity optimized by a uniform response by each plant (Weiner et al., 2010; Wu et al., 2021; Weiner, 2023).

© Burleigh Dodds Science Publishing Limited, 2025. All rights reserved.

In wheat, Weiner et al. (2001) confirmed that the attributes that result in good weed suppression in high cropping density may not be the same as those that are most advantageous in low cropping densities. The significant interaction between density and variety means that the varieties that perform best in high weed pressure at low crop density may not be those that perform best at high density.

The integration of crop competitiveness with other aspects of cultural management, such as crop density, mulch, and soil fertility, is necessary to achieve better results (Wu et al., 2021; Werle et al., 2023).

2.3 Fertility

Several studies show the effect of soil fertility on the crop–weed relationship (Kaur et al., 2018; Little et al., 2021). The studies demonstrate the complexity of interactions and responses depending on the factors involved. Although an increase in soil fertility has been associated with an increase in the degree of competition by weeds, it is essential for the crop to express its productive potential. Maintaining adequate fertility provides conditions for cultivated plants to grow faster and close soil cover earlier, thus influencing the critical periods of competition.

As with the choice of varieties, fertility management does not usually consider the issue of competition between weeds and crops. However, soil fertility effects on weed competition in more conservationist systems achieves better results when associated with various other management practices. As the popular saying goes, 'The crop is the best herbicide' meaning that, with a healthy plant community, crops can maintain greater vigor and competitive ability (Pacanoski, 2019). Another aspect is the effect on cover crops, which exert a more intense suppressive effect when under adequate conditions of fertility management.

Plants respond differently to available nutrients, which must be considered when plants are in association. In a mix of cover crops, e.g., there may be dominance/suppression of some species depending on the available nutrients. For instance, in a mixture of oats (*Avena* spp.) with vetch (*Vicia* spp.), oats will predominate if there is high nitrogen availability while, in contrast, low N availability favors vetch.

Relationships become more complex when the number of consortium species intensifies. Grasses respond well to nitrogen, radish (*Raphanus sativus*) (non-mycorrhizal) and buckwheat (*Fagopyrum esculentus*) to phosphorous, and pearl millet (*Pennisetum glaucum*) to potassium. Knowing the nutritional preferences of each species makes it easier to obtain the maximum benefits from each one and the best weed suppression.

© Burleigh Dodds Science Publishing Limited, 2025. All rights reserved.

3 Cultural methods: reduction of the presence of competing species

The seed bank is the reservoir weeds keep in the soil to ensure survival. Most seeds are dormant and can remain in the soil for prolonged periods without germinating. Dormancy is one of the main characteristics of weeds and what makes their eradication from an area difficult. The longevity of seeds in the dormancy state is variable between species. For the most common weed species, a simple seed classification proposed by Bakker et al. (1996) can be sufficient:

- 'Transitory' – species with seeds that persist in the soil for less than 1 year;
- 'Short-term persistent' – species with seeds that persist in the soil for at least 1 year but less than 5 years; and
- 'Long-term persistent' – species with seeds that persist in the soil for at least 5 years.

The decrease in the seed bank in the soil is a function of the loss of seed viability due to physiological age, the action of predators and microorganisms, or germination. The act of predators and microorganisms is a significant factor. The presence of living cover plants and their effect on soil life (increased microbial activity and presence of invertebrates) will contribute to reducing the viability of seeds in the soil and can lead to an exponential reduction in the weed seed bank (Roberts and Feast, 1973).

Maintaining soil cover with living plants, as well as using mulch and no-tillage (NT) methods, increases the concentration of seeds in the superficial layers where biological activity is more intense. This increases the presence of pathogens and predators (arthropods, fungi, and bacteria, among other organisms) that play an important role in the mortality of weed seeds (Brust and House, 1988; Menalled et al., 2000; Hakanson, 2003; Chee-Sanford et al., 2006; Meiss et al., 2010; Nikolić et al., 2020). However, the contribution that cover crops can bring to the seed bank reduction can vary depending on species and how they are managed (Davis et al., 2006).

The most apparent and immediate effect of live covers is the reduction in the presence of weeds (Baraibar et al., 2018; Osipitan et al., 2018; McKenzie-Gopsill et al., 2022). Perennial weeds are often better competitors and more challenging to control with cover crops than annual weeds because they have large nutritional reserves and rapid establishment (Hakansson, 2003; Teasdale et al., 2007). In this case, better results are obtained with cover crops with a long cycle, smothering the weeds for an extended period and depleting their nutritional reserves. However, lengthy periods without cash crops are not profitable for the farmer who needs income. Short-cycle cover crops may be less efficient but provide satisfactory results if combined with herbicides (Hakansson, 2003).

© Burleigh Dodds Science Publishing Limited, 2025. All rights reserved.

3.1 Crop rotation

Crop rotation is recognized as a fundamental practice in Conservation Agriculture. Its importance in IWM is recognized as essential in any production system. By increasing species and environmental diversity, rotations make it harder for weeds to become established. The more diverse the rotations, the more significant the impact. A crop or technique used continuously, even if it includes conservation principles, may result in weeds adapting to and proliferating in a more stable environment (Harker and O'Donovan, 2013). Guidelines and examples of rotation systems for different regions and production systems are available and can guide the implementation of a more balanced system (Taylor et al., 2008; Anderson, 2015; Martin et al., 2020; Benaragama et al., 2022).

3.2 Cover crops

The effect of cover crops on weeds occurs in two ways:

- as mulch; and
- as living soil cover.

Covering the soil with dead residues affects germination/emergence due to physical and allelopathic effects, altering the population density and floristic composition of weeds. Living soil cover suppresses both weed germination/emergence and development (Teasdale et al., 2007; Fernando and Shrestha, 2023).

Planting cover crops has made significant progress due to the necessity of producing straw to form mulch in NT systems. An example is what happened with NT in subtropical regions of Brazil. Initially winter cover crop species were used to form mulch for the cash crop the following summer. The cover crop species most planted at this period were the following grasses: black oat (*Avena strigosa*), Italian ryegrass (*Lolium multiflorum*), and rye (*Secale cereale*); forage radish (*Raphanus sativus*) among the brassicas; and vetches (*Vicia* spp.), lupins (*Lupinus* spp.), and forage pea (*Pisum sativum*) among legumes. The availability of seeds, low cost of planting, high biomass production, and the possibility of use as fodder make black oats the predominant species for use as a cover crop.

With the spread of NT system, new species (mainly summer cover crops) with rapid vegetative growth were evaluated to be inserted at intervals between crops. Crops suitable for good biomass formation and ground cover in short fallow periods were included: sunn hemp (*Crotalaria juncea*), high-density of sunflower (*Helianthus annuus*), foxtail millet (*Setaria italica*), forage sorghum

© Burleigh Dodds Science Publishing Limited, 2025. All rights reserved.

(*Sorghum bicolor*), sudangrass (*Sorghum sudanense*), pearl millet (*P. glaucum*), and buckwheat (*F. esculentum*).

In addition to these short-cycle plants, farmers also started to use longer-cycle species such as other crotalarias species (*C. breviflora*, *C. pallida*, and *C. spectabilis*, *C. ocraleuca*), giant crow's-foot grass (*Eleusine coracana*), and pigeon pea (*Cajanus cajan*). Because pearl millet has adapted well to local conditions, it has become popular. In the central tropical region of Brazil, a corn–brachiaria intercropping system was developed, allowing the use of a perennial species producing high above-ground and root biomass, with the possibility of being used as forage and for the formation of mulch.

In ecologically more sustainable production systems, cover crops have a fundamental role. They are characterized as functional plants that, in addition to the benefits in soil fertility (chemical, physical, and biological) and the management of pests and diseases, are still beneficial when considering weed substitutes. While many studies demonstrate the advantages of using cover crops, it is still necessary to make this practice more diverse to avoid the risk of cover crops becoming resistant to herbicides used to terminate them, as happened to ryegrass.

3.2.1 Mulching

The effect of mulching on weeds has been widely reported (Reberg-Horton et al., 2012; Osipitan et al., 2018; Adnan et al., 2020; Mennan et al., 2020). The studies demonstrate that changes in light availability, moisture, and temperature are the main factors that alter seed germination, emergence, and growth. Many species used for mulch formation also release allelopathic substances that interfere with weed development. The effects vary among the species. A larger volume of residue intensifies physical effects and releases more allelochemicals; residues with a high carbon:nitrogen (C:N) ratio remain in the soil longer, prolonging the suppression effect. In general, grasses are more efficient in suppressing weeds.

Planting systems that promote less soil disturbance result in better weed suppression. In annual crops, soil movement in the sowing row leads to loss of the mulching effect resulting in greater infestation in the row. The use, when possible, of discs only for cutting straw and depositing seed and fertilizer provides better coverage by the straw in the sowing row; using a chisel boot for fertilizer promotes greater movement. Planting before using a knife roller or desiccating with herbicide ('planting-green') (Ficks et al., 2023) also provides better planting row coverage.

A practice that uses the effect of mulching to suppress weeds with high efficiency is carried out in orchards (Bremer et al., 2008; Linares et al., 2008; Fidalski et al., 2010; Favarin et al., 2015; Martinelli et al., 2022; Azevedo et al.,

© Burleigh Dodds Science Publishing Limited, 2025. All rights reserved.

2020). In this system, a perennial cover crop with the potential to produce considerable biomass and a high C/N ratio is grown between rows of trees. This cover crop is then mowed periodically, and the residues are thrown into the row of plants using an ecological mower (with a circular cutting action requiring less power) (Martinelli et al., 2017). The interrow cover crop is kept fertilized to maintain constant biomass production and to prevent degradation and the development of undesirable plants.

Vegetable no-till is another system that allows better soil coverage and consequent suppression of weeds. Planting seedlings in holes allows for little soil movement and, therefore, little displacement of residues (Wildner do Prado et al., 2004; Robačer et al., 2016; Fayad et al., 2018; EMBRAPA, 2019; Fayad et al., 2019; Mennan et al., 2020). Species with higher dry matter and a high C/N ratio provide better and longer-lasting weed suppression. In general, however, this suppression is insufficient for complete control, requiring additional measures (Marchesan et al., 2016).

3.2.2 Living ground cover

Leaving no ground for weeds is important not only during the development of a crop but also in the intervals between crops. Using intercropping crops and replacing fallow periods with living crops to cover the ground prevents weeds from occupying space and replenishing the seed bank.

3.2.3 Intercropping

Plants that efficiently suppress weeds are species with fast growth and ground cover (Gerhards and Schappert, 2020). An intercropping species with rapid growth characteristics may be desirable in some systems, such as perennial crops (orchards). However, for annual crops, the intercrop crop species must have slow initial growth to avoid competition between it and the main crop (McKenzie-Gopsill et al., 2022). This will reduce the suppressive effect on weeds during the initial crop development period. Weeds present in this initial period must be controlled with another method. This may require cover crop species with a longer cycle than the main crop so that, at the end of the cash crop cycle after harvest, the cover crop would continue to vegetate and exert a suppressive effect on weeds (Hakansson, 2003).

Both slow initial growth and longer cycle than the main crop were found with signalgrass (*Brachiaria* spp.), with emphasis on *B. ruziziensis*, intercropped with corn. The low initial competitive effect (the intensity of which varies depending on the adopted system) and excellent development after the corn harvest (Melo et al., 2019; Concenço et al., 2012) make this system extremely attractive to farmers in tropical conditions.

© Burleigh Dodds Science Publishing Limited, 2025. All rights reserved.

3.3 Ley pastures

The crop–livestock farming is the closest to an ideal conservationist system. Including perennial crops during a sequence of crops allows for greater diversification (Martin et al., 2020). The system's complexity favors biodiversity and facilitates weed management (Meiss, 2010). However, more research involving several specialties is needed to understand the role of ley pastures in cropping systems (Martin et al., 2020; Schuster et al., 2020; Swanton and Vazan, 2022).

4 Organic agriculture

The importance of cultural practices for weed management assumes a more important role in organic production systems. The ban on herbicide use requires a good understanding of the relationship between crops and weeds to design efficient management systems (Liebman et al., 2001). Practices that favor the occupation of space by cultivated plants associated with the reduction of weed propagules are essential to keep the weed population at a manageable level and to facilitate complementary control by other methods (Stockdale et al., 2001; Liebman and Nichols, 2020).

A major challenge in organic agriculture is weed management in a NT system without herbicides (Reberg-Hortonn et al., 2012; Anderson, 2015; Vincent-Caboud et al., 2019; Skora Neto, 2022). The first challenge is the formation of the mulch. There are some cover crops where mechanical control with a knife roller is possible, but this depends on the development stage (Vincent-Caboud et al., 2019). The number of species suitable and the conditions for their management without using herbicides are limiting factors (Rouge et al., 2023). For some species, using a flail mower, which is more efficient on cover crops in earlier stages of development, can be an option (Skora Neto, 2022).

The second challenge is weed control during crop development. The presence of straw limits mechanical control, and manual weeding is difficult and time-consuming. In areas with low weed populations, mulch suppression may be sufficient to avoid yield losses in the crop. However, the continuity of the system can be hampered by the reproduction of the species present, with the risk of a substantial increase in weed population in the following crops. Combining a complex rotation system with occasional tillage interventions is possible (Anderson, 2015). The challenge for a total organic no-till is finding an efficient weed control method. A future possibility is the use of robotic weed control systems (Van der Weide et al., 2008; Wu et al., 2020; Taneja et al., 2023).

© Burleigh Dodds Science Publishing Limited, 2025. All rights reserved.

5 Case study

A series of studies were carried out in the southern region of Brazil to better understand fundamental aspects of the biology and behavior of the main weeds in the region, the dynamics of the relationship of weeds with cover crops and practical applications for the farmers (Skora Neto, 2022).

5.1 Weed seed longevity

Population density and survival of weed seeds were studied for 10 years in conventional tillage (CT) and NT systems. Corn was grown in the summer and cover crops in the winter (black oat with hairy vetch or blue lupine). The effect of the control undertaken during the whole cycle of corn (total control without weed reseeding) was compared to the control carried out only at the beginning of the crop life cycle, allowing reinfestation and weed reseeding (partial control with weed reseeding). The effect of the treatments was assessed by the number of plants that emerged annually, which was evaluated in the summer before undertaking weed control, and by the number of seeds in the soil seed bank, which was assessed in the winter, in the fourth and tenth year.

The weed species present in the area were alexandergrass (*Brachiaria plantaginea*) (the main weed present) (*Fig.* 1 and Table 1), southern crabgrass (*Digitaria ciliaris*), wild poinsettia (*Euphorbia heterophylla*), arrowleaf sida (*Sida rhombifolia*), and Brazil pusley (*Richardia brasiliensis*). This study shows that when reseeding was not allowed, all weed species showed **exponential population reduction**, varying in intensity according to the species. All the species studied in this chapter were characterized by a high percentage of seeds that germinate in the first year after release into the environment ('transitory').

What differentiates species from each other is the percentage of seeds that are 'persistent short-term' (1–5 years) and 'persistent long-term' (more than 5 years). Southern crabgrass and wild poinsettia are species with a low percentage of 'persistent short-term'; alexandergrass is a species with a high percentage of 'persistent short-term' and low percentage of 'persistent long-term' seeds; and Brazil pusley and arrowleaf sida are species with a high percentage of both 'persistent short-term' and 'persistent long-term' seeds.

This study also shows that the percentage of emerged plants of alexandergrass, comparing the emergence data (*Fig.* 1) and seeds in the soil (Table 1), was around 10% (roughly) of the number of seeds of the 0–5 cm layer. However, the population reduction from year to year, when seed bank replacement was not allowed, was far higher (reaching around 70% in the first years), indicating that there is **high mortality of the seeds in the soil.**

© Burleigh Dodds Science Publishing Limited, 2025. All rights reserved.

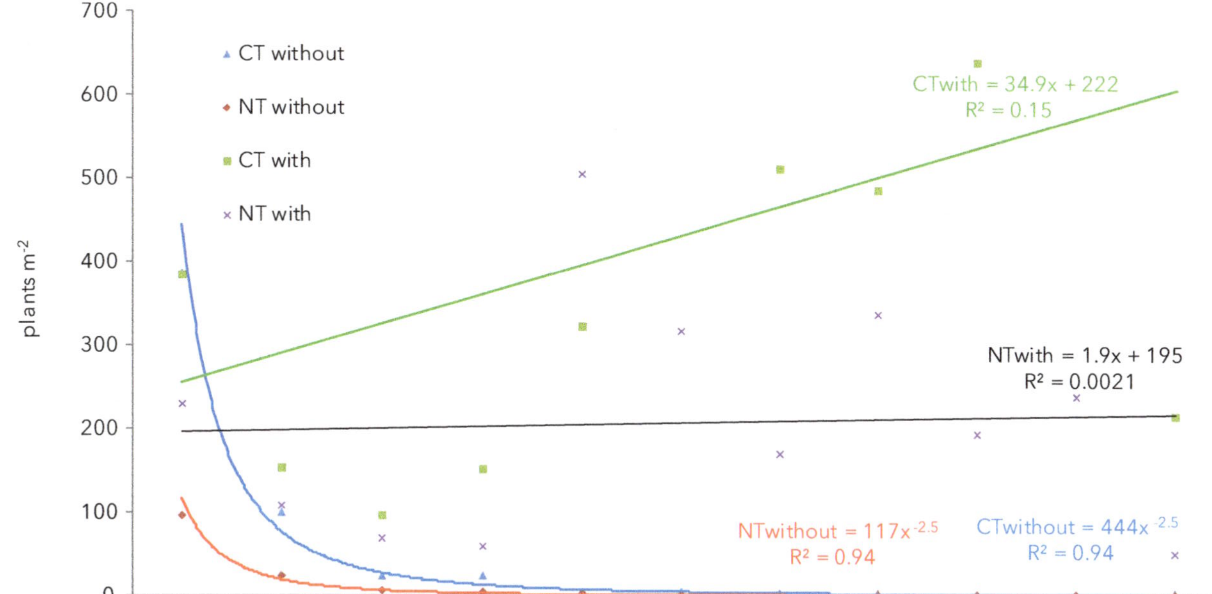

	0	1	2	3	4	5	6	7	8	9	10
CT without	386.7	100.8	25.1	24.6	5.4	5.5	2.6	2.0	0.7	1.6	2.6
NT without	97.7	24.6	7.1	4.0	3.1	0.6	1.3	0.6	0.2	0.7	0.2
CT with	385.0	153.9	97.6	152.6	322.2	1077.1	510.0	485.1	638.4	706.5	214.8
NT with	230.0	109.3	69.4	59.0	504.3	316.4	169.4	336.1	193.8	238.4	50.5

Years

Figure 1 Population density (m^{-2}) of alexandergrass (*Brachiaria plantaginea*) in conventional tillage without (CT without) and with reseeding (CT with), and no-tillage without (NT without) and with reseeding (NT with) over 10 years. Ponta Grossa, PR, Brazil.

© Burleigh Dodds Science Publishing Limited, 2025. All rights reserved.

Table 1 Number of seeds m^{-2} of alexandergrass (*Brachiaria plantaginea*) at three soil depths in corn, in conventional tillage and no-tillage, with total weed control (without reseeding) and with partial control of the weeds (with reseeding). Ponta Grossa, PR, Brazil.

Treatments			Seeds m^{-2}	
Planting system	Weeds	Depth (cm)	4^{0} YEAR	10^{0} YEAR
Conventional tillage	Total control WITHOUT reseeding	0-5	22	44
		5-10	89	22
		10-20	89	33
No-tillage		0-5	22	11
		5-10	0	22
		10-20	333	44
Conventional tillage	Partial control WITH reseeding	0-5	2,133	2,911
		5-10	2,867	3,111
		10-20	5,889	3,756
No-tillage		0-5	5,867	3,244
		5-10	2,600	1,967
		10-20	1,267	3,544

5.2 Effect of cover crops on weed population reduction

The weed population density was evaluated when cover crops substituted fallow periods. Common bean (*Phaseolus vulgaris*) was grown in spring, followed by pearl millet (*P. glaucum*) in the autumn and black oat (*Avena strigosa*) in the winter as cover crops; common bean in spring, followed by fallow in autumn and winter (without cover crops). An exponential decline in the weed population density was observed (Figs 2 and 3).

Purple nutsedge (*Cyperus rotundus*), one of the species in the area, had an exponential increase when the area was kept in fallow during the autumn and winter, and an exponential decrease in the area maintained with cash crop and cover crops throughout the year (*Fig.* 4). This study confirms the **effect of cover crops on weed population reduction** and the associated impact of desiccant herbicides and cover crops in a perennial species.

5.3 Cover crops, mulching, and herbicides

In a demonstration plot to show the change in weed density in NT, three weed control measures were adopted:

- No control;
- Using herbicides only; and
- Total control.

© Burleigh Dodds Science Publishing Limited, 2025. All rights reserved.

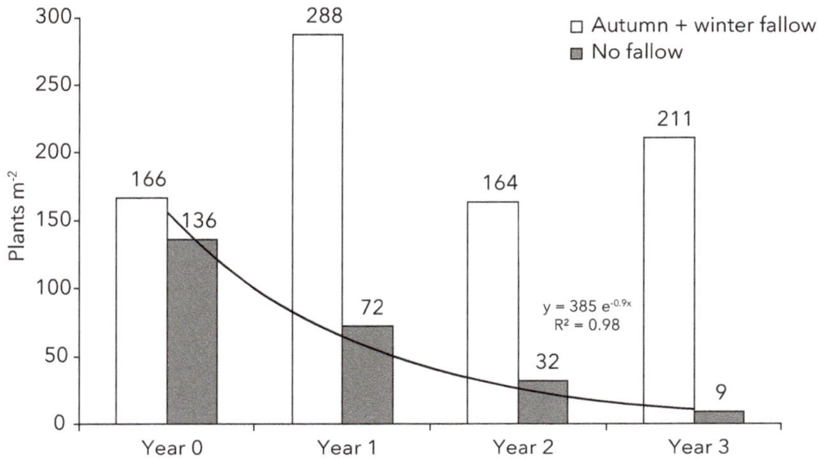

Figure 2 Effect of eliminating fallow on weed population density. Ponta Grossa, PR, Brazil.

In this third treatment, when necessary, control was performed using herbicides and weeding (hoeing) to control weed seed production and reduce the density to a level considered adequate to dispense with herbicide use. The system adopted was corn and common bean rotation as cash crops in spring/summer and cover crops in the intervals.

The initial weed flora was composed of alexandergrass (*B. plantaginea*) (77%), southern crabgrass (*D. ciliaris*) (8%), wild poinsettia (*E. heterophylla*) (9%), gallant soldier (*Galinsoga parviflora*) (5%), arrowleaf sida (*S. rhombifolia*) (0.7%), wild radish (*Raphanus raphanistrum*) (0.2%), beggarticks (*Bidens pilosa*) (0.2%), and little bell (*I. triloba*) (0.2%).

The weed density reduction was exponential using herbicide only and with total weed control. With total weed control, the weed density in the second year

Figure 3 Weed population density after three years of treatments with fallow (common bean -> autumn fallow -> winter fallow) (A) and without fallow (common bean -> pearl millet -> black oat) (B). Ponta Grossa, PR, Brazil.

© Burleigh Dodds Science Publishing Limited, 2025. All rights reserved.

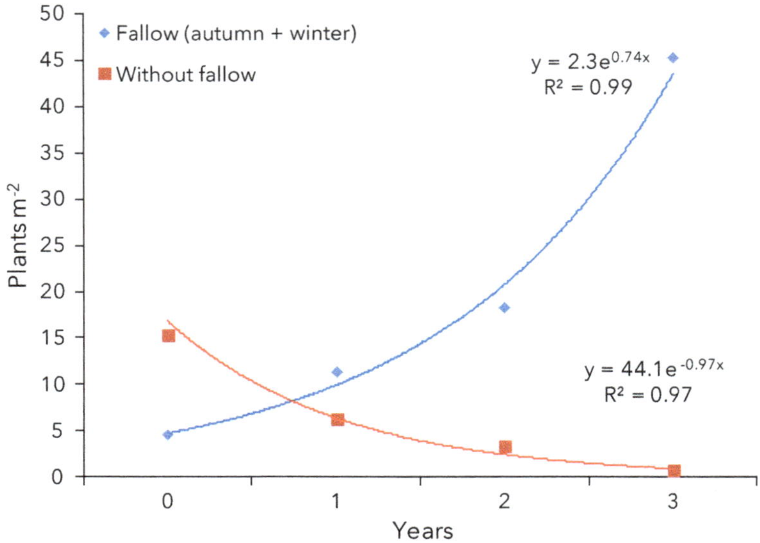

Figure 4 Evolution of population density (plants m^{-2}) of purple nutsedge (*Cyperus rotundus*) in areas with and without fallow during 3 years of treatment effect. Ponta Grossa, PR, Brazil.

was already at a level that allowed the complete elimination of herbicide from the third year. In the area with herbicide without any complementary control, such a level of weed density was reached in the third year. In the unweeded area, the density was variable, with a peak in the third year.

The study again confirms a rapid reduction in weed population density when seed production is prevented. It demonstrates that **associating herbicides with cover crops is an efficient method to obtain areas with low weed population density**. With mulching and cover crops alone, it is it not possible to get reasonable weed control and waive any control measure (*Fig*. 5).

5.4 Weed density in the period of competition

The effect of two weed densities on the common bean crop yield was evaluated to verify if the CTWR (Critical Timing for Weed Removal) changes in different weed populations. The predominant weeds were southern crabgrass (*D. ciliaris*), purple nutsedge (*C. rotundus*), alexandergrass (*B. plantaginea*), wild poinsettia (*E. heterophylla*), and little bell (*I. triloba*). The average total density in the treatment with the high population was 148 plants m^{-2} and with the low population, 13 plants m^{-2}.

© Burleigh Dodds Science Publishing Limited, 2025. All rights reserved.

	Year 0	Year 1	Year 2	Year 3
Unweeded	262	91	71	483
Herbicide	262	27	18	3.5
Herbicide + hoeing	152	17	5	2.5

Figure 5 Evolution of weed density (plants m^{-2}) in areas under different management: 1. Unweeded, 2. With herbicides, and 3. With total control. Ponta Grossa, PR, Brazil. (Demonstrative unit).

The total loss in common bean yield was 61.0% and 8.8% for high and low weed populations, respectively, when coexisting with weeds throughout the crop cycle. In addition to the effect of weed densities on the level of interference, it was also found that the period before interference was altered; for the high infestation, the CTWR was less than 15 days, and for low infestation, the CTWR was approximately 30 days (*Fig.* 6).

This study supports the view that **there are no fixed periods of coexistence with weeds for each crop**; a large number of variables (species, population

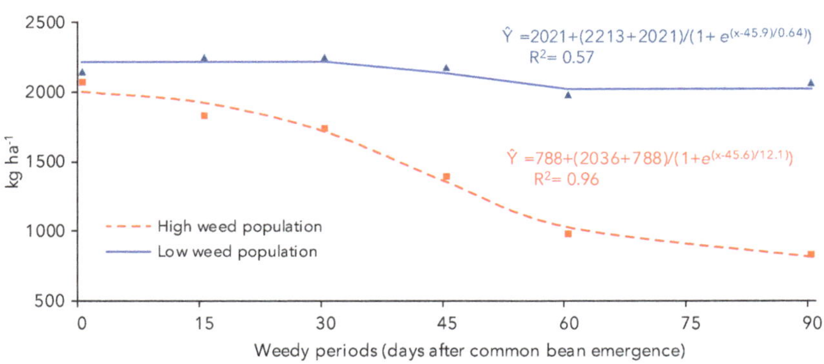

Figure 6 Interference level of different weed populations in the yield of common bean and the duration of the CTWR – Critical Timing for Weed Removal. Ponta Grossa, PR, Brazil.

© Burleigh Dodds Science Publishing Limited, 2025. All rights reserved.

density and arrangement, fertility, climate, and others) affects the duration of periods of coexistence between crops and weeds. These results may also be extrapolated to the relationship between species in intercropping systems.

5.5 Weed density compared to weed control methods

Hoeing, herbicide at a half-rate of the recommended [(fluazifop + fomesafen) + bentazon – (100 + 150) + 360 g ai ha^{-1}], and herbicide at the normal rate (sequential application – 2 × half-rate) were evaluated in two weed densities (high – 180 plants m^{-2} and low – 9 plants m^{-2}) in common bean crop. The weeds present were wild poinsettia (*E. heterophylla*), alexandergrass (*B. plantaginea*), southern crabgrass (*D. ciliaris*), purple nutsedge (*C. rotundus*), little bell (*I. triloba*), and arrowleaf sida (*Sida rhombifolia*).

In the evaluation at 60 days after crop emergence, in the lower weed population area, all the control measures provided excellent weed control; in the higher population area, the control was deficient (*Fig.* 7). In hoeing, besides the lower control, the time spent in the operation (256 h ha^{-1}) was higher than in the area with the low population (67 h ha^{-1}). Besides the efficacy of the treatments being lower, the reinfestation pressure was more intense in the area with a higher population. This study highlights **the importance of low weed density on the efficacy of weed control.**

5.6 Row and interrow weed infestation

Weed infestation in the crop row and interrow was assessed to evaluate the effect of cover crop mulching. There was a difference among cover crop species on weed density; ryegrass and black oat reduced weed emergence. Because of the low ground cover and rapid degradation, canola mulching had the same number

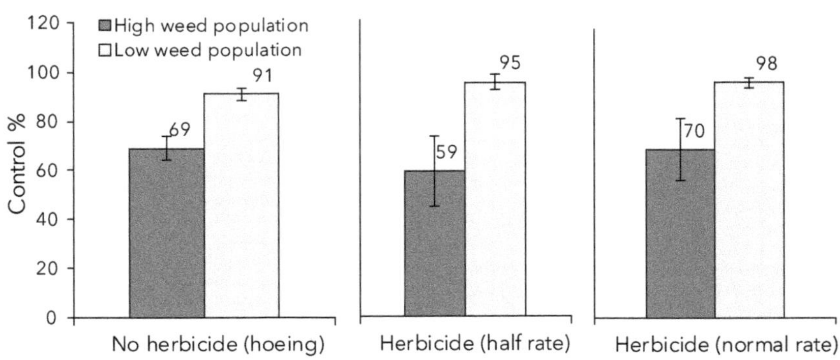

Figure 7 Percentage of weed control in high and low weed populations with hoeing and post-emergence herbicides. Ponta Grossa, PR, Brazil.

© Burleigh Dodds Science Publishing Limited, 2025. All rights reserved.

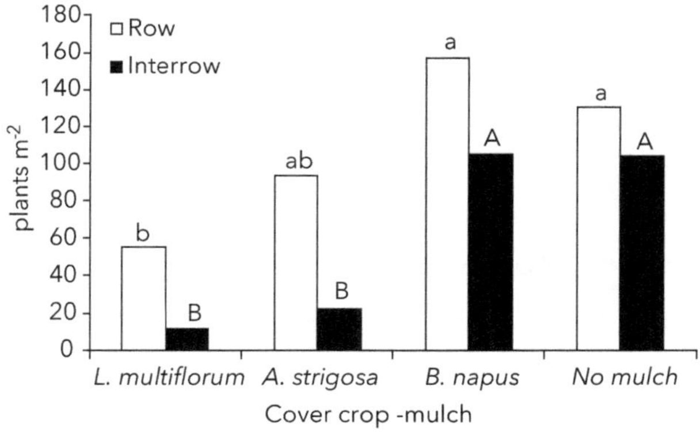

Figure 8 Weed density (plants m^{-2}) in corn row and interrow with a mulch of Italian ryegrass (*Lolium multiflorum*), black oat (*Avena strigosa*), canola (*Brassica napus*), and no mulch area. Ponta Grossa, PR, Brazil.

of weeds as the area without mulching. There was also greater weed emergence in the row than in the interrow because of soil disturbance (*Fig.* 8). This evaluation highlights the **effect of mulch and soil movement on weed emergence**.

6 Conclusion and future trends

Cultural practices such as adequate spacing and population density, adequacy of soil fertility (chemical, physical, and biological), and adapted varieties favor space occupation. They are basic principles to optimize the use of resources by crops in any cultivation system.

Cover crops and rotations are other practices that alter the life cycle and are fundamental for maintaining the weed population at a low population level. These practices are essential for proper weed management and, consequently, for conservation systems.

Isolated cultural practices are insufficient to overcome the negative effects of weeds. The proposed conservation management approach is a combination of measures ('many little hammers') (Liebman and Gallandt, 1997).

Despite studies showing the effectiveness of IWM, the adoption of these practices has been limited for a range of reasons (Moss, 2019; Bastiaans et al., 2008; Harker et al., 2017). There have been attempts to overcome resistance to adopting IWM techniques (Colas et al., 2020). The successful experience of adopting the NT system (Bolliger et al., 2006) reinforces the active role of farmers. Farmers who see the need for transformation already incorporate ecological concepts into their farming activities. Collaborating with various agents, researchers, consultants, extensionists, industry, and government seems the way to go.

© Burleigh Dodds Science Publishing Limited, 2025. All rights reserved.

IWM research must continue the search to understand the dynamics of weed interactions in agrosystems, develop equipment, and improve new control technologies (Taylor et al., 2008). There is also a need for integration with specialists in soil management, pest and disease management in the search for solutions in the broad context of Conservation Agriculture.

7 Where to look for further information

- Bastiaans, L., Paolini, R. and Baumann, D. T. (2008), 'Focus on ecological weed management: What is hindering adoption?', *Weed Res.*, 48, 481–491.
- Bhaskar, V., Westbrook, A. S., Bellinder, R.R. and DiTommaso, A. (2021), 'Integrated management of living mulches for weed control: A review', *Weed Technol.,* 35, 856–868.
- Brainard, D., Haramoto, E., Leon, R., Kells, J., Van Wychen, L., Devkota, P., Jugulan, M. and Barney, J. (2023), 'A survey o weed research priorities: key findings and future directions', *Weed Sci.,* 71(4), 330-343.
- Gage, K. L. and Schwartz-Lazaro, L. M. (2019), 'Shifting the paradigm: An ecological systems approach to weed management', *Agriculture*, 9, 179.
- Håkansson, S. (2003), *Weeds and Weed Management on Arable Land: An Ecological Approach*. Wallingford, Oxon, UK; Cambridge, MA, CABI Pub.
- Haring, S. (2021), 'A progressive approach for integrated pest management', *Weed Sci.,* 69(2), 129–131.
- Kumar, V., Obour, A., Jha, P., Liu, R., Manuchehri, M. R., Dille, J. A., Holman, J. and Stahlman, P. W. (2020), 'Integrating cover crops for weed management in the semiarid U.S. Great Plains: opportunities and challenges', *Weed Sci.*, 68, 311–323.
- MacLaren, C., Storkey, J., Menegat, A. Metcalfe, H. and Dehnen-Schmutz, K. (2020), 'An ecological future for weed science to sustain crop production and the environment. A review', *Agron. Sustain. Dev.,* 40, 24.
- McCauley, C., Legleiter, T., Herman, R., Rasoulpour, R., Schroeder, J., Pilcher, T., Meinders, K. and Wright, T. (2022), 'Sustainable weed management – What is it and how are we doing?', *Weed Technol.,* 36, 768–776.
- Liebman, M., Baraibar, B., Buckley, V., Childs, D., Christensen, S., Cousens, R., Eizenberg, H., Heijting, S., Loddo, D., Merotto Jr, A., Renton, M. and Riemens, M. (2016), 'Ecologically sustainable weed management: How do we get from proof-of-concept to adoption?', *Ecol. Appl.*, 26(5), 1352–1369.
- Martin, G., Durand, J-L., Duru, M., Gastal, F., Julier, B., Litrico, I., Louarn, G., Médiène, S., Moreau, D., Valentin-Morison, M., Novak, S., Parnaudeau, V., Paschalidou, F., Vertès, F., Voisin, A-S., Cellier, P. and Jeuffroy, M-H. (2020),

© Burleigh Dodds Science Publishing Limited, 2025. All rights reserved.

'Role of ley pastures in tomorrow's cropping systems. A review', *Agron. Sustain. Dev.,* 40, 17.

- Neve, P., Barney, J. N., Buckley, Y., Cousens, R. D., Graham, S., Jordan, N. R., Lawton-Rauh, A., Liebman, M., Mesgaran, M. B., Schut, M., Shaw, J., Storkey, J., Baraibar, B., Baucom, R. S., Chalak, M., Childs, D. Z., Christensen, S., Eizenberg, H., Fernández-Quintanilla, C., French, K., Harsch, M., Heijting, S., Harrison, L., Loddo, D., Macel, M., Maczey, N., Merotto, A., Jr, Mortensen, D., Necajeva, J., Peltzer, D. A., Recasens, J., Renton, M., Riemens, M., Sønderskov, M. and Williams, M. (2018), 'Reviewing research priorities in weed ecology, evolution and management: A horizon scan', *Weed Res.,* 58(4), 250–258.
- Vincent-Caboud, L., Casagrande, M., David, C., Ryan, M. R., Silva, E. M. & Peigne, J. (2019), 'Using mulch from cover crops to facilitate organic no-till soybean and maize production. A review', *Agron. Sustain. Dev.,* 39, 45.
- Westwood, J. H., Charudattan, R., Duke, S. O., Fennimore, S. A., Marrone, P., Slaughter, D. C., Swanton, C. and Zollinger, R. (2018), 'Weed management in 2050: Perspectives on the future of weed science', *Weed Sci.*, 66, 275–285.
- Young, S. (2020), 'A unifying approach for IWM', *Weed Sci.,* 68(5), 435–436.

8 References

Adnan, M., Asif, M., Khalid, M., Abbas, B., Hayyat, M. S., Raza, A., Khan, B. A., Hassan, M., Khan, M. A. B. and Hanif, M. S. (2020), 'Role of mulches in agriculture: A review', *Int. J. Bot. Std.*, 3, 309–314.

Anderson, R. L. (2015), 'Integrating a complex rotation with no-till improves weed management in organic farming. A review', *Agron. Sustain. Dev*., 35, 967–974.

Azevedo, F. A., Almeida, R. F., Martinelli, R., Próspero, A. G, Licerre, R., Conceição, P. M., Arantes, A. C. C., Dovis, V. L., Boaretto, R. M. and Mattos Jr, D. (2020), 'No-tillage and high-density planting for Tahiti acid lime grafted onto flying dragon trifoliate orange', *Front. Sustain. Food Syst*., 4, 108.

Bakker, J. P.; Poschlod, P.; Strykstra, R. J.; Bekker, R. M.; Thompson, K. (1996), Seed banks and seed dispersal: Important topics in restoration ecology. *Acta Botanica Neerlandica*, 45, 461-490.

Baraibar, B., Hunter, M., Schipanski, M., Hamilton, A. and Mortensen, D. (2018), 'Weed suppression in cover crop monocultures and mixtures', *Weed Sci.*, 66(1), 121–133.

Bastiaans, L., Paolini, R. and Baumann, D.T. (2008), 'Focus on ecological weed management: What is hindering adoption?', *Weed Res.*, 48, 481–491.

Benaragama, D., May, W., Gulden, R. and Willenborg, C. (2022), 'Functionally diverse flax-based rotations improve wild oat (*Avena fatua*) and cleavers (*Galium spurium*) management', *Weed Sci.,* 70(2), 220–234.

Bolliger, A.M., Magid, J., Amado, J., Skora Neto, F.S., Ribeiro, M.D., Calegari, A., Ralisch, R. and Neergaard, A. (2006), 'Taking stock of the Brazilian "zero-till revolution": A review of landmark research and farmers' practice', *Adv. Agron.,* 91, 47–110.

© Burleigh Dodds Science Publishing Limited, 2025. All rights reserved.

Bradley, K. W. (2006), 'A review of the effects of row spacing on weed management in corn and soybean', *Crop Manage.* https://doi.org/10.1094/CM-2006-0227-02-RV [Online]. Accessed August 4, 2023.

Bremer Neto, H., Victoria Filho, R., Mourão Filho, F. D. A. A., de Menezes, G. M. and Canali, E. (2008), 'Nutritional status and production of 'Pêra' sweet orange related to cover crops and mulch',*Pesq. Agrop. Bras.*, 43, 29–35.

Brust, G. and House, G. (1988), 'Weed seed destruction by arthropods and rodents in low-input soybean agroecosystems', *Am. J. Altern. Agric.*, 3(1), 19–25.

Cahill, J. F., Kembel, S. W. and Gustafson, D. J. (2005), 'Differential genetic influences on competitive effect and response in *Arabidopsis thaliana*', *J. Ecol.*, 93, 958–967.

Charles, G. W. and Taylor, I. N. (2021), 'Extending the critical period for weed control model to better include weed succession using common sunflower as a mimic weed in high-yielding cotton', *Weed Technol.*, 35(6), 1029–1037.

Chee-Sanford, J. C., Williams, M. M., Davis, A. S. and Sims, G. K. (2006), 'Do microorganisms influence seed-bank dynamics?', *Weed Sci.*, 54, 575–587.

Colas, F., Cordeau, S., Granger, S., Jeuffroy, M.-H., Pointurier, O., Queyrel, W., Rodriguez, A., Villerd, J. and Colbach, N. (2020), 'Co-development of a decision support system for integrated weed management: Contribution from future users', *Eur. J. Agron.*, 114, 126010.

Concenço, G., Ceccon, G, Fonseca, I. C., Leite, L.F., Schwerz, F. and Correia, I. V. T. (2012), 'Weeds infestation in corn intercropped with forages at different planting densities', *Planta Daninha*, 30(4), 721–728.

Davis, A., Anderson, K. I., Hallett, S. G. and Renner, K. A. (2006), 'Weed seed mortality in soils with contrasting agricultural management histories', *Weed Sci.*, 54, 291–297.

EMBRAPA (2019), *Sistema de plantio direto de hortaliças*. https://ainfo.cnptia.embrapa.br/digital/bitstream/item/210362/1/folder-spdh.pdf. Accessed August 4, 2023. (Portuguese)

Ethridge, S. R. R., Locke, A. M. M., Everman, W. J. J., Jordan, D. L. L., and Leon, R. G. G. (2022), 'Crop physiological considerations for combining variable-density planting to optimize seed costs and weed suppression', *Weed Sci.*, 70(6), 1–35.

Favarin, J. L., Tezotto, T., Pedrosa, A. W. and Paula Neto, A. (2015), 'Coffee–forage intercropping is a sustainable production system for Brazil', *Bet. Crops Plan. Food*, 99, 16–17.

Fayad, J. A., Arl, V., Comin, J. J., Mafra, A. L. and Marchesi, D. R. (2019), *Sistema de Plantio Direto de Hortaliças*, Epagri, Florianópolis, SC. (Portuguese)

Fayad, J. A., Comin, J. J., Kurtz, C. and Mafra, A. (2018), *Sistema de Plantio Direto de Hortaliças (SPDH). O cultivo da Cebola*, EPAGRI, Florianopolis, SC, Boletim didático, 146. (Portuguese)

Fernando, M. and Shrestha, A. (2023). 'The potential of cover crops for weed management: A sole tool or component of an integrated weed management system?', *Plants*, 12, 752.

Ficks, T. S., VanGessel, M. J. and Wallace, J. M. (2023), 'Cereal rye seeding rate does not affect magnitude of weed suppression when planting green within Mid-Atlantic United States', *Weed Technol.*, 36(6), 1–6.

Fidalski, J., Auler, P. A. M., Beraldo, J. M. G., Marur, C. J., de Faria, R. T. and Barbosa, G. M. (2010), 'Availability of soil water under tillage systems, mulch management and citrus rootstocks', *Rev. Bras. Ciênc. Solo,* 34, 917–924.

© Burleigh Dodds Science Publishing Limited, 2025. All rights reserved.

Fipke, M. V. and Vidal, R. A. (2019), 'Influence of density and development stage of ryegrass on glyphosate effectiveness', *Planta Daninha*, 37.

Gerhards, R. and Schappert, A. (2020), 'Advancing cover cropping in temperate integrated weed management', *Pest Manage. Sci.*, 76, 42-46.

Håkansson, S. (2003), *Weeds and Weed Management on Arable Land: An Ecological Approach.* Wallingford, Oxon, UK; Cambridge, MA, CABI Pub.

Harder, D. B., Sprague, C. L. and Renner, K. A. (2007), 'Effect of soybean row width and population on weeds, crop yield, and economic return', *Weed Technol.*, 21(3), 744–752.

Harker, K., Mallory-Smith, C., Maxwell, B., Mortensen, D. and Smith, R. (2017), 'Another view', *Weed Sci.*, 65(2), 203–205.

Harker, K. and O'Donovan, J. (2013), 'Recent weed control, weed management, and integrated weed management', *Weed Technol.*, 27, 1–11.

Heap, I., *The International Herbicide-Resistant Weed Database.* http://www.weedscience.org. Accessed July 25, 2023.

Holkem, A. S., Silva, A. L., Bianchi, M. A., Corassa, G. and Ulguim, A. R. (2022), 'Weed management in Roundup Ready® corn and soybean in Southern Brazil: Survey of consultants' perception', *Adv. Weed Sci.*, 40, e020220111.

HRAC (2023), *Best Management Practices.* https://hracglobal.com/prevention-management/best-management-practices. Accessed July 25, 2023.

Kaur, S., Kaur, R. and Chauhan, B. S. (2018), 'Understanding weed-crop-fertilizer-water interactions and their implications for weed management in agricultural systems', *Crop Prot.*, 103, 65–72.

Knezevic, Z. S., Evans, S.P., Blankenship, E., Vanacker, R. and Lindquist, J. L. (2002), 'Critical period of weed control: The concept and data analysis', *Weed Sci.*, 50, 773–786.

Kudsk, P. (ed.) (2022), *Advances in Integrated Weed Management.* Burleigh Dodds Science Publishing, Cambridge, UK.

Liebman, M. and Gallandt, E.R. (1997), 'Many little hammers: ecological approaches for management of crop-weed interactions'. In: L. E. Jackson (ed.), *Ecology in Agriculture.* Academic Press, San Diego, CA, pp. 291–343.

Liebman, M., Mohler, C.L. and Staver, C.P. (2001), *Ecological Management of Agricultural Weeds.* Cambridge University Press, Cambridge.

Liebman, M. and Nichols, V. A. (2020), 'Cropping system redesign for improved weed management: A modeling approach illustrated with giant ragweed (*Ambrosia trifida*)', *Agronomy*, 10, 262.

Linares, J., Scholberg, J., Boote, K., Chase, C. A., Ferguson, J. J. and McSorley, R. (2008), 'Use of the cover crop weed index to evaluate weed suppression by cover crops in organic citrus orchards', *HortScience*, *43*(1), 27–34.

Little, N. G., Di Tommaso, A., Westbrook, A. S., Ketterings, Q. M. and Mohler, C. L. (2021), 'Effects of fertility amendments on weed growth and weed–crop competition: A review', *Weed Sci.*, 69, 132–146.

Marchesan, E. D., Trezzi, M., Barancelli, M. V. J., Batistel, S., Pagnoncelli, F. and Vidal, R. A. (2016), 'Integration mulches with atrazine for weed management in corn', *Braz. J. Agric. Sci.*, 11, 1–7.

Marín, C. and Weiner, J. (2014), 'Effects of density and sowing pattern on weed suppression and grain yield in three varieties of maize under high weed pressure', *Weed Res.*, 54, 467–474.

© Burleigh Dodds Science Publishing Limited, 2025. All rights reserved.

Martin, A. R. and Namuth, D. M. (2017), 'Practical applications of herbicide physiology. Weed density and soil active herbicides', Lincoln: Plant and Soil Sciences eLibrary. https://passel2.unl.edu/view/lesson/61a78f20f4b1/4. Accessed August 4, 2023.

Martin, G., Durand, J. L., Duru, M., Gastal, F., Julier, B., Litrico, I., Louarn, G., Médiène, S., Moreau, D., Valentin-Morison, M. and Novak, S. (2020), 'Role of ley pastures in tomorrow's cropping systems. A review', *Agron. Sustain. Dev.*, 40, 17.

Martinelli, R., Monquero, P., Fontanetti, A., Conceição, P. and Azevedo, F. (2017), 'Ecological mowing: An option for sustainable weed management in young citrus orchards', *Weed Technol.*, 31(2), 260–268.

Martinelli, R., Rufino Jr, L. R., Alcantara-de la Cruz, R., da Conceição, P. M., Monquero, P. A. and de Azevedo, F. A. (2022), 'Glyphosate excessive use affects citrus growth and yield: The vicious (and unsustainable) circle in Brazilian orchards', *Agronomy,* 12, 453.

McCollough, M. R. and Melander, B. (2022), 'Improving upon the interrow hoed cereal system: The effects of crop density and row spacing on intrarow weeds and crop parameters in spring barley'. *Weed Sci.*, 70, 341–352.

McKenzie-Gopsill, A., Mills, A., MacDonald, A. N. and Wyand, S. (2022), 'The importance of species selection in cover crop mixture design', *Weed Sci.,* 70, 436–447.

Meiss, H. (2010), 'Diversifying crop rotations with temporary grasslands: potentials for weed management and farmland biodiversity'. *Thesis*... Agricultural Sciences. Université de Bourgogne.

Meiss, H., Lagadec, L., Munier-Jolan, N., Waldhardt, R. and Petit, S. (2010), 'Weed seed predation increases with vegetation cover in perennial forage crops', *Agric. Ecosyst. Environ.*, 138(1–2), 10–16.

Melo, T. S., Makino, P. A. and Ceccon, G. (2019), 'Weed diversity in corn with different plant arrangement patterns grown alone and intercropped with palisade grass', *Planta Daninha*, 37, e019195957.

Menalled, F., Landis, D., Lee, J., White, S. and Renner, K. (2000), *Ecology and Management of Weed Seed Predators in Michigan Agroecosystems.* MSU Extension Bulletin E-2716.

Mennan, H., Khawar Jabran, K., Zandstra, B. H. and Pala, F. (2020), 'Non-chemical weed management in vegetables by using cover crops: A review', *Agronomy*, 10(2), 257.

Miller, T. E. and Werner, P. A. (1987), 'Competitive effects and responses between plant species in a first-year old-field community', *Ecology*, 68(5), 1201–1210.

Moss, S. (2019), 'Integrated weed management (IWM): why are farmers reluctant to adopt non-chemical alternatives to herbicides?' *Pest Manage. Sci.*, 75(5), 1205–1211.

Mulugeta, D. and Boerboom, C. (2000), 'Critical time of weed removal in glyphosate-resistant Glycine max', *Weed Sci.,* 48(1), 35–42.

Nikolić, N., Squartini, A., Concheri, G., Stevanato, P., Zanin, G. and Masin, R. (2020), 'Weed seed decay in no-till field and planted riparian buffer zone', *Plants*, 9, 293.

Olsen, J., Kristensen, L., Weiner, J. and Griepentrog, H. W. (2005), 'Increased density and spatial uniformity increase weed suppression by spring wheat', *Weed Res.*, 45, 316–321.

Osipitan, O. A., Dille, J., Yared, A. and Knezevic, S. Z. (2018), 'Cover crop for early season weed suppression in crops: Systematic review and meta-analysis', *Agron. J.,* 110, 1–11.

Pacanoski, Z. (2019), 'Synergy between crop competition agronomic practices and herbicides for effective weed control in plant production', *Acta Herbol.*, 28(1), 5-15.

© Burleigh Dodds Science Publishing Limited, 2025. All rights reserved.

Place, G. T., Reberg-Horton, S. C., Dickey, D. A. and Carter Jr, T. E. (2011), 'Identifying soybean traits of interest for weed competition', *Crop Sci.*, 51, 2642–2654.

Reberg-Horton, S., Grossman, J., Kornecki, T., Meijer, A., Price, A., Place, G. and Webster, T. (2012), 'Utilizing cover crop mulches to reduce tillage in organic systems in the southeastern USA', *Renew. Agric. Food Syst.*, 27(1), 41–48.

Robačer, M., Canali, S., Kristensen, H. L., Bavec, F., Mlakar, S. G., Jakop, M. and Bavec, M. (2016), 'Cover crops in organic field vegetable production', *Sci. Hortic.*, 208, 104–110.

Roberts, H. A. and Feast, P. M. (1973), 'Emergence and longevity of seeds of annual weeds in cultivated and undisturbed soil', *J. Appl. Ecol.*, 10(1), 133–143.

Rosset, J. D. and Gulden, R. H. (2020), 'Cultural weed management practices shorten the critical weed-free period for soybeans grown in the Northern Great Plains', *Weed Sci.*, 68(1), 79–91.

Rouge, A., Adeux, G., Busset, H., Hugard, R., Martin, J., Matejicek, A., Moreau, D., Guillemin, J.-P. and Cordeau, S. (2023), 'Carry-over effects of cover crops on weeds and crop productivity in no-till systems', *Field Crops Res.*, 295, 108899.

Ruberti, I., Sessa, G., Ciolfi, A., Possenti, M., Carabelli, M. and Morelli, G. (2012), 'Plant adaptation to dynamically changing environment: The shade avoidance response',*Biotechnol. Adv.*, 30(5), 1047–1058.

Schuster, M. Z., Gastal, F., Doisy, D., Charrier, X., Moraes, A., Mediene, S. and Barbu, C. (2020), 'Weed regulation by crop and grassland competition: critical biomass level and persistence rate', *Eur. J. Agron.*, 113, 125963.

Skora Neto, F. (2022), *Manejo sustentável de plantas daninhas: fundamentos para um sistema de plantio direto sem herbicida*. Londrina, PR, IDR-Paraná. (Portuguese)

Stockdale, E., Lampkin, N., Hovi, M., Keatinge, R., Lennartsson, M., Macdonald, D., Padel, S., Tattersall, F., Wolfe, M. S. and Watson, C. (2001), 'Agronomic and environmental implications of organic farming systems', *Adv. Agron.*, 70, 261–327.

Swanton, C. and Vazan, S. (2022), 'Influence of agronomic practices on the persistence of weed seedbanks'. In Upadhyaya, M. K., Clements, D. R. and Shrestha, A. (eds.),*Persistence Strategies of Weeds*. John Wiley & Sons Ltd., Hoboken, New Jersey, USA. Chapter 10.

Taneja, A., Nair, G., Joshi, M., Sharma, S., Sharma, S., Jambrak, A. R., Roselló-Soto, E., Barba, F. J., Castagnini, J. M., Leksawasdi, N. and Phimolsiripol, Y. (2023), 'Artificial intelligence: Implications for the agri-food sector', *Agronomy*, 13(5), 1397.

Taylor, E., Renner, K. and Sprague, C. (2008), *Integrated Weed Management: Fine Tuning the System*, Extension bulletin E-3065. Michigan State University, East Lansing, MI.

Teasdale, J. R. (1995), 'Influence of narrow row/high corn population (*Zea mays*) on weed control and light transmittance', *Weed Technol.*, 9, 113–118.

Teasdale, J. R., Brandsaeter, L. O., Calegari, A., Skora Neto, F., Upadhyaya, M. K. and Blackshaw, R. E. (2007), 'Cover crops and weed management'. In M. K. Upadhyaya and R. E. Blackshaw (eds.), *Non-chemical Weed Management: Principles, Concepts and Technology*. CAB International, Cambridge, MA, pp. 49–64.

Trezzi, M. M., Balbinot Jr, A. A., Benin, G., Debastiani, F., Patel, F. and Miotto Jr, E. (2013), 'Competitive ability of soybean cultivars with horseweed (*Conyza bonariensis*)', *Planta Daninha*, 31, 543–550.

Van der Schans, D. A., Bleeker, P. O., Molendijk, L. P. G., Plentinger, M. C., van der Weide, R. Y., Lotz, L. A. P., Bauermeister, R., Total, R. and Baumann, D. T. (2006), *Practical Weed Control in Arable Farming and Outdoor Vegetable Cultivation Without Chemicals.*

© Burleigh Dodds Science Publishing Limited, 2025. All rights reserved.

(PPO publication; no. 352). Applied Plant Research. https://edepot.wur.nl/24553. Accessed August 4, 2023.

Van der Weide, R. Y., Bleeker, P. O., Achten, V. T. J. M., Lotz, L. A. P., Fogelberg, F. and Melander, B. (2008), 'Innovation in mechanical weed control in crop rows', *Weed Res.*, 48, 215–224.

Vincent-Caboud, L., Casagrande, M., David, C., Ryan, M. R., Silva, E. M. and Peigne, J. (2019), 'Using mulch from cover crops to facilitate organic no-till soybean and maize production. A review', *Agron. Sustain. Dev.*, 39, 45.

Weiner, J. (2023), 'Weed suppression by cereals: Beyond 'competitive ability', *Weed Res.*, 63, 133–138.

Weiner, J., Andersen, S. B., Wille, W. K. M., Griepentrog, H. W. and Olsen, J. M. (2010), 'Evolutionary agroecology: The potential for cooperative, high density, weed-suppressing cereals', *Evol. Appl.*, 3, 473–479.

Weiner, J. and Freckleton, R. P. (2010), 'Constant final yield', *Annu. Rev. Ecol. Evol. Syst.*, 41, 173–192.

Weiner, J., Griepentrog, H.-W. and Kristensen, L. (2001), 'Suppression of weeds by spring wheat (*Triticum aestivum*) increases with crop density and spatial uniformity', *J. Appl. Ecol.*, 38, 784–790.

Werle, I., Noguera, M. M., Karaikal, S. K., Carvalho-Moore, P., Kouame, K. B., De Lima, G. H.B., Roberts, T. L. and Roma-Burgos, N. (2023), 'Integrating weed-suppressive cultivar and cover crops for weed management in organic sweet potato production', *Weed Sci.*, 71(3), 255–264.

Wildner do Prado, L., de Freitas, V. H. and McGuire, M. (2004), 'Use of green manure/cover crops and conservation tillage in Santa Catarina, Brazil'. In M. Eilitta, J. Mureithi and R. Derpsch (eds.), *Green Manure/Cover Crop Systems of Smallholders: Experiences from Tropical and Subtropical Regions*. Kluwer Academic, Dordrecht, pp. 1–36.

Wu, X., Aravecchia, S., Lottes, P., Stachniss, C. and Pradalier, C. (2020), 'Robotic weed control using automated weed and crop classification', *J. Field Robot.*, 37, 322–340.

Wu, Y., Xi, N., Weiner, J. and Zhang, D.-Y. (2021), 'Differences in weed suppression between two modern and two old wheat cultivars at different sowing densities', *Agronomy*, 11, 253.

Yelverton, F. H. and Coble, H. D. (1991), 'Narrow row spacing and canopy formation reduces weed resurgence in soybean (Glycine max)', *Weed Technol.*, 5, 169–174.

© Burleigh Dodds Science Publishing Limited, 2025. All rights reserved.

Chapter 4

Allelopathic weed control in Conservation Agriculture systems

Omer Farooq and Sami Ul-Allah, Bahauddin Zakariya University, Pakistan; Amar Matloob, Muhammad Nawaz Shareef University of Agriculture, Pakistan; Mubshar Hussain, Bahauddin Zakariya University, Pakistan; and Muhammad Farooq, Sultan Qaboos University, Oman

1 Introduction

Allelopathy is a naturally occurring biological process involving the release of chemicals from the plant roots that can affect both biological and agricultural systems. The Austrian plant physiologist Hans Molìsch first coined the term 'allelopathy' (Molísch, 1937), and defined it as 'any direct or indirect harmful effect by one plant (including microorganisms) on another through the release of chemicals that escape into the environment'. It encompasses both direct (e.g. root exudates) and indirect (e.g. volatiles) chemical interactions between plants. Such chemical compounds are usually secondary metabolites or byproducts of the major metabolic pathways in plants.

Conservative Agriculture (CA) is a farming approach consisting of practices intended to promote sustainable and environmentally friendly crop production

http://dx.doi.org/10.19103/AS.2024.0145.05
© Burleigh Dodds Science Publishing Limited, 2025. All rights reserved.

and to improve long-term land productivity (Farooq and Siddique, 2015a, 2015b). Reduced tillage is one of the major components of CA (Farooq and Siddique, 2015a). However, weed control continues to be a major challenge in CA. During the early stages of adopting CA, it was anticipated that there would be an increase in weed density (Farooq et al., 2011a; Sosnoskie et al., 2006).

Modern agriculture is heavily dependent upon the use of chemical herbicides for weed management. Non-judicious use of chemical herbicides, with similar modes of action for a long time, may also lead to the development of herbicide-resistant weed biotypes (Busi et al., 2021). For instance, rigid ryegrass (*Lolium rigidum* Gaud.) and wild oat (*Avena fatua* L.) have developed resistance against PS-II-inhibitors, acetolactate synthase (ALS) inhibitors, and Acetyl-CoA Carboxylase inhibitors (Boutsalis et al., 2012; Broster et al., 2013), while broadleaf weeds, like kochia (*Bassia scoparia* (L.) A.J. Scott), horseweed (*Erigeron canadensis* L.) and Russian thistle (*Salsola australis* R. Br.), etc., have developed resistance against glyphosate (Dhanda et al., 2023; Kumar et al., 2017; VanGessel, 2001). The use of allelopathy as a component of the weed management package may help address the issues of herbicide resistance (Farooq et al., 2011b; Jabran et al., 2015) and weed management in the CA systems (Cordeau, 2022; Nawaz et al., 2019a).

The idea of using allelopathic crops for weed control was proposed by Putnam and Duke (1974). This included the development of weed-suppressive crop genotypes and later intercrops, cover crops, and rotational sequence of crops (Putnam and Duke, 1978). Moreover, allelopathic crop mulches (Minhas et al., 2023a) were also included in the weed control programs through allelopathy (Farooq et al., 2011b).

Weed management in CA is one of the main hurdles in its wide-scale adoption. In this regard, the phenomenon of allelopathy can be employed for weed management in the CA system (Farooq et al., 2017; Nawaz et al., 2019a; Jabran and Farooq, 2013). In this chapter, various methods of utilizing allelopathy for effective control of weeds are discussed. These methods are not only limited to incorporating allelopathic crops into intercropping and rotation patterns but also extend to mulching with allelopathic crops, using allelopathic crops as cover crops, and combining allelopathic crop water extracts (WEs) with reduced doses of herbicides.

2 Weeds in Conservation Agriculture systems

Conservation agriculture relies on three key practices:

- no or minimum tillage;
- permanent crop residue cover; and
- diversified crop rotation.

© Burleigh Dodds Science Publishing Limited, 2025. All rights reserved.

Weeds behave differently in CA compared to conventional agriculture. For instance, immediately after conversion to CA, weed density can increase but later weed pressure can then decrease in more established CA systems (Farooq et al., 2020). Several examples indicate that weed pressure is reduced in more established CA systems compared with conventional agriculture in a variety of crops/cropping systems in different agro-ecological zones of the world (Table 1).

However, weeds remain one of the major challenges in the adoption of CA. In conventional agriculture systems, weeds are controlled by uprooting and deep burying through tillage (Swanton et al., 2000). In addition, in conventional agriculture systems, previous crop residues are usually not present at the time of sowing of the next crop which allows the application of pre-emergence

Table 1 The impact of adopting Conservation Agriculture on weed flora compared to conventional agriculture in different regions of the world

Region	Crop	Changes in weed flora	References
Mexico	Wheat	81% reduction in weed biomass	Fonteyne et al., 2020
Mexico	Maize	91% reduction in weed biomass	Fonteyne et al., 2020
Indo-Gangetic Plains	Wheat	Reduction in weed density and biomass	Sharma et al., 2023
Indo-Gangetic Plains	Rice	Reduction in the emergence of flat sedge (*Cyperus iria* L.) (62%), jungle rice (*Echinochloa colonum* (L.) Link) (82–90%), and total weeds (81–83%)	Mishra et al., 2022
India	Pigeon pea-castor system	Increase in density and biomass of dicots (*Tridax procumbens* L and *Euphorbia hirta* L.) and a marginal decrease in monocots (*Bulbostylis barbata*, *Digitaria sanguinalis* L.)	Pratibha et al., 2021
India	Rice	Lower weed seed density	Nandan et al., 2020
France	Wheat	Increase in species richness, total weed abundance and α-functional diversity, but no change in species diversity and species evenness over time	Derrouch et al., 2021
France	Wheat	Shift in the weed community toward perennial and monocotyledon species	Trichard et al., 2013
Italy	Italian ryegrass	Decrease in weed cover	Nikolić et al., 2022
India	Soybean	Lower number of grassy weeds	Blaise et al., 2015
India	Rice-wheat	No shift	Singh et al., 2010
India		Higher competition from common vetch (*Vicia sativa* L.) and less from goosefoot (*Chenopodium album* L.)	Sharma et al., 2013

© Burleigh Dodds Science Publishing Limited, 2025. All rights reserved.

herbicides. In a CA system, such residues may limit the amount of applied herbicide reaching the soil surface and decrease the effectiveness of herbicide application (Streit et al., 2003). It is difficult to calculate how much of a negative effect on herbicide efficacy is caused by crop residues. Rich soil organic matter content in CA may also influence the efficacy of applied herbicides (Chauhan et al., 2006).

Several studies have indicated a shift in weed flora in CA systems, especially toward perennial weeds (Ranaivoson et al., 2016; Thierfelder et al., 2018; Derrouch et al., 2021) and certain annuals like Russian thistle and Kochia (Vencill and Banks, 1994). Derrouch et al. (2021) reported results that, after 20 years of CA adoption, weed diversity and evenness did not change, but species richness, functional diversity, and total abundance increased. Based on a 7-year study, Pratibha et al. (2021) recorded a change in weed behavior under conventional, reduced, and zero tillage systems. In conventional and reduced tillage systems, annual weed species were dominant, while in zero tillage a shift toward perennials was detected. Due to these changes, farmers need to adjust weed management strategies.

3 The use of allelopathy in weed management in Conservation Agriculture systems: intercropping and crop rotation

Allelopathy is a relatively new area of research that involves plant-produced bioactive compounds, called allelochemicals, that can act as natural herbicides (Shahzad et al., 2021; Kostina-Bednarz et al., 2023; Table 2). Crop plants having the allelopathic potential to suppress weed growth could be used in CA systems. Historically, several crops such as sorghum (*Sorghum bicolor* (L.) Moench), buckwheat (*Fagopyrum esculentum* Moench), sunflower (*Helianthus annuus* L.), *Brassica* spp. (*B. hirta*, *B. juncea*, *B. nigra*, and *B. napus*), oat (*Avena sativa* L.), wheat (*Triticum* spp.), maize (*Zea mays* L.), black walnut (*Juglans nigra* L.), velvet bean (*Mucuna deeringianum* Borr. Merr), rye (*Secale cereale* L.), and barley (*Hordeum vulgare* L.) have been widely reported to suppress different weed species (Cheema et al., 2013; Scavo and Mauromicale, 2021). About 2000 plant species from 39 different families are reported to have allelopathic properties (Li et al., 2019). These plants with allelopathic properties can be employed in CA systems as intercrops, cover crops, mulch, crop rotation, and/or WEs for weed management (Table 3). This section looks at intercropping and rotations.

3.1 Intercropping

Intercropping is a significant component of CA systems for managing weeds. A substitute for chemical weed control, intercropping refers to

© Burleigh Dodds Science Publishing Limited, 2025. All rights reserved.

Table 2 Allelopathic plants and their potent allelochemicals

Plant species	Allelochemicals	References
Alfalfa (*Medicago sativa* L.)	Catechin, gallic acid, ferulic acid, protocatechuic acid, p-coumaric acid, p-hydroxybenzoic acid, salicylic acid, sinapic acid, syringic acid, vanillin, vanillic acid	Nakahisa et al., 1994; Xuan et al., 2003
Barley (*Hordeum vulgare* L.)	Benzoxazinones, gramine, hordenine, scopoletin	Hanson et al., 1981; Liu and Lovett, 1993; Lovett et al., 1994; Baghestani et al., 1999
Brassica spp.	Glucosinolates (phenylethyl glucosinolate; 4-pentenyl, 2-hydroxybutenyl, 3-butenyl, and 2-hydroxypentenyl glucosinolate, allyl isothiocyanate, isothiocyanate, oxazolidinethione)	Vaughn and Boydston, 1997; Eberlein et al., 1998; Vaughn, 1999
Cereal rye (*Secale cereale* L.)	Benzoxazinoids, phenolic acids (coumaric acid, ferulic acid, 4-hydroxybenzoic acid, sinapic acid, syringic acid, vanillic acid, sinapic acid)	Otte et al., 2020; Rice, 2022
Cotton (*Gossypium hirsutum* L.)	Gossypol	Hedin and McCarty, 1994
Crimson clover (*Trifolium incarnatum* L.)	Biochainin, calycosin, formononetin, isoflavones, phenolics, and prunetin	Lou et al., 2016
Hairy vetch (*Vicia villosa* Roth)	Cyanamide	Fujii, 2003; Kamo et al., 2008
Neem (*Azadirachta indica* A. Juss)	Benzoic acid, gallic acid, ferulic acid, p-coumaric acid, p-hydroxybenzoic acid, vanillic acid	Xuan, 2004
Rice (*Oryza sativa* L.)	Phenolic acids, momilactones	Rimando et al., 2001; Dilday et al., 2001; Kong et al., 2002; Kato-Noguchi et al.,2002; Kato-Noguchi, 2003
Sorghum (*Sorghum bicolor* (L.) Moench)	Sorgoleone, vanillic acid, p-hydroxybenzoic acid, p-hydroxybenzaldehyde, p-coumaric acid, ferulic acid, gallic acid, synrigic acid	Chang et al., 1986; Nimbal et al., 1996; Séne et al., 2001
Sunflower (*Helianthus annuus* L.)	Flavanoids, phenolics, terpenoids, sesquiterpenes, steroids,	Macías et al., 1999
Walnut (*Juglans regia* L.)	Juglone (5-hydroxy-1,4-napthoquinone)	Rietveld, 1983
Wheat (*Triticum aestivum* L.)	Phenolics (ferulic acid, p-coumaric acid, p-hydroxybenzoic acid, hydroxamic acid, vanillic acid, syringic acid) benzoxazinones (dimboa (2,4-dihydroxy-7-methoxy-1,4-benzoxazin-3-one)) sterols, ketosteroids	Guenzi et al., 1967; Lodhi et al., 1987; Salomonsson et al., 1978; Bohidar et al., 1986; Niemeyer, 1988; Gaspar et al., 1999; Wu et al., 2000a, 2001

© Burleigh Dodds Science Publishing Limited, 2025. All rights reserved.

Table 3 Allelopathic weed control in Conservation Agriculture system

Methods/techniques	Allelopathic crops	Main crops	Weeds	Reduction (%)	References
Soil mixing and mulching	Sorghum	Wheat	Total weeds	75	Minhas et al., 2023a
	Sunflower			71	
	Cotton			54	
	Mungbean			57	
	Rye	Maize	Grass and Broadleaf weeds	66	Tabaglio et al., 2013
Intercropping	Wheat + Bitter vetch	Wheat	Seasonal weeds	59	Amini et al., 2021
	Wheat + Grass pea			48	
Crop rotation	Sorghum	Wheat	Total weeds	54	Minhas et al., 2023b
	Sunflower			56	
	Sorghum	Wheat	Sweet clover	76	Shahzad et al., 2016
	Rice		Broadleaf weeds	74	
	Maize	Barley	Grassy weeds	55	Naeem et al., 2021
	Sorghum			54	
Cover crops	Rape	–	Seasonal weeds	67	Mennan and Ngouajio, 2012
	Field mustard	–		61	
	Oriental mustard	–		73	
Allelopathic water extract	A balanced mixture of water extracts of mulberry, sorghum, eucalyptus, and sunflower	Barley	Total weeds	70	Naeem et al., 2022
	Sorghum	Rice	Total weeds	74	Farooq et al., 2017

© Burleigh Dodds Science Publishing Limited, 2025. All rights reserved.

growing a crop along with one, two, or more crops with different growing behaviors concurrently on the same land. In general, intercrops establish a canopy and use seedbeds, and competition for resources (including canopy interception of light) to limit weed growth. Adding an allelopathic crop in intercropping can suppress weeds by limiting available resources and by releasing allelochemicals in the shared environment (Shahzad et al., 2016; Khamare et al., 2022). Intercropping can help maximize land/fertilizer/water use efficiency to improve crop productivity and economic returns (Naeem et al., 2021).

An allelopathic crop-based intercropping system has proved to be a viable choice for managing weeds in CA systems. For instance, in a study on weed control in maize and wheat crops grown in a rotation under the CA system, a weed management package, including soybean (*Glycine max* L.) intercropping in wheat and brassica intercropping in maize, recorded the lowest weed population, total weed density, weed species richness, and evenness (Kumar et al., 2023). In another study, Zerihun et al. (2014) recorded 29–52% less time needed for weeding by intercropping various crops under a CA system compared to a conventional system. Intercropping of allelopathic crops in maize caused a significant reduction in various broad and narrow-leaved weeds (Nawaz et al., 2014). Amini et al. (2021) also reported that intercropping bitter vetch (*Lathyrus linifolius* (Reichard) Bässler) with wheat can be a suitable strategy for controlling weeds in CA.

3.2 Crop rotation

Crop rotation is an arrangement of growing crops one after another on a piece of land to maintain soil productivity and sustainability. However, it is also a vital tool for managing associated weeds among non-chemical weeding strategies. This strategy becomes more beneficial when a crop with allelopathic potential is added to this rotational arrangement because such crops release several allelochemicals in soil that help to eliminate the weeds both in standing and following future crops (Fonteyne et al., 2020; Naeem et al., 2021). A well-designed rotation has advantages such as maintaining or improving soil structure and enhancing organic matter content as well as controlling weeds (Farooq et al., 2020; Minhas et al., 2023b).

Minhas et al. (2023b) studied different wheat-based cropping systems, where wheat was grown followed by cotton (*Gossypium hirsutum* L.), sorghum, mungbean (*Vigna radiata* L.), rice (*Oryza sativa* L.), and sunflower under conventional and CA systems. They reported that minimum weed density and biomass were recorded in the sorghum–wheat rotation system under conventional and CA systems at all three different locations. This weed inhibition pattern was followed by sunflower. In another study, wheat was

© Burleigh Dodds Science Publishing Limited, 2025. All rights reserved.

planted after harvesting rice, cotton, mungbean, and sorghum under different tillage systems including zero tillage, conventional tillage, deep tillage, and bed sowing. The strongest inhibitory effects on the density and biomass of weeds were observed in zero-tilled wheat under a sorghum–wheat rotation (Shahzad et al., 2016).

4 The use of allelopathy in weed management in Conservation Agriculture systems: mulching and cover crops

4.1 Mulching

Mulch is a material, natural or synthetic, applied as a layer on the soil surface to protect it from erosion, conserve soil moisture, improve soil structure, and control weeds (Scavo and Mauromicale, 2021). The word mulch is derivative of the Old French word *muche* which means 'soft, rotten wood' first used in the sixteenth century. Initially, mulching was used in small nurseries and small home gardens to conserve soil moisture and then shifted to large commercial nurseries for the same purpose (Sportelli et al., 2022; Biswas et al., 2022). It has now become an important component of non-chemical weed management (Fonteyne et al., 2020). Mulching of allelopathic plant parts has attracted particular attention for weed control in CA systems and organic agriculture (Chauhan et al., 2012; Minhas et al., 2023a).

Different allelopathic mulches, e.g. cotton stalk mulch, sorghum straw mulch, rice straw mulch, and sunflower straw mulch, can both help manage weeds and improve crop yield. In comparison among different mulches, sorghum straw mulch showed the highest capacity to inhibit weeds (Minhas et al., 2023a). Mulches of allelopathic plant materials either hamper weed seed germination or prevent the growth of weed seedlings as a result of release of allelochemicals (Ali et al., 2018; Table 2). For instance, the use of sorghum straw as a surface-applied mulch in maize caused a 37% decrease in weed density. Sorghum mulching in rice and cotton crops caused a reduction in weed density by 50% and 60%, respectively (Abbas et al., 2021). In another study, Xuan et al. (2005) recorded a more than 70% reduction in major weeds of rice (barnyard grass (*Echinochloa crus-galli* (L.) P. Beauv.), purple nutsedge (*Cyperus rotundus* L.), flat sedge (*Cyperus difformis* L.), and jungle rice (*Echinochloa colonum* (L.) Link)) by applying allelopathic crop mulches (1–2 t ha^{-1}) with subsequent 20% improvement in rice yield. Diverse allelochemicals are present in all the allelopathic crops causing weed suppression (Table 2).

© Burleigh Dodds Science Publishing Limited, 2025. All rights reserved.

4.2 Use of cover crops

One of the key components of CA is to keep the soil covered and this can be achieved using cover crops. Benefits of using these crops include preserving and enhancing soil fertility, as well as controlling weeds and other pests (Jabran et al., 2015). The weed control mechanism of allelopathic cover crops is similar to the case of allelopathic crop mulching. The main driving force is the diversity of allelochemicals present in such crops that can suppress the weeds.

Allelopathic crops that can be used as cover crops include oilseed rape (*Brassica napus* L.), rye (*Secale cereale* L.), crimson clover (*Trifolium incarnatum* L.), red clover (*Trifolium pratense* L.), buckwheat, brown mustard (*Brassica juncea* (L.) Czern.), cowpea (*Vigna unguiculata* (L.) Walp.), ryegrass (*Lolium rigidum* Gaud.), and hairy vetch (*Vicia villosa* Roth) (Haramoto and Gallandt, 2004; Farooq et al., 2011b; Jabran et al., 2015; Altieri et al., 2011; Mirsky et al., 2013; Antichi et al., 2022; Table 2).

Cover crops are mostly used in annual cropping systems (Fernando and Shrestha, 2023), since annual cover crops suppress weeds for a limited period of the growing season (Bybee-Finley et al., 2022; Smith et al., 2020). Cover crops inhibit weed growth through competition for resources like space, water, light, and nutrient and particularly by releasing a diversity of allelochemicals (Mahé et al., 2022). The use of cover crops for weed control in CA has been very effective. For example, the use of rye as a cover crop under a zero-tillage system caused significant suppression of weeds throughout the season, resulting in successful soybean production (Bernstein et al., 2014). However, rye cultivars differ in their allelochemical contents and thus their ability to suppress weeds (Tabaglio et al., 2013). In another study, Norsworthy et al. (2011) recorded a substantial reduction in the density of Palmer amaranth (*Amaranthus palmeri* L.), crowfootgrass (*Eleusine indica* L.), and morning glory (*Ipomoea lacunose* L.) when wheat and rye were grown as cover crops in cotton under a CA system.

Certain cover crops, such as oat and hairy vetch, can also decrease weed seeds in the soil. For instance, a 30–70% reduction in the seeds of some weeds (like *Amaranthus retroflexus* L., *Datura stramonium* L., *Eleusine indica*, and *Digitaria sanguinalis* L.) in the top soil layer was noted with the use of oat and hairy vetch as cover crops (Dube et al., 2012). Crabgrass (*Digitaria sanguinalis* (L.) Scop.) and barnyard grass are troublesome weeds of soybean, particularly in the CA system. However, using barley as a cover crop in soybean fields caused a significant reduction in the density of both weeds in CA (Kobayashi et al., 2004).

The members of the Brassicaceae family are rich in allelochemicals, suggesting a role for oilseed rape and white mustard (*Sinapis alba*) as cover crops in controlling weeds (Haramoto and Gallandt, 2004). The root exudates of brassica contain glucosinolates as major allelochemicals. Under natural

© Burleigh Dodds Science Publishing Limited, 2025. All rights reserved.

conditions, these are decomposed into various other compounds such as isothiocyanates (Haramoto and Gallandt, 2004; Halkier and Gershenzon, 2006). These are biologically active compounds and have an inhibitory impact on the emergence and subsequent growth of plants exposed to them. These compounds are helpful in season-long control over morning glory, palmer amaranth, and yellow nutsedge (*Cyperus esculentus* L.). Soybean can be grown in a standing rye cover crop and reduces the density of associated weeds without causing any harm to soybean under a CA system (Bernstein et al., 2014).

5 The use of allelopathy in weed management in Conservation Agriculture systems: allelopathic water extracts

The concept of utilizing WEs of mature plant foliage, roots, or even their residues is very old (Guenzi and McCalla, 1962; Lehle and Putnam, 1982). The use of WEs of allelopathic plants for management of pests, including weeds, has gained much attention in recent years (Cheema and Khaliq, 2000; Farooq et al., 2011b, 2017; Carrubba et al., 2020; Scavo et al., 2020; Naeem et al., 2022; Ain et al., 2023).

It has been found that the application of sorghum WEs caused a significant reduction in weed number and biomass in dry direct-seeded rice, compared to a control (Farooq et al., 2017). Shahzad (2016) also reported a significant reduction in weed density and biomass (compared to a control) through application of WEs of sorghum, sunflower, mulberry (*Morus alba* L.), and eucalyptus (*Eucalyptus obliqua* L'Hér.) in wheat sown under different wheat-based cropping systems using conventional and conservation tillage systems. Naeem (2021) also reported similar results for barley sown in different barley-based cropping systems under conventional and CA systems. Certain weeds like Bonpland's croton (*Croton bonplandianus* Baill.), an annual weed mostly found in Asia, are allelopathic in nature. Water extracts of Bonpland's croton have strong herbicidal effects on different weeds like sweet clover (*Melilotus alba* L.), common vetch (*Vicia sativa* L.), and toothed medick (*Medicago hispida* Gaertn) (Sisodia and Siddiqui, 2010) and can be used for weed control in CA systems.

The application of allelopathic plant WEs has shown encouraging results in many studies. Application is both environmentally friendly and economical, but suppression of weed population is lower than when using chemical weed control. The idea of utilizing allelopathy in combination with lower rates of herbicides was postulated by Kebede (1994) and Einhellig (1996). Research has shown that herbicide usage can be reduced by 50–75% in major crops such as cotton, rice, wheat, and maize if combined with allelopathic WEs of sorghum, sunflower, and brassica (Cheema et al., 2004; Jamil et al., 2009).

© Burleigh Dodds Science Publishing Limited, 2025. All rights reserved.

Other researchers have found similar results (Iqbal et al., 2008; Farooq et al., 2011b, 2017).

Khaliq et al. (2013) reported that the application of sorghum WE in combination with a 50% reduced dose of herbicide (iodo+mesosulfuron) (at early post-emergence – 30 days after sowing) had the same level of reduction in weed density and weed biomass as achieved by the recommended dose of herbicide. However, the idea of using allelopathic WEs in combination with reduced doses of commercial herbicides has not yet been widely tested under CA. Most of the literature available for using allelopathic WEs in combination with reduced herbicide doses is tested under conventional tillage systems where it has provided effective control of weeds and improved the crop yield.

6 Case study: weed management in a Conservation Agriculture rice–wheat system

The rice–wheat system is one of the major cropping systems of the Indo-Gangetic plains (Nawaz et al., 2019b). In a conventional system, rice seedlings are transplanted in muddy flooded fields and the field is kept flooded during most of the growing season. This method is efficient in controlling weeds due to the flooded environment. However, this system requires significant amounts of water (Farooq et al., 2011c; Kumar and Ladha, 2011). Increasing water shortages exacerbated by climate change are threatening the productivity as well as the sustainability of transplanted rice in Southeast Asia (Bhatt et al., 2016). Puddling consumes about 30% of the total water input in flooded transplanted rice (Aslam et al., 2002; Gopal et al., 2010). After the harvest of rice, primary and secondary tillage practices are undertaken to prepare fine seedbed for wheat sowing. This extends the turnaround time between rice harvest and wheat planting and delays wheat sowing, in particular after the harvest of late-maturing basmati rice varieties (Farooq et al., 2011c; Matloob et al., 2020; Nawaz et al., 2019b).

The CA approach, consisting of direct-seeded rice (DSR)-zero-tillage wheat, is an attractive alternative to the conventional rice–wheat system (Nawaz et al., 2019b). However, weed infestation is a major constraint limiting productivity and wide-scale adoption of CA (Nawaz et al., 2019b). For example, weed infestation may cause up to 75–80% reduction in paddy yield (Khaliq et al., 2012a, 2012b, 2013). Fields of DSR are dominated by diverse and difficult-to-control weeds, while the wheat fields established under zero-tillage can be dominated by perennial grasses (Kumar and Ladha, 2011; Nawaz and Farooq, 2016; Nawaz et al., 2019b). With modifications in growing environments and establishment techniques such as water-saving methods of rice cultivation, adoption of permanent raised beds, and zero tillage systems, existing weed control practices became ineffective.

© Burleigh Dodds Science Publishing Limited, 2025. All rights reserved.

Chemical control has therefore gained popularity as a potential solution to weed proliferation. Herbicides are now considered a vital tool to protect crops against weeds. However, sole reliance on herbicides for controlling weeds needs to be discouraged due to the associated trade-off in shifts in weed flora, the evolution of herbicide-resistant biotypes, seed bank buildup, and risks of environmental pollution. Endemic rice and wheat herbicides targeting the ALS enzyme are increasingly ineffective due to the evolution of herbicide resistance in target weed species (Heap, 2023). Tackling this challenge necessitates a holistic approach in CA (Nawaz and Farooq, 2016; Bhatt et al., 2016; Nawaz et al., 2019b).

Farooq et al. (2017) reported the potential of the combined application of sorghum mulch and sorghum WE in reducing weed infestation in DSR. Farooq et al. (2022) also reported that the inclusion of sorghum in a rice–wheat cropping system improved productivity, especially by suppressing weeds. In another comprehensive review, Ali et al. (2018) found allelopathic weed management options like brown manuring, mulching, and intercropping of allelopathic crops to avert weed infestation in DSR. A better understanding of the cause-and-effect relationship between weed eco-biology in response to altered soil moisture regimes, stand establishment techniques, and soil physico-chemical properties is needed to combat weeds and devise eco-efficient weed management practices. The synergistic interaction of cultural practices that provide crops with a competitive advantage over weeds can contribute significantly to limiting and optimizing more limited application of herbicides.

7 Improving the allelopathic potential of crops

The diversity of allopathic traits, characterization of their genetic basis, and effective breeding strategies are crucial aspects of breeding weed-resistant crops (Fragasso et al., 2013). The allelopathic potential of crops is influenced by both genetic and environmental factors (Jabran et al., 2015; Aslam et al., 2017). The available germplasm of crops can be initially screened for their allelopathic activity and subsequent improvement can be accomplished via breeding (Kim and Shin, 2003; Worthington and Reberg-Horton, 2013). In this context, there are significant genetic differences in allelopathic activity for a crop like wheat (Wu et al., 2000a, 2000b, 2000c, 2000d; Rizvi et al., 2004; Mahmood et al., 2013), such variation being positively correlated with total phenolic contents (Wu et al., 2003). Dilday et al. (1998) found that a rice hybrid derived from allelopathic accessions recorded fewer weed species in the field.

Conventional breeding has been employed to improve the allelopathic potential of spring wheat (Bertholdsson, 2010). After preliminary screening, a Tunisian cultivar with high allelopathic potential (Mohan 73) and a better yielding Swedish cultivar with low allelopathic potential (Zebra) were crossed.

© Burleigh Dodds Science Publishing Limited, 2025. All rights reserved.

The product of this cross was used as a baseline material for a breeding program focused on achieving highly allelopathic wheat cultivars. In the F6 and F7 generations, three breeding lines (with an average improved allelopathy of 20%) and one line (same phenotype as the high allelopathic lines) with an unimproved allelopathy activity were tested together with the low allelopathic Swedish parent. Further field scale testing revealed a 19% reduction in weed biomass by highly allelopathic lines; however, grain yield was 9% less owing to lower early biomass compared to the control.

Modern wheat cultivars can inhibit root growth of perennial ryegrass (15–25%) due to high allelopathic potential (Bertholdsson, 2007). Bertholdsson (2012) has proposed crossing wheat and rye to improve allelopathic activity. These authors found that lines with substitution of 1R or 2R manifested the highest allelopathic activity against mustard. Bertholdsson (2012) included wheat–rye translocation lines listed in the Swedish National List and tested cultivars against herbicide-resistant and susceptible black-grass (*Alopecurus myosuroides* Huds.) using bioassays followed by 2-year field studies. A wheat–rye hybrid 'Dinaro' and wheat cultivar 'Nimbus' exhibited the highest allopathic suppression of black grass. The black grass biomass was reduced to almost half in experimental plots of these wheat cultivars as compared to the cultivars with low allelopathic activity.

In China, crosses between the local rice cultivars N9S and PI 312777 resulted in a promising F8 generation line to produce the first commercially acceptable cultivar capable of weed suppression (Kong et al., 2011). A breeding program to develop allelopathic rice was also launched in Korea (Ma et al., 2006). Single seed descent method was employed to produce F4 and F5 generations by crossing allelopathic, 'Kouketsumochi' with non-allelopathic but high-yielding 'Donginbyeo'. These were then tested for their allopathic potential against weeds in field conditions. From the study, K21 was selected as a promising line since it combined the allopathic activity of the strong allelopathic parent and recorded yield and agronomic performance equivalent to the non-allelopathic parent. Lee et al. (2008) also confirmed the inhibitory effects of root exudates of K21 on barnyard grass seedling growth. Allelopathic traits can be incorporated into an elite restorer line while developing three-line hybrid rice (Lin et al., 2000). These authors used simultaneous back-crossing and selfing techniques to develop hybrid rice with allelopathic potential. Three lines of rice viz. Koukesumochi, Rexmont, and IR24 were used as the donors of allopathic, non-allelopathic and restoring genes, respectively. The study found a significant positive heterotic effect on rice allelopathy. Interestingly, specific rice hybrid developed through this approach demonstrated higher suppressive effects on barnyard grass.

Functional genomics is being employed to identify genes responsible for allelochemical biosynthesis, and molecular breeding can pave the way

© Burleigh Dodds Science Publishing Limited, 2025. All rights reserved.

to improve the allelopathic potential of field crops (Belz, 2007; Jabran et al., 2015). A better understanding of variation in allelochemical production and release patterns in response to season, and genotype and their respective QTLs can be a step toward weed-suppressive wheat cultivars possessing improved allelopathic traits (Rector, 2008; Trezzi et al., 2016; Table 4). In wheat and rice, QTLs maps associated with allelopathic traits have been developed (Jensen et al., 2001; Wu et al., 2003).

Genes governing the weed-suppressive ability of wheat were found on chromosomes 1A, 2B, and 5D via QTL identification (Zuo et al., 2012). Analysis of QTLs can be done using restriction fragment length polymorphism, amplified fragment length polymorphism, and microsatellite markers to unravel the genetic basis of allelopathy (Scavo and Mauromicale, 2021). Based upon an extensive research program, Olofsdotter et al. (2002) suggested marker-assisted selection based on quantitative trait loci to improve the allelopathic potential of recombinant inbred lines of rice. However, genes involved in the biosynthesis of hydroxamic acids in wheat have not been identified to date (Trezzi et al., 2016). Identification of genes and corresponding enzymes can elucidate the major transcriptional and post-transcriptional mechanisms (Duke et al., 2001; Weston et al., 2013).

Belz (2007) postulated that, whether through conventional breeding, molecular approaches, or transgene technologies, research in rice has the potential to make it the first annual crop where weed-suppressive commercial

Table 4 Different types of QTLs/molecular markers identified related to allopathic effects identified in different crops

Name of crop	Type of marker/QTL	Number of markers/QTL linked with allelopathy	Chromosome number	References
Rice (*Oryza sativa* L.)	RFLP/QTLs	05	3, 5, 6, 7, 12	Ebana et al., 2001
	SSR/QTLs	09	1, 2, 3, 4, 5, 8, 9, 12	Lee et al., 2005
	SSR markers	02	4, 5	El-Naby et al., 2019
Wheat (*Triticum aestivum* L.)	SSR/QTLs	02	2B	Wu e al., 2003
	SSR/QTLs	02	2B	Wu, 2005
Soybean (*Glycine max* L.)	RPAD/ISSR/QTLs	07 RPAD, 11 SSR	Throughout the genome	Hala and Maissa, 2018
Sorghum (*Sorghum bicolor* L.)	SSR/QTLs	11	Throughout the genome	Shehzad and Okuno, 2020

QTL - Quantitative Trait Loci; FLP - Restriction Fragment Length Polymorphism; RAPD - Random Amplified Polymorphic DNA; SSR - Simple Sequence Repeats; ISSR - Inter-Simple Sequence Repeats Random amplified polymorphic DNA.

© Burleigh Dodds Science Publishing Limited, 2025. All rights reserved.

cultivars with improved allelopathic potential will be available. Concurrently, a high-yielding (5.95 t ha^{-1}) and weed-suppressive (26–29%) allelopathic rice cultivar (Huagan-3) has been released in China (Kong et al., 2011). Strong allelopathic activity of rice line 'PI 312777' against barnyard grass was presumably because of the enhanced activation of genes such as phenylalanine ammonia lyase (PAL) and cytochrome P450 as increased abundance of their transcription factors was noticed at low nitrogen supply (Song et al., 2008). For this rice line, Fang et al. (2009) also noticed that up-regulation of genes related to antioxidant proteins and phenylpropanoid metabolism increased in allelopathic potential. Besides gene expression, the RNA interference (RNAi) approach has also been utilized to silence PAL-related genes to demonstrate the strong correlation between higher expression of PAL in the phenylpropanoid metabolism and inhibitory effects of allelopathic rice on weeds (Xu et al., 2012). Although the genome-based investigation of allelopathy is an emerging area, its inclusion in specific breeding programs remains limited to date.

8 Conclusion

Allelopathy can contribute to weed management in CA systems. Incorporating allelopathic crops into the rotation, using them as intercrops or cover crops, or applying mulches of allelopathic crops, can provide long-term benefits for weed control. Similarly, the application of allelopathic crop WEs can provide better weed control in all major crops. Mixing these allelopathic crop WEs with 50–67% lower doses of herbicides can provide the same level of reduction in weed infestation as achieved by the recommended dose of commercial herbicides. Breeding allelopathic crops with weed-suppressive abilities is also promising. Innovative breeding methods are required to improve allelopathic potential in ideotypes having desirable agronomic characteristics. Conventional breeding and modern genomics in a harmonized manner can pave the way toward this goal. A breeding-oriented framework with diverse genetic resources is required for important crops to improve their genetic architecture related to allelopathy.

9 Where to look for further information

For further information on the subject of this chapter, the following references are suggested:

- Bajwa, A. A. (2014). Sustainable weed management in conservation agriculture. Crop Protection, 65, 105–113.
- Cordeau, S. (2022). Conservation agriculture and agroecological weed management. Agronomy, 12(4),867.

© Burleigh Dodds Science Publishing Limited, 2025. All rights reserved.

- Farooq, M., Flower, K., Jabran, K., Wahid, A., & Siddique, K. H. M. (2011). Crop yield and weed management in rainfed conservation agriculture. Soil and Tillage Research, 117, 172–183.
- Farooq, M., Jabran, K., Cheema, Z. A., Wahid, A., & Siddique, K. H. M. (2011). The role of allelopathy in agricultural pest management. Pest Management Science, 67(4), 493–506.
- Farooq, M., & Siddique, K. H. M. (2015). Conservation Agriculture. Springer International Publishing, Switzerland. Singh, V., Barman, K., Singh, R., & Sharma, A. (2015). Weed management in conservation agriculture systems. In M. Farooq & K. H. M. Siddique (Eds.), Conservation Agriculture (pp. 55–72). Springer, Cham.https://doi.org/10.1007/978-3-319-11620-4_3

10 References

Abbas, T., et al. (2021), Ways to use allelopathic potential for weed management: a review, *Int. J. Food Sci. Agric.*, 5: 492–498.

Ain, Q., et al. (2023), Allelopathy: an alternative tool for sustainable agriculture, *Physiol. Mol. Biol. Plants*, 29: 495–511.

Ali, N., et al. (2018), Different approaches in direct seeded rice system to avert weed infestation: Review, *Int. J. Sci. Engin.* Res., 9: 1538–1558.

Altieri, M. A., et al. (2011), Enhancing crop productivity via weed suppression in organic no-till cropping systems in Santa Catarina, *Brazil. J. Sustain. Agri.*, 35: 855–869.

Amini, E., et al. (2021), Evaluation the effect of wheat intercropping with bitter vetch and grass pea on weed diversity and density under different soil tillage systems, *J. Iran. Plant Prot. Res.*, 35: 231–250.

Antichi, D., et al. (2022), Targeted timing of hairy vetch cover crop termination with roller crimper can eliminate glyphosate requirements in no-till sunflower, *Agron. Sustain. Develop.*, *42*(5): 87.

Aslam, F., et al. (2017), Allelopathy in agro-ecosystems: a critical review of wheat allelopathy-concepts and implications, *Chemoecology*, 27: 1–24.

Aslam, M., et al. (2002), Water-saving strategies for irrigated rice, *J. Drain. Water Manage.*, 6: 25–36.

Baghestani, A., et al. (1999), Determination of allelochemicals in spring cereal cultivars of different competitiveness, *Weed Sci.*, 47: 498–504.

Belz, R. G. (2007), Allelopathy in crop/weed interactions – an update, *Pest Manag Sci.*, 63: 308–326.

Bernstein, E. R., et al. (2014), Weed community dynamics and suppression in tilled and no-tillage transitional organic winter rye-soybean systems, *Weed Sci.*, 62: 125–137.

Bertholdsson, N. O. (2007), Varietal variation in allelopathic activity in wheat and barley and possibilities to use this in breeding, *Allelopathy J.*, 19: 193–202.

Bertholdsson, N. O. (2010), Breeding spring wheat for improved allelopathic potential, *Weed Res.*, 50: 49–57.

Bertholdsson, N. O. (2012), Allelopathy–A tool to improve the weed competitive ability of wheat with herbicide-resistant black-grass (*Alopecurus myosuroides* Huds.), *Agronomy*, 2: 284–94.

© Burleigh Dodds Science Publishing Limited, 2025. All rights reserved.

Bhatt, R., et al. (2016), Sustainability issues on rice–wheat cropping system, *Int. Soil Water Cons. Res.*, 4: 64–74.

Biswas, T., et al. (2022), Impact of mulching and nutrients on soil water balance and actual evapotranspiration of irrigated winter cabbage (*Brassica oleracea* var. capitata L.), *Agric. Water Manage.*, 263: 107456.

Blaise, D., et al. (2015), The response of weed community in soybean with conventional and conservation tillage systems on rainfed Vertisols, *Arch. Agron. Soil Sci.*, 61: 1289–1301.

Bohidar, K., et al. (1986), Effects of hydroxamic acids on the resistance of wheat to the aphid *Sitobion avenae*, *Ann. Appl. Biol.*, 109: 193–98.

Boutsalis, P., et al. (2012), Incidence of herbicide resistance in rigid ryegrass (*Lolium rigidum*) across southeastern Australia, *Weed Technol.*, 26: 391–398.

Broster, J. C., et al. (2013), Herbicide resistance levels in annual ryegrass (*Lolium rigidum* Gaud.) and wild oat (*Avena* spp.) in southwestern New South Wales, *Plant Prot.*, 28: 126.

Busi, R., et al. (2021), Herbicide resistance across the Australia an continent, *Pest Manage. Sci.*, 77: 5139–5148.

Bybee-Finley, A. K., et al. (2022), Finding the right mix: a framework for selecting seeding rates for cover crop mixtures, *Ecol. Appl.*, 32: e02484.

Carrubba, A., et al. (2020), Use of plant water extracts for weed control in durum wheat (*Triticum turgidum* L. Subsp. durum Desf.), *Agronomy*, 10: 364.

Chang, M., et al. (1986), Chemical regulation of distance – Characterization of the 1st natural host germination stimulant for *Striga asiatica*. *J. Am. Chem. Soc.*, 108: 7858–7860.

Chauhan, B. S., et al. (2012), Ecology and management of weeds under conservation agriculture: a review, *Crop. Prot.*, 38: 57–65.

Chauhan, B. S., et al. (2006), Tillage system effects on weed ecology, herbicide activity and persistence: a review, *Aus. J. Experi. Agri.*, 46: 1557–1570.

Cheema, Z. A. and Khaliq, A. (2000), Use of sorghum allelopathic properties to control weeds in irrigated wheat in a semi arid region of Punjab, *Agric. Ecosys. Environ.*, 79: 105–112.

Cheema, Z. A., et al. (2013), Application of allelopathy in crop production: success story from Pakistan. In: Cheema, Z. A., et al. (eds.), *Allelopathy: Current Trends and Future Applications*. Heidelberg, Germany, Springer-Verlag, pp. 113–143.

Cheema, Z. A., et al. (2004), Weed control in maize (*Zea mays* L.) through sorghum allelopathy, *J. Sust. Agric.*, 23: 73–86.

Cordeau, S. (2022), Conservation agriculture and agroecological weed management, *Agronomy*, 12: 867.

Derrouch, D., et al. (2021), Weed community diversity in conservation agriculture: Post-adoption changes, *Agric. Ecosyst. Environ*. 312: 107351.

Dhanda, S., et al. (2023), Synergistic interactions of 2,4-D, dichlorprop-p, dicamba, and halauxifen/fluroxypyr for controlling multiple herbicide-resistant kochia (*Bassia scoparia* L.), *Weed Technol.* doi:10.1017/wet.2023.48

Dilday, R. H., et al. (2001), Allelopathic potential in rice germplasm against ducksalad, redstem and barnyardgrass, *J. Crop Prod.*, 4: 287–301.

© Burleigh Dodds Science Publishing Limited, 2025. All rights reserved.

Dilday, R. H., et al. (1998), Allelopathic activity in rice for controlling major aquatic weeds. In: Olofosdotter, M. (ed.), *Allelopathy in Rice*. International Rice Research Institute, Los Banos, Philippine, pp. 7–26.

Dube, E., et al. (2012), Winter cover crops and fertiliser effects on the weed seed bank in a low-input maize-based conservation agriculture system, *South Afri. J. Plant Soil*, 29: 195–197.

Duke, S. O., et al. (2001), Strategies for using transgenes to produce allelopathic crops, *Weed Sci.*, 15: 826–834.

Ebana, K., et al. (2001), Analysis of QTL associated with the allelopathic effect of rice using water-soluble extracts, *Breed. Sci.*, 51: 47–51.

Eberlein, C. V., et al. (1998), Gluco- sinolate production by five field-grown *Brassica napus* cultivars used as green manures, *Weed Technol.*, 12: 712–718.

Einhellig, F. A. (1996), Interactions involving allelopathy in cropping systems: allelopathy in cropping systems, *Agron. J.*, 88: 883–893.

El-Naby, A., et al. (2019), Field evaluation and molecular detection of allelopathic potential for some rice genotypes to improve weed management and grain yield of broadcast-seeded rice, *J. Plant Prod.*, 10: 21–30.

Fang, C. X., et al. (2009), Analysis of gene expressions associated with increased allelopathy in rice (*Oryza sativa* L.) induced by exogenous salicylic acid, *Plant Growth Regul.*, 57: 163–172.

Farooq, M., et al. (2017), Using sorghum to suppress weeds in dry seeded aerobic and puddled transplanted rice, *Field Crops Res.*, 214: 211–218.

Farooq, M., et al. (2011b), The role of allelopathy in agricultural pest management. *Pest Manage. Sci.*, 67: 493–506.

Farooq, M., et al. (2022), Brown manuring and sorghum allelopathy affects weeds, soil health, and paddy yield. *Agron. J.*, 114: 3040–3051.

Farooq, M., et al. (2011c), Rice direct seeding: Experiences, challenges and opportunities, *Soil Tillage Res.*, 111: 87–98.

Farooq, M. and Siddique, K. H. M. (2015a), Conservation agriculture: concepts, brief history and impacts on agricultural systems. In: Farooq, M. and Siddique, K. H. M. (eds.), *Conservation Agriculture*. Springer International Publishing, Switzerland, pp. 3–17.

Farooq, M. and Siddique, K.H.M. (2015b), *Conservation Agriculture*. Springer International Publishing Switzerland.

Farooq, M., et al. (2011a), Crop yield and weed management in rainfed conservation agriculture, *Soil Tillage Res.*, 117: 172–183.

Farooq, M., et al. (2020), Crop and cropping systems management practices and benefits in conservation agriculture systems. In: Kasssm, A. (ed.), *Advances in Conservation Agriculture Volume 2: Practice and Benefits*, Sawston, Cambridge, Burleigh Dodds Science Publishing Limited.

Fernando, M. and Shrestha, A. (2023), The potential of cover crops for weed management: a sole tool or component of an integrated weed management system?, *Plants*, 12: 752.

Fonteyne, S., et al. (2020), Rotation, mulch and zero tillage reduce weeds in a long-term conservation agriculture trial, *Agronomy*, 10: 962.

Fragasso, M., et al. (2013), Durum wheat and allelopathy: toward wheat breeding for natural weed management, *Front. Plant Sci.*, 4: 375.

© Burleigh Dodds Science Publishing Limited, 2025. All rights reserved.

Fujii, Y. (2003),Allelopathy in the natural and agricultural ecosystems and isolation of potent allelochemicals from Velvet bean (*Mucuna pruriens*) and Hairy vetch (*Vicia villosa*), *Biol. Sci. Space*, 17: 6–13.

Gaspar, E. M. M. S. M., et al. (1999), Triterpenoids and other potentially active compounds from wheat straw: isolation, identification, and synthesis. In: Cutler, H. G. and Cutler, S. J. (eds.), *Biologically Active Natural Products*, CRC Press, Boca Raton, USA, pp. 69–80.

Gopal, R., et al. (2010), Direct dry seeded rice production technology and weed management in rice based systems, Technical Bulletin. *International Maize and Wheat Improvement Center*, New Delhi India, p. 28.

Guenzi, W. D. and McCalla, T. M. (1962), Inhibition of germination and seedling development by crop residues, *Soil Sci. Soc. America J.*, 26: 456–458.

Guenzi, W. D., et al. (1967), Presence and persistence of phytotoxic substances in wheat, oat, corn and sorghum residues. *Agron. J.*, 59: 163–165.

Hala, M. A. and Maissa, M. M. (2018), Genetic fidelity testing of soybean plants under allelopathic stress of eucalyptus ground leaves through RAPD and ISSR markers, *Int. J. Bot.*, 14: 14–23.

Halkier, B. A. and Gershenzon, J. (2006), Biology and biochemistry of glucosinolates, *Annu. Rev. Plant Biol.*, 57: 303–333.

Hanson, A. D., et al. (1981), Gramine in barley forage-effects of genotype and environment, *Crop Sci.*, 21: 726–730.

Haramoto, E. R. and Gallandt, E. R. (2004), Brassica cover cropping for weed management: a review, *Renewable Agric. Food Sys.*, 19: 187–198.

Heap, I. (2023), Current Status of the International Herbicide-Resistant Weed Database, Available at https://www.weedscience.org/home.aspx. (Accessed on 11 November 2023).

Hedin, P. A. and McCarty Jr., J. C. (1994), Effects of several commercial plant growth regulator formulations on yield and allelochemicals of cotton (*Gossypium hirsutum*), *J. Agric. Food Chem.*, 42: 1355–1357.

Iqbal, J. and Cheema, Z. A. (2008), Purple nutsedge (*Cyperus rotundus* L.) management in cotton with *combined* application of sorghum water extract and s-metolachlor, *Pak. J. Bot.*, 40: 2383–2391.

Jabran, K. and Farooq, M. (2013), Implications of potential allelopathic crops in agricultural systems. In: Cheema, Z. A., et al. (eds.), *Allelopathy: Current Trends and Future Applications*. Springer-Verlag, Heidelberg, Germany, pp. 349–388.

Jabran, K., et al. (2015), Allelopathy for weed control in agricultural systems, *Crop Prot.*, 72: 57–65.

Jamil, M., et al. (2009), Alternative control of wild oat and canary grass in wheat fields by allelopathic plant water extracts, *Agron. Sustain. Develop.*, 29: 475–482.

Jensen, L. B., et al. (2001), Locating genes controlling allelopathic effects against barnyardgrass in upland rice, *Agron. J.*, 93: 21–26.

Kamo, T., et al. (2008) Limited Distribution of Natural Cyanamide in Higher Plants: Occurrence in *Vicia villosa* Subsp. varia, V. cracca, and Robinia pseudo-acacia. *Phytochemist*, 69: 1166–1172.

Kato-Noguchi H. and Ino T. (2003) Rice seedlings release momilactone B into the environment, *Phytochemistry*, 63: 551–554.

Kato-Noguchi H., et al. (2002) Isolation and identification of a potent allelopathic substance in rice root exudates, *Physio. Plant.*, 115: 401–405.

© Burleigh Dodds Science Publishing Limited, 2025. All rights reserved.

Kebede, Z. (1994), Allelopathic chemicals: their potential uses for weed control in agro ecosystems. Review articles Chemical Ecology EN570. Colorado State University, Colorado. Retrieved from http://www.colostate.edu/Depts/Entomology/Courses/en570/papers_1994/kebede.html

Khaliq, A., et al. (2012a), Optimizing seeding density and tank mixture of herbicides help reduce yield losses in dry seeded fine rice, *Pak. J. Weed Sci. Res.*, 18: 167–181.

Khaliq, A., et al. (2012b), Seeding density and herbicide tank mixture furnishes better weed control and improve the growth, yield and quality of direct seeded fine rice, *Int. J. Agric. Biol.*, 14: 499–508.

Khaliq, A., et al. (2013), Supplementing herbicides with manual weeding improves weed control efficiency, growth and yield of dry seeded rice, *Int. J. Agric. Biol.*, 15: 191–199.

Khamare, Y., et al. (2022), Allelopathy and its application as a weed management tool: a review, *Front. Plant Sci.*, 13: 1034649.

Kim, K. U. and Shin, D. H. (2003), The importance of allelopathy in breeding new cultivars, In: Labrada, R. (ed.), *Weed Management for Developing Countries*. FAO Plant Production and Protection Paper No. 120 (addendum1), pp. 195–210.

Kobayashi, H., et al. (2004), Effects of winter barley as a cover crop on the weed vegetation in a no-tillage soybean, *Weed Biol. Manage.*, 4: 195–205.

Kong, C. H., et al. (2011), Breeding of commercially acceptable allelopathic rice cultivars in China, *Pest. Manage. Sci.*, 67: 1100–1106.

Kong, C., et al. (2002), Using specific secondary metabolites as markers to evaluate allelopathic potentials of rice varieties and individual plants. *Chin. Sci. Bull.*, 47: 839–843.

Kostina-Bednarz, M., et al. (2023), Allelopathy as a source of bioherbicides: challenges and prospects for sustainable agriculture, *Rev. Environ. Sci. Bio/Technol.*, 22: 471–504.

Kumar, S., et al. (2023), Weed phyto-sociology and diversity in relation to conservation agriculture and weed management strategies under Northwestern Indian Himalayas, *J. King Saud Uni. Sci.*, 35: 102728.

Kumar, V. and Ladha, J. K. (2011), Direct seeding of rice: recent developments and future needs, *Adv. Agron.*, 111: 297–413.

Kumar, V., et al. (2017), Glyphosate resistant Russian thistle (*Salsola tragus*) identified in Montana and Washington, *Weed Technol.*, 31: 238–251.

Lee, H. W., et al. (2008), Allelopathic effect of the root exudates of K21, a potent allelopathic rice, *Weed Biol. Manage.*, 8: 85–90.

Lee, S. B., et al. (2005), QTLs and molecular markers associated with rice allelopathy, In: *Proceedings of the 4th World Congress on Allelopathy*. International Allelopathy Society, Wagga Wagga, NSW, Australia, pp. 505–507.

Lehle, F. R. and Putnam, A. R. (1982), Quantification of allelopathic potential of sorghum residues by novel indexing of Richards' function fitted to cumulative cress seed germination curves, *Plant Physiol.*, 69: 1212–1216.

Li, Z. R., et al. (2019), Allelopathy in sustainable weeds management, *Allelopathy J.*, 48: 109–138.

Lin, W., et al. (2000), Hybrid rice with allelopathy. In: Kim, K. U. and Shin, D. H. (eds.), *Rice Allelopathy, Proceedings of the International Workshop in Rice Allelopathy*, 17–19 August 2000, Institute of Agricultural Science and Technology, Kyungpook National University, Taegu, Korea, pp. 49–56.

© Burleigh Dodds Science Publishing Limited, 2025. All rights reserved.

Liu, D. L. and Lovett, J. V. (1993), Biologically active secondary metabolites of barley. I. Developing techniques and assessing allelopathy in barley. *J. Chem. Ecol.*, 19: 2217–2230.

Lodhi, M. A. K., et al. (1987), Allelopathy in agroecosystems: wheat phytotoxicity and its possible roles in crop rotation. *J. Chem. Ecol.*, 13: 1881–1891.

Lou, Y., et al. (2016), Interactions between allelochemicals and the microbial community affect weed suppression following cover crop residue incorporation into soil, *Plant Soil*, 399: 357–371.

Lovett, J. V., et al. (1994), Biologically active secondary metabolites of barley, IV: Hordenine production by different barley lines, *J. Chem. Ecol.*, 20: 1945–1954.

Ma, H. J., et al. (2006), Allelopathic potential of K21, selected as a promising allelopathic rice, *Weed Biol. Manage.*, 6: 189–196.

Macías, F. A., et al. (1999), Allelopathic studies in cultivar species, 14: allelochemicals from sunflower leaves cv. Peredovick. *Phytochem.*, 52: 613–621.

Mahé, I., et al. (2022), Deciphering field-based evidences for crop allelopathy in weed regulation, A review, *Agron. Sus. Develop.*, *42*: 50.

Mahmood, K., et al. (2013), Allelopathic activity of Pakistani wheat genotypes against wild oat (*Avena fatua*), *Pak. J. Agric. Sci.*, 50: 169–176.

Matloob, A., et al. (2020), Challenges and prospects for weed management in Pakistan: a review, *Crop Prot.*, 134: 104724.

Mennan, H. and Ngouajio, M. (2012), Effect of brassica cover crops and hazelnut husk mulch on weed control in hazelnut orchards, *Hort. Technol.*, 22: 99–105.

Minhas, W. A., et al. (2023a), The influence of different crop mulches on weed infestation, soil properties and productivity of wheat under conventional and conservation production systems, *Plants*, 12: 9.

Minhas, W. A., et al. (2023b), Weed infestation and productivity of wheat crop sown in various cropping systems under conventional and conservation tillage, *Front. Plant Sci.*, 14: 1176738.

Mirsky, S., et al. (2013), Overcoming weed management challenges in cover crop–based organic rotational no-till soybean production in the eastern United States, *Weed Technol.*, 27: 193–203.

Mishra, J. S., et al. (2022), Tillage and crop establishment effects on weeds and productivity of a rice-wheat-mungbean rotation, *Field Crops Res.*, 284: 108577.

Molísch, H. (1937), Der Einfluss einer Pfalnze auf die Andere-allelopathie. Fischer, Jena, Germany.

Naeem, M. (2021), Weed spectrum and management studies in barley-based cropping systems under conventional and conservative tillage practices, PhD Dissertation, Department of agronomy, Bahauddin Zakariya University, Multan, Pakistan.

Naeem, M., et al. (2022), The impact of different weed management systems on weed flora and dry biomass production of barley grown under various barley-based cropping systems, *Plants*, 11: 718.

Naeem, M., et al. (2021), Weed flora composition of different barley-based cropping systems under conventional and conservation tillage practices, *Phytoparasitica*, 49: 751–769.

Nakahisa, K., et al. (1994), Study on the allelopathy of alfalfa (*Medicago sativa* L.). 2. Isolation and identification of allelopathic substance in alfalfa, *Jpn. J. Crop Sci.*, 63: 278–284.

© Burleigh Dodds Science Publishing Limited, 2025. All rights reserved.

Nandan, R., et al. P., (2020), Viable weed seed density and diversity in soil and crop productivity under conservation agriculture practices in rice-based cropping systems, *Crop Prot.*, 136: 105210.

Nawaz, A. and Farooq, M. (2016), Weed management in resource conservation production systems in Pakistan, *Crop Prot.*, 85: 89–103.

Nawaz, A., et al. (2014), Role of allelopathy in weed management. In: Chauhan, B. Mahajan, G. (eds.), *Recent Advances in Weed Management*. Springer, New York, pp. 39–61.

Nawaz, A., et al. (2019a), Ecological management of agricultural pests through allelopathy. In: Merillon, J. M. and Ramawat, K. (eds.), *Co-Evolution of Secondary Metabolites. Reference Series in Phytochemistry*. Springer, Cham, pp. 543–574.

Nawaz, A., et al. (2019b), Rice–wheat cropping systems in South Asia: issues, options and opportunities, *Crop Pasture Sci.*, 70: 395–427.

Niemeyer, H. M. (1988), Hydroxamic acids (4-hydroxy-1,4-benzoxazin-3-one), defense chemicals in the gramineae, *Phytochemistry* 27: 3349–3358.

Nikolić, N., et al. (2022), Opportunities from unmanned aerial vehicles to identify differences in weed spatial distribution between conventional and conservation agriculture, *Sustainability* 14: 6324.

Nimbal, C. I., et al. (1996), Phytotoxicity and distribution of sorgoleone in grain sorghum germplasm, *J. Agric. Food Chem.*, 44: 1343–1347.

Norsworthy, J. K., et al. (2011), Evaluation of cereal and Brassicaceae cover crops in conservation-tillage, enhanced, glyphosate-resistant cotton, *Weed Technol.*, 25: 6–13.

Olofsdotter, M., et al. (2002) Improving crop competitive ability using allelopathy-an example from rice, *Plant Breed.*, 121: 1–9.

Otte, B. A., et al. (2020), Phenolic acids released to soil during cereal rye cover crop decomposition, *Chemoecol.*, 30: 25–34.

Pratibha, G., et al. (2021), Weed shift and community diversity in conservation and conventional agriculture systems in pigeonpea-castor systems under rainfed semi-arid tropics, *Soil Tillage Res.*, 212: 105075.

Putnam, A. R. and Duke, W. O. (1974), Biological suppression of weeds: Evidence for allelopathy in accessions of cucumber, *Science*, 185: 370–72.

Putnam, A. R. and Duke, W. O. (1978), Allelopathy in agroecosystems, *Ann. Rev. Phytopathol.*, 16: 431–451.

Ranaivoson, L., et al. (2016), Agro-ecological functions of crop residues under conservation agriculture. A review, *Agron. Sustain. Develop.*, 37: 1–17.

Rector, B. G. (2008), Molecular biology approaches to control of intractable weeds: new strategies and complements to existing biological practices, *Plant Sci.*, 175: 437–448.

Rice, C. P., et al. (2022), Benzoxazinoids in roots and shoots of cereal rye (*Secale cereale*) and their fates in soil after cover crop termination, *Chemoecology*, 32: 117–128.

Rietveld, W. J. (1983), Allelopathic effects of juglone on germination and growth of several herbaceaous and woody species, *J. Chem. Ecol.*, 9: 295–308.

Rimando, A. M., et al. (2001), Searching for rice allelochemicals, *Agron. J.*, 93: 16–20.

Rizvi, S. J. H., et al. (2004), A discipline called allelopathy, *in Allelopathy: Basic and Applied Aspects, ed.* by Rizvi SJH and Rizvi V. Chapman and Hall, London, UK, pp. 1–10.

Salomonsson, A. C., et al. (1978), Quantitative determination by GLC of phenolic acids as ethyl derivatives in cereal straw, *J. Agric. Food Chem.*, 26: 830–835.

© Burleigh Dodds Science Publishing Limited, 2025. All rights reserved.

Scavo, A. and Mauromicale, G. (2021), Crop allelopathy for sustainable weed management in agroecosystems: Knowing the present with a view to the future, *Agronomy*, 11: 2104.

Scavo, A., et al. (2020), Leaf extracts of cultivated cardoon as potential bioherbicide, *Sci. Horti.*, 261: 109024.

Séne, M. N., et al. (2001), Phenolic compounds in a Sahelian sorghum (*Sorghum bicolor*) genotype (CE145-66) and associated soils, *J. Chem. Ecol.*, 27: 81–92.

Shahzad, M. (2016), Weed dynamics and management studies in wheat-based cropping systems under conventional and conservative tillage practices, PhD Dissertation, Department of agronomy, Bahauddin Zakariya University, Multan, Pakistan.

Shahzad, M., et al. (2016), Weed spectrum in different wheat-based cropping systems under conservation and conventional tillage practices in Punjab, Pakistan, *Soil Tillage Res.*, 163: 71–79.

Shahzad, M., et al. (2021) The impact of different crop rotations by weed management strategies' interactions on weed infestation and productivity of wheat (*Triticum aestivum* L.), *Agronomy*, 11: 2088.

Sharma, A. R., et al. (2013), Weed management in conservation agriculture systems – problems and prospects. In: *Proceedings of 24th Asian-Pacific weed Science Society Conference*, 22–25 October 2013, Bandung, pp. 31–35.

Sharma, T., et al. (2023), Long-term conservation agriculture influences weed diversity, water productivity, grain yield, and energy budgeting of wheat in north-western Indo-Gangetic plains, *Sustainability* 15: 7290.

Shehzad, T. and Okuno, K. (2020), Genetic analysis of QTLs controlling allelopathic characteristics in sorghum, *PLoS One*, 15: e0235896.

Singh, V. P., et al. (2010), Current status of zero tillage in weed management, *Indian J. Weed Sci.*, 42: 1–9.

Sisodia, S. and Siddiqui, M. B. (2010), Allelopathic effect by aqueous extracts of different parts of *Croton bonplandianum* Baill. on some crop and weed plants, *J. Agri. Ext. Rural Develop.*, 2: 22–28.

Smith, R. G., et al. (2020), Are cover crop mixtures better at suppressing weeds than cover crop monocultures? *Weed Sci.*, 68: 186–194.

Song, B., et al. (2008), Allelopathic enhancement and differential gene expression in rice under low nitrogen treatment, *J. Chem. Ecol.*, 34: 688–695.

Sosnoskie, L. M., et al. (2006), Weed seed bank community composition in a 35 years old tillage and rotation experiment, *Weed Sci.*, 54: 263–273.

Sportelli, M., et al. (2022), Innovative living mulch management strategies for organic conservation field vegetables: Evaluation of continuous mowing, flaming, and tillage performances, *Agronomy*, 12: 622.

Streit, B., et al. (2003), Weed populations in winter wheat as affected by crop sequence, intensity of tillage and time of herbicide application in a cool and humid climate, *Weed Res.*, 43: 20–32.

Swanton, C. J., et al. (2000), Influence of tillage type on vertical seed bank distribution in a sandy soil, *Can. J. Plant Sci.*, 80: 455–457.

Tabaglio, V., et al. (2013) Allelopathic cover crop of rye for integrated weed control in sustainable agroecosystems, *Italian J. Agron.*, 8: e5.

Thierfelder, C., et al. (2018), Complementary practices supporting conservation agriculture in southern Africa: a review, *Agron. Sustain. Develop.*, 38: 16.

© Burleigh Dodds Science Publishing Limited, 2025. All rights reserved.

Trezzi, M. M., et al. (2016), Allelopathy: driving mechanisms governing its activity in agriculture, *J. Plant Interactions*, 11: 53–60.

Trichard, A., et al. (2013), Identification of weed community traits response to conservation agriculture, *Agric. Ecosyst. Environ.*, 179: 179–186.

VanGessel, M. J. (2001), Glyphosate-resistant horseweed from Delaware. *Weed Sci.*, 49: 703–705.

Vaughn, S. F. (1999), Glucosinolates as natural pesticides. In: Cutler, H. G. and Cutler, S. J. (eds.), *Biologically Active Natural Products*, CRC Press, Boca Raton, USA, pp. 81–92.

Vaughn, S. F. and Boydston, R. A. (1997), Volatile allelochemicals released by crucifer green manures, *J. Chem. Ecol.*, 23: 2107–16.

Vencill, W. K. and Banks, P. A. (1994), Effects of tillage systems and weed management on weed populations in grain sorghum (*Sorghum bicolor*), *Weed Sci.*, 42: 541–547.

Weston, L. A., et al. (2013), Sorghum allelopathy –from ecosystem to molecule, *J. Chem. Ecol.*, 39: 142–153.

Worthington, M. and Reberg-Horton, C. (2013), Breeding cereal crops for enhanced weed suppression: optimizing allelopathy and competitive ability, *J. Chem. Ecol.*, 39: 213–231.

Wu, H. (2005), Molecular approaches in improving wheat allelopathy, In: *Proceedings of the 4th World Congress on Allelopathy*, pp. 11–14.

Wu, H., et al. (2000a), Allelochemicals in wheat (*Triticum aestivum* L.): variation of phenolic acids in root tissues, *J. Agric. Food Chem.*, 48: 5321–5325.

Wu, H., et al. (2000b), Evaluation of seedling allelopathy in 453 wheat (*Triticum aestivum*) accessions against annaul rye grass (*Lolium rigidum*), *Aust. J. Agric. Res.*, 51: 937–944.

Wu, H., et al. (2000c), Laboratory screening for allelopathic potential of wheat (*Triticum aestivum*) accessions against annual ryegrass (*Lolium rigidum*), *Aust. J. Agric. Res.*, 51: 259–266.

Wu, H., et al. (2000d), Distribution and exudation of allelochemicals in wheat (*Triticum aestivum*), *J. Chem. Ecol.*, 26: 2141–2154.

Wu, H., et al. (2001), Allelochemicals in wheat (*Triticum aestivum* L.): variation of phenolic acids in shoot tissues, *J. Chem. Ecol.*, 27: 125–135.

Wu, H., et al. (2003), Quantitative trait loci and molecular markers associated with wheat allelopathy, *Theor. Appl. Genet.*, 107: 1477–1481.

Xu, M., et al. (2012), Genetic evidence for natural product-mediated plant-plant allelopathy in rice (*Oryza sativa*), *New Phytol.*, 193: 570–575.

Xuan, T. D. (2004), Utilization of allelopathy in higher plants for control of paddy weeds. Ph.D. Thesis, School of Agricultural Sciences, Kagoshima University, Kagoshima, Japan, 167p.

Xuan, T. D., et al. (2003), Correlation between inhibitory exhibition and suspected allelochemicals in alfalfa (*Medicago sativa* L.), *Plant Prod. Sci.*, 6: 165–171.

Xuan, T. D., et al. (2005), Biological control of weeds and plant pathogens in paddy rice by exploiting plant allelopathy: an overview, *Crop Prot.*, 24: 197–206.

Zerihun, A., et al. (2014), Conservation agriculture: maize-legume intensification for yield, profitability and soil fertility improvement in maize belt areas of western Ethiopia, *Int. J. Plant Soil Sci.*, 3: 969–985.

Zuo, S. P., et al. (2012), Genetic basis of allelopathic potential of winter wheat based on the perspective of quantitative trait locus, *Field Crops Res.*, 135: 67–73.

© Burleigh Dodds Science Publishing Limited, 2025. All rights reserved.

Chapter 5

Thermal weed control in Conservation Agriculture

Bernhard Streit and Martin V. Bauer, Bern University of Applied Sciences, Switzerland

1 Introduction

Conservation Agriculture (CA), characterized by minimal soil disturbance, permanent soil cover (e.g. through cover cropping), and enhanced biodiversity through crop rotations and alternative cropping layouts, has gained prominence as a sustainable farming approach. Weed management within this framework is crucial, and the limitations of traditional herbicide-dependent methods have spurred interest in alternative strategies. Although permanent soil cover offers weed-suppressing benefits, it does not completely eliminate weed growth (Dorn et al., 2015). Herbicides are often promoted as a solution for reducing weed pressure both before and during crop growth. While non-chemical measures exist, alternatives to herbicides for in-crop weed control in CA systems are limited.

Maintaining significant quantities of plant residues on the soil surface is essential for suppressing weed emergence and stabilizing soil structure against erosion. These plant residues are critical to the success of CA (Chauhan et al., 2012). However, traditional implements for within-crop mechanical weed control (such as cultivators, hoes, or harrows (pulled by a tractor)) are inefficient and impractical when large quantities of plant residues are present (Erenstein, 2003). Other tools potentially suitable for mechanical weed control

http://dx.doi.org/10.19103/AS.2024.0145.08
© Burleigh Dodds Science Publishing Limited, 2025. All rights reserved.

where there are crop residues, such as rotary hoes or cultivators, are generally not efficient enough (Shirtliffe and Johnson, 2012; Podolsky et al., 2016; Zinati et al., 2017).

Thermal weed control techniques, utilizing heat to manage weed populations and described in this chapter, have emerged as a viable and environmentally friendly option. In commercial applications today, thermal weed control methods are typically used prior to crop emergence, especially when the crop is a poor competitor to weeds (Perruzzi et al., 2004; Seaman, 2016). This is often the case for crops like carrots, which face competition from broad-leaved weeds such as *Bidens pilosa* or *Amaranthus* spp. and grass weeds like *Echinochloa crus-galli*. Thermal weed control is also used when the crop is more resistant to heat than the surrounding weeds (Mutch et al., 2008; Ulloa et al., 2012), such as in post-emergent maize with major weeds like *Cyanodon dactylon* or *Parthenium hysterophorus*.

2 Thermal weed control: general aspects

Physical weed control methods are generally based on thermal or mechanical effects (Vincent et al., 2014). Thermal energy can be induced by hot matter or radiation, while mechanical energy is transmitted through gases, liquids, or solids. Mechanical methods primarily control weeds by rupturing the plant's cell structure through a combination of uprooting and soil burial, followed by tissue desiccation (Clout, 2009; Laguë et al., 2014). However, since mechanical energy cannot be transferred to plants without impacting the soil (violation of the "contact-free" criterion), it will not be discussed in this chapter.

The aim of thermal weed control is to have a lethal effect on the weed. In thermal weed control, heat is transferred to the plants from an on-site energy source, causing a localized temperature increase. Heat transfer occurs through three main physical processes (and subsequent material absorption) (Bertram and Meyer, 1997):

- convection (including advection);
- conduction; and
- radiation.

Based on general physical principles, we can deduce that:

- wind will affect convection and will cool the plant with ambient air;
- plant and soil moisture content increases heat capacity and conductivity;
- wet leaves increase heat capacity and may hinder heat transfer to the plant by generating latent heat through vaporization;
- soil moisture increases the likelihood of regrowth; and

© Burleigh Dodds Science Publishing Limited, 2025. All rights reserved.

- layers of leaf hair or wax or lignification protects the vital plant tissue and may reduce the efficacy of thermal weed control methods.

Several studies have shown that dose-response effects depend on the species being treated. For young plants, thermal energy can be applied to areas such as cotyledons, the stem, the roots, the first leaves, or the entire plant. Species with protected growth points at or below soil level are less vulnerable or unaffected by thermal methods.

Due to a correlation between the applied energy dose and the ratio of directly to indirectly affected tissue (i.e. cellular heat conduction), small plants at early growth stages are more sensitive to thermal treatment than older, larger plants (Parish, 1989; Johnson et al., 1989; Ascard, 1995; Ulloa et al., 2012; Marx et al., 2012a). Leaf morphology also influences the effectiveness of thermal treatment; for instance, leaves with relatively thick epidermides are more resistant to heat than those with thinner structures (Hansson and Ascard, 2002; Leon and Ferreira, 2008).

While annual and biennial weeds are generally well controllable by thermal methods, perennial weeds are more resistant due to their underground rhizomes and roots (Cisneros and Zandstra, 2008; Stepanovic, 2013). Repeated thermal weed control can lead to the dominance of monoculture weeds, such as *Taraxacum* spp. in vineyards. Both thermal and chemical weed control methods are typically applied at similar development stages of the weeds.

In general, there is an increased risk of fire when using heat against weeds. This risk is particularly high when the soil is covered with dead, dry plant residues, a common condition in CA systems. As a result, the use of this alternative technology is somewhat limited. For this reason, this chapter focuses on selective methods of thermal weed control that minimize the risk of wildfires. It is also assumed that active fire extinguishing equipment is an integral part of any thermal weed control system.

3 Flame weeding

3.1 Surface applications

Flaming has been widely researched using both tractor-drawn implements and handheld devices (Ascard, 1998; Rasmussen et al., 2011; Ghantous et al., 2012; Hoyle et al., 2012; Rask et al., 2012; Raffaelli et al., 2013). The primary disadvantages of flaming are its low selectivity and the need for repeated application (Ascard, 1995). To improve energy efficiency, models have been developed for optimizing flame weeding implements (Bertram and Meyer, 1997; Andersen, 1997). Selectivity in flame weeding has been studied in relation to crops that can tolerate heat without major loss in yield (biological selectivity)

© Burleigh Dodds Science Publishing Limited, 2025. All rights reserved.

(Sivesind, 2010; Peruzzi et al., 2017; Abou Chehade et al., 2018). However, in most cases, flame tools have been used non-selectively on entire surfaces maintaining a safe distance from crops to prevent damage. In situations where only living plants are present on the soil surface and the risk of fires is low, flaming the entire surface can achieve a similar effect against annual weeds as non-selective herbicides.

Infrared radiation, an important component of thermal weed control, uses specialized equipment that emits infrared wavelengths to selectively target and heat weed cells, causing cellular damage and inhibiting growth. Like flames, infrared radiation exploits the differential heat sensitivity between weeds and crops and is typically used on larger surfaces.

3.2 Spot flaming

For greater efficiency and to minimize collateral damage, the flame design must ensure low spatial heat distribution in the horizontal plane. Andersen (1997) found that a hydrogen burner delivers a more consistent flame than propane burners. This consistency offers an advantage by making heat dispersion more controllable, allowing for closer contact with crop plants (and plant residues) without damaging them. Ideally, the lethal energy is focused on an area significantly smaller than the overall surface of the weed plant being treated.

Current selective burner systems, such as the Robovator (F. Poulsen Engineering ApS.), do not yet allow for individual weed plants to be selectively eliminated without harming nearby crops or plant residues. In CA, this can increase the risk of fire due to the large quantities of (sometimes dry) plant residues around the weed plants. Moreover, time-dependent factors must be considered: conventional flaming should be applied when there is a difference in growth (or resistance) between the crop and the weeds. This means either the crop has not yet grown or it is big enough to survive the thermal treatment and random heat exposure (Raffaelli et al., 2011).

4 Electrical weed control

The use of electrical energy for weed eradication dates back to the late 1800s, with numerous patents registered in the United States from 1890 onward (Sharp, 1892). A modern approach involves a tractor-driven apparatus developed by Lasco Inc. The LW5 Lightning Weeder was designed to eliminate resilient weeds in row crops after conventional chemical interventions. Manufactured in Vicksburg, Mississippi, this device was introduced in 1980, demonstrating the ongoing evolution of electrical weed control technology in agriculture (Vigneault and Benoît, 2001)(Fig. 1).

© Burleigh Dodds Science Publishing Limited, 2025. All rights reserved.

Figure 1 Superficial flaming of residual weeds after sugar beet harvest.

Electric currents in plants cause dielectric losses, leading to a temperature increase that damages cell structures (Broadhurst et al., 1987). These currents can be induced through direct contact with an electric power source or by capacitive coupling in an electric field (Table 1). The equivalent electrical circuit can be described as a high-voltage current generator with serial resistance from the machine and the soil (Rs), parallel resistance from obstacles (Rp), and resistance from the vegetation (Rv) (Fig. 2).

Rp represents the surroundings of the plant that may conduct current parallel to the plant and depends on objects which are close to the target – in the worst case, an arable crop. Rs depends on the internal resistance of the machine, as well as on the resistance of the soil and roots. Effectiveness is highest when Rv is much smaller than Rp and much larger than Rs, as the current will follow the path with lowest resistance. Electrical resistance is calculated by R = p l, where l is the length of the conductor, A is the cross-sectional area of the conductor, and p is the electrical resistivity of the material. The cross-section of soil, always larger than that of the roots, explains why bulbous roots like those of weed beets tend to be damaged as the current flows through a substantial part of the tissue before leaving the root. Dry soil increases root damage, while plants with extensive root systems, such as couch grass, are often not killed (Reed, 2009), similar to grass-clover stands in fodder crops (Ruf et al., 2023). Power frequency also affects resistance, requiring matching frequencies for high weed-damaging.

Recent advancements have led to the development and commercialization of devices such as Electroherb or Xpower (Zasso, Indaiatuba, Brazil) (Slaven et al., 2023). However, controlling green target plants with these devices

© Burleigh Dodds Science Publishing Limited, 2025. All rights reserved.

Table 1 Comparison of methods for electrothermal weed control

Method	*DC and AC (low freq.)*	*High freq. (13 MHz)*	*Microwave (2.45 GHz)*	*Microwave (>10 GHz)*	*Thermal IR*	*IR Laser*
Effect on roots	Noticeable	Difficult to induce electr. field	Moderate	Poor	None	none
Effect on stalks and leaves	Good	Good	Good	Good	Good if not shaded	Good if not shaded
Hazard potential	Low	Low	Very high	High	Moderate	High
Energy consumption	Low	Low	Very high	High	Very high	Moderate
Development costs	Moderate	High	High	High	Low	High

Source: Adapted from Mätzler (1993).

© Burleigh Dodds Science Publishing Limited, 2025. All rights reserved.

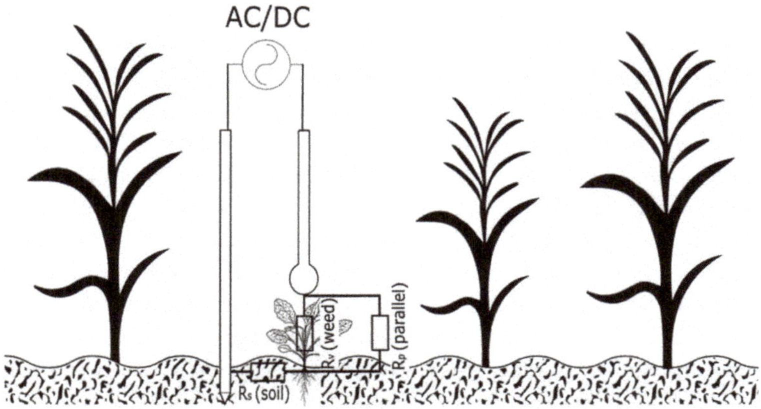

Figure 2 Electric diagram of electric weed control. Source: Adapted from Mätzler (1993).

Figure 3 Use of a Xpower device (Zasso, Indaiatuba, Brazil) to non-selectively control annual weeds.

requires significant time and energy. Practical observations report driving speeds of around 3 km/h and diesel consumption exceeding 20 l/ha when using 3-meter-wide equipment powered by a 100 kW tractor (Fig. 3).

Other alternative applications, such as microwaves or capacitive coupling, have also been tested experimentally. Studies on the use of microwave radiation have employed the industrial frequency of 2.45 GHz, resulting in wavelengths

© Burleigh Dodds Science Publishing Limited, 2025. All rights reserved.

of 12.25 cm. Magnetrons operating at this frequency are readily available at low cost from household microwave ovens. However, unless new ways of applying microwave radiation are developed, the long wavelength of 12.25 cm will not be capable of targeting seedlings and small plants in early growth stages. Regardless of focusing capabilities (which implies resolution), less than 7% of 2.45 GHz radiation is absorbed by a leaf with a thickness of 0.2 mm at an ambient temperature of 20°C, meaning that most of the energy is lost to the surroundings, potentially causing unintended damage (Mätzler, 1993). A benefit of microwave application is that it is unaffected by wind, as the radiation penetrates all parts of the plants simultaneously, unlike traditional heating process where heat is conducted from the outside toward the center (Diprose et al., 1984).

Experiments have also been conducted on electrical applied to individual plants. The findings may lead to promising applications in robotic weed control in the near future, potentially overcoming the non-selectivity and inefficiency of large-area electrical applications.

5 Hot water or steam used for weed control

Water, in either its liquid or gaseous state (steam), is used to transfer thermal energy to plants, leading to a temperature increase that damages the plant's cell structure (Kristoffersen et al., 2008; Rask et al., 2013). In general, a negative exponential relationship has been observed between lethal temperature and exposure time (Sutcliffe, 1977; Levitt, 1980).

When higher temperatures are applied for weed control, the heat sum required to kill the plants is lower (Storeheier, 1994). Structural changes in cells are more pronounced when cellular temperature rises rapidly, as seen in flaming, compared to slower, gradual temperature changes, such as in hot water treatment (Daniell et al., 1969; Astatkie et al., 2014). Repeated treatments are needed for perennial weeds, as hot water primarily affects above-ground plant parts (Hansson and Ascard, 2002). Due to these limitations, there are currently no known commercial applications of hot water or steam for weed control on larger areas.

6 Laser-based weed control

Laser-based weed control represents an emerging technology in precision agriculture (Andreasen et al., 2022), leveraging the principles of selective photothermolysis for targeted, non-chemical weed suppression (Mathiassen et al., 2006). Laser weeding is potentially well-suited for CA due to its high spatial resolution, which allows for the selective treatment of emerging weeds between plant residues and crop. Within this chapter, the underlying scientific principles, technological advancements, and potential applications of laser systems in weed management will be explored.

© Burleigh Dodds Science Publishing Limited, 2025. All rights reserved.

Laser systems designed for weed control typically consist of a laser source, optics for beam delivery and focusing, and control mechanisms for precision targeting. The complexity of laser weed control arises from the interdependence of the following factors:

- generation of a suitable laser beam and spot;
- absorption of the laser energy by plant tissues; and
- enabling technologies required for weed localization and beam positioning.

Key variables for effective laser weed control are energy density, spot size, and wavelength. Laser light has a single wavelength and is coherent. However, the most relevant characteristic for weeding applications is its low divergence, allowing for high spatial energy densities. In general,

- the cost of the laser unit increases as the wavelength decreases at constant power;
- energy conversion efficiency within solid-state lasers decreases with shorter wavelength;
- light absorption by plant tissues is low at the most affordable wavelength (1 micron); and
- building a safe device becomes more difficult as the wavelength increases.

Relationships between absorbance potential of plant tissue and laser wavelengths are shown in Fig. 4.

Laser spots of industrial applications are typically in the millimeter range or smaller but, for weed control, larger spot sizes are advantageous. Smaller spot sizes make it difficult to target weed plants accurately, while larger spot sizes require higher energy consumption to maintain constant energy density. Tissue damage is related to energy density. If the plant is not killed by the process, the risk of adaption arises, potentially leading to increased resistance.

CO_2-lasers are the most suitable for generating large spot sizes at the required energy densities. This may be the main reason Carbon Robotics chose this technology for the only commercial laser weeding product to date. However, CO_2 lasers have a relatively low power conversion efficiency, around 10%, compared to solid-state lasers, which can approach 50%. This means that, with a CO_2 laser-based weeding tool, powered by a diesel generator, more than 95% of the fuel is lost in conversion (assuming approximately 35% fuel-to-electricity efficiency). Another reasons for choosing 10 micron wavelength (as used in CO_2 lasers) is the high absorption rate in plant tissue and the easier implementation of safety measures compared to solid-state lasers.

© Burleigh Dodds Science Publishing Limited, 2025. All rights reserved.

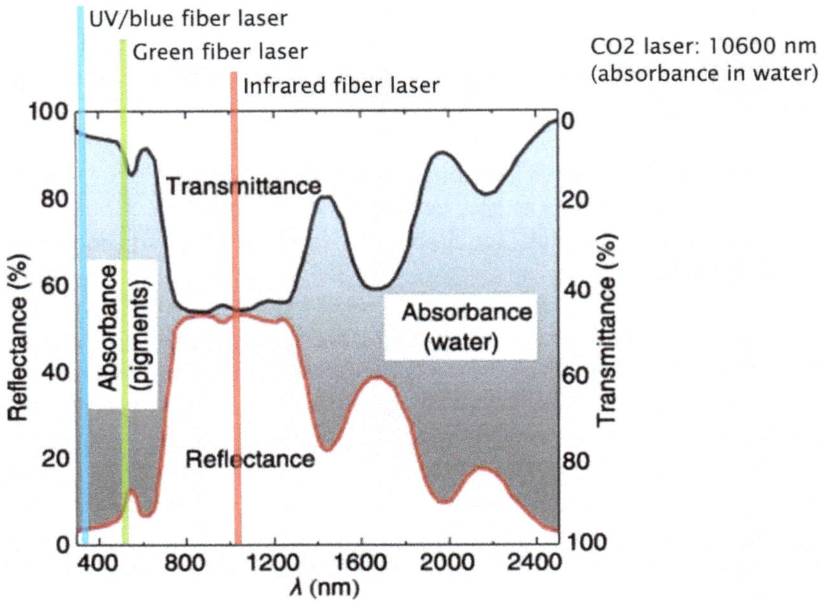

Figure 4 Typical absorbance of plant tissue and popular solid state laser wavelengths based on Gao and Ye (2017).

Various research groups and startups are working on weeding equipment based on solid-state lasers, but no commercial solution is available at the time of writing. These research efforts face the challenge of finding an affordable laser source that can generate the required spot size at the nessecary energy density. Optimizing spot size to reduce cost could compromise effectiveness and increase the accuracy requirement for weed localization and targeting, thus increasing system complexity. Eye safety is also a significant concern with decreasing wavelenths.

Effective weed control requires precise energy delivery to the targeted plants. Adaptive optics and advanced beam steering technologies help optimize laser intensity, minimizing collateral damage to non-target crops and reducing overall energy consumption of the systems. To meet these requirements in small and efficient control systems, precise detection of the target organisms and their vulnerable points (e.g. the meristem) is essential (Wöltjen et al., 2008). It is likely that these technological challenges will be addressed in the near future as advancements continue.

7 Conclusion

Thermal weed control methods, including flaming, electrical weed control, hot water or steam treatments, and laser technology, present a viable and

© Burleigh Dodds Science Publishing Limited, 2025. All rights reserved.

environmentally sustainable alternative to chemical herbicides. These methods address both weed management and soil conservation concerns. The effectiveness of thermal weed control varies across weed species, with younger, smaller weeds being more susceptible to thermal treatment than older, larger ones. However, the challenge of selectively targeting weeds without harming adjacent crops or causing unintended environmental impacts, such as increased fire risk, remains a significant issue.

Technological advancements in equipment and application techniques are critical for improving the selectivity and efficiency of thermal weed control methods. For instance, the development of precision targeting systems could greatly reduce the risk of damaging non-target plants. Additionally, integrating thermal weed control with other non-chemical methods within a holistic weed management framework could enhance the sustainability and resilience of CA systems.

Future research should focus on optimizing thermal application techniques to minimize energy consumption and reduce the environmental footprint. Furthermore, exploring the integration of thermal weed control methods with digital agriculture technologies, such as drones and AI-driven precision agriculture tools, could open new opportunities for targeted weed management at a larger scale.

As the global agricultural sector continues to shift toward sustainable practices, the role of thermal weed control in CAis likely to expand. However, the successful adoption of these technologies will depend on their cost-effectiveness, ease of use, and compatibility with existing agricultural practices. In conclusion, while challenges remain, the potential of thermal weed control to contribute to sustainable and productive agriculture systems is significant, warranting continued research, development, and adoption efforts.

8 References

Abou Chehade, L., et al. (2018) Effects of flame weeding on organic garlic production. *Horttechnology*, 28(4), 502–508.

Andersen, J. (1997) Experimental trials and modelling of hydrogen and propane burners for use in selective flaming. *Biological Agriculture & Horticulture*, 14, 207–219.

Andreasen, C., et al. (2022) Laser weeding with small autonomous vehicles: friends or foes? *Frontiers in Agronomy*, 4, 841086.

Ascard, J. (1995) *Thermal weed control by flaming*. Alnarp: Biological and Technical Aspects, Sveriges lantbruksuniversitet.

Ascard, J. (1998) Flame weeding. Effects of burner angle on weed control and temperature patterns. *Acta Agriculturae Scandinavica, Section B - Soil & Plant Science*, 48, 248–254.

Astatkie, T., et al. (2014) Effectiveness of hot water, infrared and open flame thermal units for controlling weeds. *Biological Agriculture & Horticulture*, 25, 1–12.

© Burleigh Dodds Science Publishing Limited, 2025. All rights reserved.

Bertram, A. and Meyer, J. (1997) Modellierung der Wärmeübertragung bei der thermischen Unkrautregulierung als Basis für die gerätechnische Optimierung. *Agrartechnische Forschung*, 3, 86–98.

Broadhurst, M.G., et al. (1987) The dielectric properties of biological tissue (Crassula portulacea) from 10–2 to 109 Hz. *Journal of Molecular Liquids*, 36, 65–73.

Chauhan, B.S., et al. (2012) Ecology and man- agement of weeds under conservation agriculture: a review. *Crop Protection*, 38, 57–65.

Cisneros, J.J. and Zandstra, B.H. (2008) Flame weeding effects on several weed species. *Weed Technology*, 22, 290–295.

Clout, M.N. (2009) *Invasive species management: a handbook of principles and techniques*. Oxford: Oxford University Press.

Daniell, J.W., et al. (1969) Effect of sublethal and lethal temperature on plant cells. *Plant Physiology*, 44, 1684–1689.

Diprose, M.F., et al. (1984) The effect of externally applied electrostatic fields, microwave radiation and electric currents on plants and other organisms, with special reference to weed control. *The Botanical Review*, 50, 171–223.

Dorn, B., et al. (2015) Weed suppression by cover crops. Comparative on-farm experiments under integrated and organic conservation tillage. *Weed Research*, 55, 586–597.

Erenstein, O. (2003) Smallholder conservation farming in the tropics and sub-tropics. A guide to the development and dissemination of mulching with crop residues and cover crops. *Agriculture, Ecosystems & Environment*, 100, 17–37.

Gao, Y. and Ye, H. (2017) Bionic membrane simulating solar spectrum reflection characteristics of natural leaf. *International Journal of Heat and Mass Transfer*, 114, 115–124.

Ghantous, K.M., et al. (2012) Handheld flame cultivators as a management option for woody weeds. *Weed Technology*, 26, 371–375.

Hansson, D. and Ascard, J. (2002) Influence of developmental stage and time of assessment on hot water weed control. *Weed Research*, 42, 307–316.

Hoyle, J.A., et al. (2012) Weed control using an enclosed thermal heating apparatus. *Weed Technology*, 26, 699–707.

Johnson, D.W., et al. (1989) Response of monocot and dicot weed species to fresnel lens concen- trated solar radiation. *Weed Science*, 37, 797–801.

Kristoffersen, P., et al. (2008) Non-chemical weed control on traffic islands: a comparison of the efficacy of five weed control techniques. *Weed Research*, 48, 124–130.

Laguë, C., et al. (2014) Thermal control in plant protection. In Vincent, C., et al. (Eds.), *Physical control methods in plant protection*. Berlin Heidelberg: Springer, pp. 35–46.

Leon, R.G. and Ferreira, D.T. (2008) Interspecific differences in weed susceptibility to steam injury. *Weed Technology*, 22, 719–723.

Levitt, J. (1980) *Responses of plants to environmental stresses. Vol. I: chilling, freezing, and high temperature stresses*, 2nd edn. New York: Academic Press.

Marx, C., et al. (2012a) Design and application of a weed damage model for laser-based weed control. *Biosystems Engineering*, 113, 148–157.

Mathiassen, S.K., et al. (2006) The effect of laser treatment as a weed control method. *Biosystems Engineering*, 95, 497–505.

Mätzler, C. (1993) *Elektrothermische Vegetationskontrolle*. Bern, Switzerland: Machbarkeitsstudie.

© Burleigh Dodds Science Publishing Limited, 2025. All rights reserved.

Mutch, D.R., et al. (2008) *Flaming as a method of weed control in organic farming systems.* East Lansing, MI: Michigan State University Extension.

Parish, S. (1989) Weed control - testing the effects of infrared radiation. *Agricultural Engineer*, 44(2), 53–55.

Perruzzi, A., et al. (2004) Physical weed control in organic carrot production. In Cloutier, D.C. and Ascard, J. (Eds.), *Proceedings 6th EWRS workshop on Physical And Cultural Weed Control*. Norway: European Weed Research Society EWRS, pp. 24–38.

Peruzzi, A., et al. (2017) Machines for non-chemical intra-row weed control in narrow and wide-row crops: a review. *Journal of Agricultural Engineering*, 48, 57.

Podolsky, K., et al. (2016) A comparison of reduced tillage implements for organic wheat production in Western Canada. *Agronomy Journal*, 108, 2003.

Raffaelli, M., et al. (2011) Physical weed control in processing tomatoes in Central Italy. *Renewable Agriculture and Food Systems*, 26, 95–103.

Raffaelli, M., et al. (2013) Dvelopment of machines for flaming weed control on hard surfaces. *Applied Engineering and Agriculture*, 29, 663–673.

Rask, A.M., et al. (2012) Response of Lolium perenne to repeated flame treatments with various doses of propane. *Weed Research*, 52, 131–139.

Rask, A.M., et al. (2013) Determining treatment frequency for controlling weeds on traffic islands using chemical and non-chemical weed control. *Weed Research*, 53, 249–258.

Rasmussen, J., et al. (2011) Punch planting, flame weeding and delayed sowing to reduce intra-row weeds in row crops. *Weed Research*, 51, 489–498.

Reed, J. (2009) Desk study: electrical weed control in field vegetables. Final report, Great Britain.

Ruf, T., et al. (2023) Field fodder conversion using electricity – negative effects on earthworms and changes in labile carbon fractions. *Soil and Tillage Research*, 232, 105746.

Seaman, A. (2016) *Production guide for organic carrots for processing*. Geneva, NY: New York State Agricultural Experiment Station.

Sharp, A.A. (1892) *Vegetation-exterminator.* Patent US492635 A.

Shirtliffe, S.J. and Johnson, E.N. (2012) Progress towards no-till organic weed control in western Canada. *Renewable Agriculture and Food Systems*, 27, 60–67.

Sivesind, E. (2010) *Selective flame weeding in vegetable crops*. Montreal: McGill University.

Slaven, M.J., et al. (2023) Exploring the potential of electric weed control: a review. *Weed Science*, 71(5), 403–421.

Stepanovic, S.V. (2013) *Positioning an innovative flame-weeding technology into crop production*. MSc, University of Nebraska, Nebraska.

Storeheier, K.J. (1994) Basic investigations into flaming for weed control. *Acta Horticulturae*, 372, 195–204.

Sutcliffe, J. (1977) *Plants and temperature*. London: Edward Arnold.

Ulloa, S.M., et al. (2012) Weed control and crop tolerance to propane flaming as influenced by the time of day. *Crop Protection*, 31, 1–7.

Vigneault, C. and Benoit, D.L. (2001) Electrical weed control: theory and applications. In: Vincent, C., et al. (Eds.), *Physical control methods in plant protection*. Berlin, Heidelberg: Springer Science & Business Media.

© Burleigh Dodds Science Publishing Limited, 2025. All rights reserved.

Vincent, C., et al. (eds) (2014) *Physical control methods in plant protection*. Berlin Heidelberg: Springer.
Wöltjen, C., et al. (2008) Plant growth depression by selective irradiation of the meristem with CO2 and diode lasers. *Biosystems Engineering*, 101(3), 316–324.
Zinati, G.M., et al. (2017) High-residue cultivation timing impact on organic no-till soybean weed management. *Weed Technology*, 31, 320–329.

© Burleigh Dodds Science Publishing Limited, 2025. All rights reserved.

Chapter 6

Using herbicides in Conservation Agriculture systems: the state of the art

Per Kudsk, Aarhus University, Denmark

1 Introduction

Conservation Agriculture (CA) has received increasing interest worldwide as a cropping system that addresses the future challenge of feeding an increasing world population while minimizing the negative environmental and climate impacts of agriculture and preserving soil fertility (Reicosky, 2015). CA is based on three principles: no tillage, continuing plant/residue cover on the soil surface and a diverse crop rotation (Hobbs, 2007; FAO, 2011). There is a general agreement that the use of cover crops/green mulches and diverse crop rotations tend to minimize weed problems (Adeux et al., 2019; Andert et al., 2016; Guinet et al., 2023; Teasdale et al., 2007, Haramoto & Gallandt, 2004) although the effect of cover crops on weed management recently was questioned (Adeux et al., 2021). In contrast, the adoption of no-tillage practices has often been associated with increasing weed problems (Derksen et al., 1993; Melander et al., 2013; Vullioud et al., 2006) although omitting tillage ensures that weed seeds remain on the soil surface, a situation expected to minimize germination and increase seed mortality (Cordeau et al., 2020). Nonetheless, several studies have reported higher weed pressure under no-tillage than under ploughing and no-tillage systems can promote problematic weed species such

http://dx.doi.org/10.19103/AS.2024.0145.09
© Burleigh Dodds Science Publishing Limited, 2025. All rights reserved.

as annual grass species and perennial weed species (Armengot et al., 2016; Cardina et al., 2002; Thomas et al., 2004; Trichard et al., 2013). Nichols et al. (2015) reviewed the literature on CA and weed dynamics and concluded that implementing no-till without crop rotation can result in severe weed problems with grass weeds and perennial weeds becoming more dominant but also found that greater crop diversity resulted in easier weed management.

There is an ongoing discussion on whether CA, adopting all three principles, will exacerbate problems with weeds and hence increase the need for weed control including herbicides. Friedrich and Kassam (2012) argued that there is no need for additional input of chemicals in CA if all three principles are implemented simultaneously. A long-term study conducted in Mexico supported this assumption showing that weed biomass was significantly lower in CA than conventional tillage both without herbicide use and with annual herbicide applications (Fonteyne et al., 2020). In contrast, a higher herbicide use has been found among no-tillage farmers (not all practicing CA) in France and Germany compared to farmers doing conventional tillage (Adeux et al., 2022; Andert et al., 2018). Recently, a simulation study, using the model FlorSys and data from 395 arable cropping systems in France, suggested that herbicide use, expressed as the Treatment Frequency Index, will increase if tillage frequency is reduced (Colbach & Cordeau, 2022).

In organic farming, where farmers do not have access to effective herbicides, it is noteworthy that very few farmers have adopted CA (Peignè et al., 2016) suggesting that organic farmers perceive weed management without herbicides to be difficult in a CA setting. Légère et al. (2013) found that while competitive crops like barley and red clover could successfully be grown organically in a CA setting, the cultivation of the less competitive crops like maize and soybeans failed due to problems with weeds.

This chapter will only deal with herbicide performance in CA systems. Omitting ploughing and shifting to no-tillage practices can also affect herbicide fate in the soil, which could indirectly affect herbicide performance and selectivity but effects on herbicide fate will not be covered in this chapter. Instead, the reader is referred to earlier reviews by Locke & Bryson (1997) and Alletto et al. (2010).

2 Herbicide use in Conservation Agriculture: an overview

Before CA was introduced, minimum tillage, no-tillage and conservation tillage were popular among farmers. These tillage practices are not CA, as they do not incorporate all three principles of CA but, in particular conservation tillage and no-tillage, include some of the features of CA (no-tillage and plant cover or crop residues). Hence, research on herbicide use and performance in CA builds

© Burleigh Dodds Science Publishing Limited, 2025. All rights reserved.

upon the results of earlier research on herbicide use in no-tillage, conservation tillage and other systems of reduced tillage. The number of papers focussing specifically on herbicide performance in CA is limited and this chapter therefore also considers results originating from no-tillage and conservation tillage studies.

In the 1970s and 1980s, many studies were published examining the performance of residual herbicides in maize, wheat and soybean grown in different soil cultivation systems. With the introduction of glyphosate-tolerant varieties, farmers shifted to CA because glyphosate provided farmers with a highly effective, broad-spectrum, foliar-applied in-crop herbicide where crop residues no longer were a concern. Also, large weed plants surviving the previous crop could be controlled not only before crop sowing but also after. Fernandez-Cornejo et al. (2013) showed how the adoption of glyphosate-tolerant crops was closely related to the adoption of CA in soybean in the USA and a similar observation was made by Trigo and Cap (2003) concerning the cultivation of herbicide-tolerant soybean and the adoption of no-till in Argentina. As a result of the injudicious use of glyphosate, glyphosate resistance became a widespread problem in glyphosate-tolerant crops forcing farmers to combine the use of glyphosate with residual herbicides to control glyphosate-resistant weed biotypes and continue to practice no-tillage systems (Baek et al., 2021).

In Europe, conservation tillage was introduced later and it has not been as widely adopted as in North America (Derpsch et al., 2010), which can be attributed to the fact that genetically modified herbicide-tolerant crops are not authorised for cultivation in Europe. Nonetheless, glyphosate is also a key herbicide in CA in Europe where it is used pre-emergence to terminate cover crops and kill weeds surviving the previous crop or germinating after harvest. In some countries, glyphosate is also authorized for pre-harvest use to control primarily perennial weeds (Antier et al., 2020). Glyphosate was recently re-authorized in the EU following years of controversy on whether to re-authorize the herbicide or not (Kudsk & Mathiassen, 2020; Finger et al., 2023). One of the strongest agronomic arguments for re-authorization of glyphosate was that without this herbicide, it would be very difficult to continue practicing CA (Neve et al., 2024), which several EU countries had been subsidizing for many years primarily to reduce soil erosion. Furthermore, the increased costs for farmers practicing CA would have been significant if glyphosate had been banned (Pardo & Martinez, 2019).

In conventional agriculture, synthetic herbicides are an integral part of the cropping system and often applied by default in contrast to fungicides and insecticides where the use often is based on an assessment of the need for disease and pest control (Jørgensen et al., 2019). The primary concerns associated with herbicide performance in CA are:

© Burleigh Dodds Science Publishing Limited, 2025. All rights reserved.

- the higher content of organic matter in the uppermost soil layer that may reduce the efficacy of residual herbicides and
- the occurrence of crop residues on the soil surface that may intercept the herbicide reducing deposition on the soil and weeds.

The next two sections review the ways soil organic matter and plant cover/crop residues may affect herbicide efficacy.

3 Herbicide performance in Conservation Agriculture: effects of soil organic matter

Sorption of herbicides to soil particles reduces the availability to uptake by plants and can reduce the efficacy of residual herbicides. The affinity of an herbicide to the soil particles is determined primarily by the content of soil organic matter and the physical-chemical properties of the herbicide (Takeshita et al., 2019) although soil parameters like clay content, soil pH and cation exchange capacity also influence herbicide sorption (Harrison et al., 1976; Peter & Weber, 1985a). With a few exceptions like glyphosate, paraquat and diquat, herbicides with low water solubility generally bind much stronger to the soil particles than water-soluble herbicides. The affinity of a herbicide to the soil is expressed by the sorption constant, K_d (ratio between the concentration of the soil particles and in the soil solution) and K_d values for most residual herbicides increase with increasing content of soil organic matter (Pedersen et al., 1995).

Several studies have shown that the performance of residual herbicides is related to soil organic matter content (e.g. Rahman & Matthews, 1979; Peter & Weber, 1985a, 1985b, 1985c). Blumhorst et al. (1990), applying a dose range of five herbicides on six soils with varying content of soil organic matter, clay and silt and different cation exchange capacity and soil pH, found a strong correlation between soil organic matter content and herbicide efficacy ($r = 0.66$ to 0.84) but an even stronger correlation to soil humic matter content, considered to be the fraction of the soil organic matter with the highest sorptive capacity (Kozak et al., 1983). Gonese & Weber (1998) derived herbicide rate equations based on soil organic matter and humic matter and found a good correlation to the recommended dose rates.

In conventionally tilled soils, the distribution of soil organic matter tends to be uniform in the tilled layer (typically 20–25 cm for ploughed soils), whereas a strong stratification is often observed in fields with no tillage. Unger (1991) found a higher content of soil organic matter in the upper 0–1 cm than in the layers below in two no-till plots (7 and 9 years of no-till), whereas no differences were found at different depths in the plot that had received stubble mulch tillage (0–8 cm depth). Martínez et al. (2016) analyzed soil organic matter in soil samples from a long-term experiment in Switzerland where no-till had been

© Burleigh Dodds Science Publishing Limited, 2025. All rights reserved.

practiced for two decades and found a strong stratification in the no-till plots while no differences were detected in the ploughed plots in the 0–25 cm layer. Soil organic matter content in the uppermost soil layer of the ploughed plot was only 50% of that in the no-till plot.

An example of the effect of soil organic matter content on herbicide performance is shown in Fig. 1. Prosulfocarb, a residual herbicide with low water solubility, was applied pre- and post-emergent on the grass weed species *Apera spica-venti*. The soil in the pots was collected at two depths, 0–2.5 cm and 0–20 cm in a field where no-tillage had been practiced for many years. The soil samples were mixed, and filled into the pots and seeds of *A. spice-venti* were sown. The 0–20 cm treatment simulated the effect of ploughing the field. The efficacy of prosulfocarb was considerably lower in the 0–2.5 cm soil sample than in the 0–20 cm soil sample, irrespectively of the time of application. The lower efficacy in the 0–2.5 cm soil layer correlated with a significantly higher K_d value of 57 compared to 28 for the 0–20 cm soil layer. In the same experiment, a much smaller difference in efficacy were observed with flupyrsulfuron-methyl, a herbicide binding less strongly to the soil (K_d values of 1.6 for 0–2.5 cm and 0.7 for 0–20 cm).

4 Herbicide performance in Conservation Agriculture: effects of crop residues on herbicide interception and performance

Another feature of CA that can potentially influence herbicide performance is that plant residues will be present at the soil surface when herbicides are

Figure 1 Effect of increasing doses of prosulfocarb (0, 100, 200, 400, 800, 1600 and 3200 g a.i./ha) against *Apera spica-venti* grown in soil samples collected from 0-2.5 cm and 0-20 cm soil depth in a no-tilled field. Prosulfocarb was applied pre-emergent and post-emergent at the 2-3 leaf stage. The 0-20 cm soil depth mimics the effect if the field had been ploughed prior to herbicide application.

© Burleigh Dodds Science Publishing Limited, 2025. All rights reserved.

applied. Plant residues can either be stubble and residues from a harvested crop (in this chapter referred to as crop residues) or residues from a cover crop terminated chemically or mechanically.

Plant residues can intercept spray droplets reducing the amount of herbicide deposited on the soil surface and weeds and, thus, the herbicide dose the weeds are exposed to. Besides intercepting spray droplets, plant residues can also bind herbicides making the herbicide unavailable until it is washed off, desorbed or released by other means. The sorption capacity of crop residues can be much higher than that of the soil due to a higher carbon content compared to soils. Boyd et al. (1990) found that sorption of nonionic organic compounds on maize residues was inversely related to water solubility and that the sorption capacity of maize residues was 35 to 60 times higher than that of soil with 1.9% organic matter highlighting that plant residues on the soil surface could influence herbicide availability and efficacy. Similar results were reported for cyanazine and herbicide-killed Italian ryegrass (*Lolium multiflorum*) plant material (Reddy et al., 1997) and fluometuron and fresh plant material of a native weed population, hairy vetch (*Vicia villosa*) and wheat (Gaston et al., 2001).

4.1 Amount, orientation and origin of plant residues

Banks and Robinson (1982) found that increasing the amount of wheat straw (0, 2250, 4500 and 9000 kg/ha) reduced the amount of metribuzin deposited on the soil by ca. 70%, 88% and 98–99%, respectively. A similar experiment with oryzalin and 0, 2250, 4500, 6720 and 9000 kg/ha wheat straw revealed reductions in the amount of herbicide reaching the soil of 53%, 73%, 84% and 80%, respectively (Banks & Robinson, 1984). The same authors reported similar results for acetochlor, alachlor and metolachlor in a study using 0 to 6720 kg/ha wheat straw (Banks & Robinson, 1986). Crutchfield et al. (1986) found, averaged over two doses of metolachlor, a 32%, 60%, 73% and 85% reduction in the concentration of metolachlor in the upper 10 cm of the soil following application in a field with 1700, 3400, 5100 and 6800 kg/ha wheat straw. Johnson et al. (1989) found significant reductions in alachlor soil deposition following application to a field with up to 10000 kg/ha maize straw. Wheat straw amounts of 840, 1680, 3360 and 6720 kg/ha intercepted 16%, 43%, 69% and 89% of the herbicides imazethapyr, imazaquin and chlorimuron applied to maize (Schmitz et al., 2001). More recently, Khalil et al. (2018) reported similar results from a lab study using a bioassay to assess herbicide activity. Increasing the amount of crop residues from 2000 to 4000 kg/ha reduced soil deposition of prosulfocarb, pyroxasulfone and trifluralin. In some of the abovementioned studies, interception was related to ground cover of the residues and the results of these studies suggest

© Burleigh Dodds Science Publishing Limited, 2025. All rights reserved.

that a curve-linear relationship exists between soil deposition and ground cover. A linear relationship was observed until around 70–80% ground cover whereafter an increase in ground cover has a diminishing effect on soil deposition compared to at lower levels of ground cover resulting in a flattening out of the curve at high ground cover levels. Khalil et al. (2018), using a bioassay and not chemical analysis, found in contrast to previous studies a linear relationship between herbicide activity and ground cover of residues of five crop species.

Following harvest, crop residues will be present either on the soil surface, sometimes called flat or horizontal crop residues, or as stubble, which is sometimes referred to as standing or vertical crop residues. Ghadiri et al. (1984), applying atrazine to a field with 3400 kg/ha flat and 3000 kg/ha standing wheat residues, respectively, found that 61% of the herbicide was intercepted by the crop residues with the two fractions of residues retaining the same amount (30% and 31%). In a laboratory setup, Wolf et al. (2000) measured spray deposition on *Amaranthus hybridus* and *Setaria faberi* plants grown in plastic cones. The target plants were placed in a rack in between rows of plastic cones containing wheat culms collected in the field and cut to 25 cm height. The distance between the rows of wheat culms was 18 cm while the density was 800 culms/m^2 mimicking the conditions in a winter wheat field. The wheat culms (standing residue) intercepted up to 52% of the applied imazethapyr. Khalil et al. (2019) examined pyroxasulfone interception at various combinations of standing (0, 0.1, 0.2 and 0.3 m height equivalent to ca. 600, 1000 and 1600 kg/ha) and flat wheat residues (0, 1000, 2000 and 4000 kg/ha). Spray deposition was measured as spray coverage using water-sensitive paper. Flat wheat residues had a more significant effect on pyroxasulfone interception than standing wheat residues. Spray coverage with 1000 kg/ha flat and standing wheat residue was 10.3% and 15.4%, respectively. Cutting height only had a minor effect on spray deposition on the soil surface. The results of the published studies are not conclusive regarding the relative importance of flat and standing crop residues but it has been demonstrated that both can have a significant influence on the amount of herbicide reaching the soil surface.

Spray interception can also be influenced by the origin of crop residues. Khalil et al. (2018) found that barley and wheat residues intercepted more herbicide than equivalent amounts of oilseed rape, chickpea and lupin residues. They attributed this to differences in ground cover with barley and wheat providing a higher ground cover than the legume species. A similar observation was made by Lamoreaux et al. (1993), studying weed control with three herbicides with increasing amounts of residues of maize, oat and soybean residues (1120 to 8960 kg/ha). Ground cover by soybean residues was lower and weed control was better.

© Burleigh Dodds Science Publishing Limited, 2025. All rights reserved.

4.2 Age and moisture content of plant residues

Another aspect to consider concerning crop residues and herbicide interception is the moisture content and age of the crop residues at the time of application. Erbach and Lovely (1975) found no influence of wetting maize residues before or after application on the activity of alachlor. Khalil et al. (2018) hypothesized that herbicides intercepted by crop residues will bind stronger to wet than dry residues and found that this was the case for prosulfocarb and trifluralin but not for pyroxasulfone. The evidence on the role of residue moisture at the time of herbicide application are inconclusive and likely also depends on the physical-chemical properties of the herbicide.

Dao (1991) found that lignin accounted for the majority of the capacity of wheat straw to adsorb metribuzin and one of its metabolites, S-ethyl-metribuzin, with the cellulose fraction only contributing marginally to the overall sorption. When crop residues age, the lignin content increases as cellulose degrades faster than lignin and, accordingly, Dao (1991) found an increased sorption of metribuzin and S-ethyl metribuzin to aged compared to fresh wheat straw. Khalil et al. (2018) compared fresh and 1-year-old crop residues using a bioassay and found small and insignificant differences, although the lignin content was higher in aged residues of oilseed rape, chickpea and lupin, while no differences were observed for wheat and barley.

Besides crop residues, cover crops terminated chemically or mechanically may also be present on the soil surface when spraying residual herbicides in CA. As shown by Reddy et al. (1997) and Gaston et al. (2001) (see above), fresh cover crop plant material also has a high affinity to bind herbicides. Reddy et al. (1995) found that aged cover crop residues of rye and hairy vetch (up to 11 weeks) had a higher sorption capacity of chlorimuron-ethyl than fresh plant material (54% and 40%, respectively). Similarly, Cassigneul et al. (2015), using s-metolachlor and glyphosate and the fungicide epoxiconazole and residues of four cover crop species, found that sorption on aged residues (up to 56 days old) increased by 1.6- to 4.7-fold depending on pesticide and cover crop.

In summary, ageing tends to increase the sorption capacity of both crop residues and cover crops. On the other hand, the amount of plant material will decrease over time negating the effect of the increasing sorption capacity.

4.3 The role of precipitation and irrigation

Precipitation or irrigation following herbicide application can wash off herbicide intercepted by crop residues or a cover crop and transport it into the soil. Several studies have looked into the effect of natural or simulated rainfall on herbicide release from crop/plant residues. Banks and Robinson (1984) reported that irrigation with 1.5 to 13 mm of water immediately after herbicide application

© Burleigh Dodds Science Publishing Limited, 2025. All rights reserved.

increased the concentration of oryzalin in the soil following application to a field with up to 9000 kg/ha wheat residues. At crop residue levels above 2250 kg/ha, the soil concentration of oryzalin was markedly lower than without wheat residues, irrespectively of the amount of water. In a greenhouse study, Ghadiri et al. (1984) applied 12.5, 25 or 50 mm of water every second day for 12 days to wheat residues sprayed with atrazine. The loss of atrazine was monitored by collecting the water and determining the amount of atrazine in the wheat residues after the final application of water. Following the application of 300 mm rain (6 times 50 mm), 18% of the atrazine was still present in the wheat residues. The first application of water applied immediately after atrazine application removed ca. 60% of the total amount washed off. Banks and Robinson (1986) found that 13 mm simulated rain washed off ca. 15% to 20% of the acetochlor, alachlor and metolachlor retained on wheat residues, irrespectively of the amount of wheat residues.

More recently, Carbonari et al. (2016), in a greenhouse experiment, examined the effect of simulated rainfall volume (0 to 100 mm) and timing (1 to 60 days after application) on the release of sulfentrazone from sugarcane residues (5000 to 20000 kg/ha). Sulfentrazone release was measured by collecting the water running through the sugarcane residues. Increasing the volume of rain above 20 mm did not increase the amount of herbicide washed off from the straw, regardless of the time after application. The maximum amount of sulfentrazone collected varied between 60% and 80%, depending on the amounts of residues. Significantly more sulfentrazone was washed off the residues when rain occurred 1 day after application compared to 30 and 60 days after application (77%, 62% and 42%). Whether this reduction in release was due to degradation or an increased intensity in sorption was not studied.

Khalil et al. (2018), using a bioassay, studied the effect of simulated rainfall on the release and activity of prosulfocarb, pyroxsulam and trifluralin from fresh and 10 months aged barley, wheat, oilseed rape, chickpea and lupin residues. They found that rainfall generally increased the performance of the herbicides and that the effect was greatest with chickpeas and lowest for barley residues. No significant differences were found between fresh and aged residues. In another study, Khalil et al. (2019), using a bioassay, examined the effect of rainfall amounts (5, 10 and 20 mm), the number of rain events (4 x 5mm, 2 x 10 mm and 1 x 20 mm) and intensity (5, 10 and 20 mm/h) on the release and performance of prosulfocarb, pyroxsulam and trifluralin sprayed onto wheat residues. Herbicide performance increased with increasing amounts of rain whereas rain intensity had no influence on herbicide activity. One rain event was more effective than several events.

Rainfall or irrigation after herbicide application is required to wash off herbicide intercepted by plant residues. It can be assumed that most of the herbicides deposited on plant residues will have no or little effect on

© Burleigh Dodds Science Publishing Limited, 2025. All rights reserved.

germinating weeds if not washed off into the soil. Furthermore, herbicide adsorbed to plant residues could be more vulnerable to photodegradation and in some cases, volatilization may also reduce the amount of herbicide (Weinhold & Gish, 1994). In contrast, herbicides adsorbed to plant residues may not be exposed to microbial or chemical degradation to the same extent as herbicides in the soil. The dynamics of residual herbicides in CA can be assumed to be complicated and weather, in particular precipitation, will be a significant parameter determining both the dynamics and performance of a herbicide. In summary, the results of the published studies suggest that even very high amounts of rain will not wash off all herbicides intercepted by crop or plant residues.

4.4 Herbicide chemistry and formulation and application technique

Herbicide chemistry and formulation can influence herbicide interception and sorption to plant residues as well as wash off by rainfall or irrigation. Banks and Robinson (1986) found that marginally more metolachlor than acetochlor and alachlor was retained by wheat straw following 13 mm of simulated rain. The three herbicides belong to the group of chloroacetanilides and have similar physical-chemical properties but metolachlor has a slightly higher water solubility. Lamoreaux et al. (1993) consistently found better weed control of dimethenamid than metolachlor and alachlor with high levels of crop residues and low amounts of simulated rain. Dimethenamid is also a chloroacetanilide herbicide but water solubility is higher and Log P (logarithm of the partition coefficient (P), which is a measure of lipophilicity) is lower than that of metolachlor and alachlor, which could explain the observed differences. Khalil et al. (2018), using an *L. multiflorum* bioassay, only found complete control with trifluralin with no crop residues after 5 mm of simulated rainfall, whereas prosulfucarb and pyroxasulfone provided full control with 0 and 1000 kg/ha and 0, 1000, 2000 and 4000 kg/ha crop residues, respectively, suggesting that pyroxasulfone was adsorbed less strongly to the crop residues than trifluralin and prosulfocarb. These results were confirmed in a subsequent experiment examining the effect of increasing rainfall amounts and intensities (Khalil et al., 2019). The response of the three herbicides, which belong to different chemical groups, to wash-off by water corresponds with their physical-chemical properties with trifluralin and pyroxasulfone having the highest and lowest Log P value of the three herbicides, respectively. In contrast, Schmitz et al. (2001) found that the same amount of imazethapyr and imazaquin was retained on wheat straw following 10 mm of rain despite the fact that imazaquin is significantly more water soluble and has a lower log P than imazethapyr. Overall, the available results suggest

© Burleigh Dodds Science Publishing Limited, 2025. All rights reserved.

that the susceptibility of nonionic herbicides to wash off can largely be related to their physical-chemical properties.

Johnson et al. (1989) compared weed control with an emulsifiable concentrate, a granular and a micro-encapsulated formulation of alachlor, a dry flowable and a granular formulation of cyanazine and an emulsifiable concentrate and a granular formulation of metolachlor in four tillage systems. Better weed control was achieved with granular formulations in the no-tillage system whereas no differences were observed between the formulations in the mouldboard plough, chisel plough or ridge tillage systems. The difference in performance in the no-tillage system was attributed to a two- to three-fold lower interception of the granular formulation by the maize residues. Increasing amounts of rain after herbicide application decreased the differences between the formulations.

Another source of loss of herbicide following application to the soil surface covered by plant residues is volatilization. Wienhold and Gish (1994) found that volatilization from experimental starch-encapsulated formulations of atrazine and alachlor was lower than from a commercial formulation and suggested that encapsulated formulation could be a good option for no-tillage systems. In summary, results from the literature have revealed that the amount of herbicide retained by plant residues depends on the herbicide formulation, however, the significance of formulation diminished with increasing amounts of post-application rain.

Borger et al. (2013) examined the effect of spray volume and droplet size on soil deposition (using water-sensitive paper) and pre-emergence control of *Lolium rigidum* with trifluralin and pyroxasulfone in a no-tillage system wheat stubble field (578–785 kg/ha standing and 995 to 1482 kg/ha flat wheat residues, respectively). Increasing the spray volume from 30 to 150 L/ha increased soil deposition from 5% to 32% and *L. rigidum* control from 53% to 78%. Reducing droplet size improved spray deposition but had no impact on herbicide performance. Wolf et al. (2000) examined spray deposition of imazethapyr on *A. hybridus* and *S. faberi* in a simulated post-emergence weed control setting using the set-up with weed plants and wheat culms in plastic cones (see section 3.2.1). Using hydraulic nozzles, a reduction in spray deposition of 9% to 38% was observed at a speed of 8 km/h increasing to 36% to 52% at 16 km/h due to a more horizontal trajectory. Droplet bouncing studies showed that the wheat culms and *A. hybridus* retained even very large droplets of 800 μm, whereas *S. faberi* had a higher affinity for smaller droplets. Electrostatic charging significantly increased spray retention of a fine hydraulic spray on the weed plants. The two studies clearly showed that application technique can have a significant influence on the effect of both residual and foliar-applied herbicides in CA and other no-tillage systems.

© Burleigh Dodds Science Publishing Limited, 2025. All rights reserved.

4.5 Plant residues and herbicide performance: summary

The results presented in the previous sections have shown that the higher soil organic carbon content as well as the presence of large amount of standing and flat plant residues in a CA system can potentially reduce the performance of particularly residual herbicides. However, neither results of field studies nor experiences from practice have unequivocally shown that herbicide performance of residual herbicides is lower when applied in no-tillage systems.

A confounding factor in some of the studies is that weed density can vary significantly between tillage systems. Buhler and Daniel (1988) reported a 10-times higher density of *S. faberi* in a no-tillage system compared with a tilled systems whereas the density of *Abutilon theophrasti* was lower and found a lower effect on *S. faberi* under no-tillage than conventional tillage whereas the opposite was seen for *A. theophrasti*. Similar results were reported by Buhler and Oplinger (1990) regarding the density of *S. faberi* and *A. theophrasti* and control of *S. faberi* and while the density of *Chenopodium album* was not influenced by tillage system. Chauhan and Opeña (2012) found a much higher density of four weed species, two grass and two broadleaf weed species, in a no-tillage rice system compared to a conventional system and the effect of oxadiazon followed by various post-emergence herbicides was lower in the no-tillage system.

Rainfall is another important parameter determining herbicide efficacy in no-tillage systems with large amounts of plant residues on the soil surface. Johnson et al. (1989) concluded that increasing amounts of post-application rainfall decreased the differences in the performance of herbicide formulation observed in the no-tillage system. Borger et al. (2013), studying the effect of spray volume on the control of *L. rigidum* in a no-tillage system, found a higher efficacy in one location in spite of a high weed density and attributed this to higher-than-average rainfall in that year.

Weed species respond differently to herbicides and in many studies, only one herbicide dose was applied. Hence, it can be expected that the presence of plant residues will affect weed species differently depending on how much the dose can be reduced without a concomitant reduction in herbicide performance. The effect of dose was clearly demonstrated by Erbach and Lovely (1975), who found that maize residues had no effect on the performance of atrazine and alachlor applied at the recommended dose but had a detrimental effect when herbicide rates were reduced. Differences in susceptibility to the herbicides used in the study may also explain why Chauhan and Abugho (2012), in a screenhouse experiment, found different responses of *Echinocloa crus-galli*, *Echinocloa colona*, *Dactylotenium aegyptium* and *Cyperus iria* to oxadiazon and pendimethalin applied to pots covered with what was equivalent to 6000 kg/ha rice residues.

© Burleigh Dodds Science Publishing Limited, 2025. All rights reserved.

Finally, plant residues itself may provide some measure of weed control masking the effect of lower effect of the herbicides. Crutchfield et al. (1986) concluded that although wheat residues reduced the metolachlor concentration in the soil compared to no wheat residues, it was not necessary to increase the metolachlor dose because the wheat residues provided some level of weed control. Similarly, Johnson et al. (1989) concluded that maize residues provided some measure of weed control and that binding of residual herbicides to plant residues may not be the cause of weed control failures in no-tillage systems. In contrast, Teasdale et al. (2003) found that the combination of atrazine and alachlor and a *V. villosa* cover crop provided less control than expected considering the effect of the herbicides and cover crops used alone suggesting some level of antagonism between herbicides and the cover crop. On the other hand, the combination of herbicide and cover crop delayed the emergence of weeds thereby reducing the soil concentration of metolachlor required to control the weeds. Applying residual herbicides pre-planting, the highest level of weed control was observed when the cover crop was terminated 21 days before sowing compared to 7 days before sowing, whereas the 7 days before sowing timing was better if the residual herbicides were applied post-emergent showing a clear interaction between biomass accumulation and herbicide interception/performance (Whalen et al., 2019).

5 Recent trends in Conservation Agriculture affecting herbicide use

With the advent of genetically modified glyphosate and glufosinate-tolerant crops in North and South America, the adoption of no-tillage systems including CA exploded. Weed management practices became much simpler because the herbicide application window became much wider, particularly for glyphosate. The first genetically modified glyphosate-resistant crops were introduced in the United States in 1996. The agricultural use of glyphosate increased almost 10-fold from 1995 to 2014 (Benbrook, 2016) and residual herbicides were used much less than before the introduction of genetically modified glyphosate tolerant crops (Young, 2006). As a result of the injudicious use of glyphosate and probably also incorrect use of glyphosate, applying low glyphosate doses to large weed plants, glyphosate resistance quickly became a major problem in North and South America. Since the first glyphosate-resistant *L. rigidum* biotype was found in Australia in 1996, another 50 weed species have been found to be resistant to glyphosate (Heap, 2024) of which many were detected in genetically modified glyphosate-tolerant crops. Besides an increased number of weed species evolving resistance to glyphosate, Baek et al. (2021) also speculated that cultivation of genetically modified glyphosate tolerant crops and the widespread use of glyphosate may also have contributed to

© Burleigh Dodds Science Publishing Limited, 2025. All rights reserved.

an expansion of the areas infested by weed species like *Conyza canadensis*, *Amaranthus tuberculatus* and *Amaranthus palmeri*, which were already resistant to glyphosate or subsequently evolved glyphosate resistance. Initially, many farmers responded to glyphosate resistance by increasing the glyphosate dose but later they began to use residual herbicides in combination with glyphosate, reversing the trend observed from 1995 to 2004, resulting in an increase in the use of residual herbicide use such as PPO inhibitors in the United States (Dayan et al., 2018). Van Deynze et al. (2022), in a modelling study, found that while the first cases of glyphosate-resistant weeds will have little impact on the adoption of no-tillage and conservation tillage in soybean production, the adoption would fall by 7.6% and 3.9%, respectively, when eight glyphosate-resistant weed species are identified, i.e. more widespread glyphosate resistance eventually can reduce the adoption of CA. Recently, genetically modified soybean and cotton varieties with resistance to both glyphosate and dicamba have been released in the United States, as a tool to overcome the problem with glyphosate resistant weeds, but the widespread use of dicamba has led to significant problems with off-target drift damaging crops susceptible to dicamba (Soltani et al., 2020; Travlou et al., 2024) questioning the sustainability of this strategy.

Unsuccessful weed control in any no-tillage practice can lead to a fast build-up of large populations of particularly grass weeds (Nichols et al., 2015). One way of resolving a problem with large weed populations or resistant weed biotypes is occasional tillage bringing the weed seeds down to a soil depth from where they cannot germinate. Occasional tillage has been fiercely debated among CA practitioners (Cordeau et al., 2020). The debate centres around the potential benefits regarding not only weed management but also soil acidification and nutrient stratification versus the potential loss of benefits such as control of wind and water soil erosion, reduced energy costs and ecosystem services in general (reviewed by Blanco-Canqui & Wortmann, 2020). However, lower productivity due to high weed infestation levels may be an argument for adopting occasional tillage (Díaz-Zorita, M, 2004; Van den Putte, 2010). Kettler et al. (2000) studied the effect of occasional mouldboard ploughing on the control of *Bromus tectorum*, winter wheat yield and soil quality indicators in a no-tillage wheat-fallow system. Compared to continuous no-tillage, the *B. tectorum* populations were reduced by 97% and 41% in the first and third year after ploughing while winter wheat yields were up by 30% and 9%, respectively. In a study in France, Cordeau et al. (2020) compared three types of fallow management: ploughing, reduced tillage and, no-tillage with glyphosate on four fields after 17 years of no-ploughing of which the last 7 years were CA. Tillage promoted the density, richness and composition of the weed flora in the succeeding winter wheat crop and crop yield increased with tillage intensity (11% for reduced tillage and 31% for ploughing). Blanco-Canqui and

© Burleigh Dodds Science Publishing Limited, 2025. All rights reserved.

Wortman (2020), reviewing the literature on the influence of occasional tillage on ecosystem services with no-tillage, concluded that occasional tillage does not reduce the amount of soil organic matter, but it reduces vertical stratification of soil organic matter and nutrients (Martinez et al., 2016; Unger 1991). Based on the available information, It can be concluded that occasional tillage can be an effective tool for solving certain weed problems but 5 years after ploughing, soil organic matter distribution in the 0–30 cm soil layer is still different from before ploughing, which may enhance the performance of residual herbicides (Kettler et al., 2000).

Recently, a new practice referred to as planting green has emerged in the cooler and more humid regions of the United States. In the planting green system, farmers practicing conservation tillage and no-tillage do not terminate their cover crops 1 to 2 weeks before sowing the succeeding crop, as recommended to avoid competition with the crop, but delay it until the crop is sown (Reed et al., 2019). Allowing the cover crops to grow for longer will lead to more biomass accumulation and, hence, a better suppression of weeds in the succeeding crop (Dang et al., 2015). The increasing problem with herbicide-resistant weeds is also a reason for the interest in the planting green system (Dang et al., 2015). Grint et al. (2022) found that an autumn cereal rye cover crop terminated in the spring at the time of sowing of maize and soybean reduced weed density by 31% and weed biomass by 61% compared to no cover crop in Wisconsin in the Upper Midwest region of the United States where biomass accumulation from cover crops can be low. Moreover, higher biomass may also reduce soil moisture leading to better conditions for crop establishment under humid conditions and, in the long term, higher biomass may lead to a better soil structure increasing soil organic matter, improving water holding capacity and water infiltration (Basche et al., 2016; Blanco-Canqui et al., 2015; Karlen et al., 1994).

As mentioned above, the reliance on residual herbicides has increased significantly due to the increasing problems with resistant weed biotypes and an open question has been how the planting green approach will affect the performance of residual herbicides applied around the time of crop sowing. Nunes et al. (2023a) examined the fate of flumioxazin + pyroxasulfone applied pre-emergence at soybean sowing under no-tillage without a cover crop and with a cereal rye cover crop terminated ca. 2 weeks before sowing and at sowing. Herbicide concentrations in the soil 0, 7 and 21 days after application were lower with a cereal rye cover crop than without a cover crop but the time of termination of the cover crop had no influence. Effects on *Amaranthus* spp. was not affected by the presence of the cereal rye cover crop. In another planting green experiment with a cereal rye cover crop and soybean, Nunes et al. (2023b) found that the presence of a cereal rye cover crop reduced the deposition of sulfentrazone + *S*-metolachlor on the soil surface and the

© Burleigh Dodds Science Publishing Limited, 2025. All rights reserved.

concentration in the soil. The cover crop was terminated either chemically with glyphosate or mechanically with a roller crimper. With mechanical cover crop termination soybean yields were reduced in all three years of experimentation whereas this was only the case in one of the three years with chemical cover crop termination. The planting green approach seems to be a promising strategy for suppressing weeds in regions where biomass accumulation of cover crops is limited by climatic conditions.

6 Conclusion

Primary tillage like ploughing has always been an integral part of weed management to 'clean up' after a crop and prepare the land for the next crop. Abandoning primary tillage and also mechanical weed control mean that farmers practicing CA/no-tillage will have to step up on the use of other preventive weed control methods such as crop rotation, cover crops and optimum crop establishment and integrate these tools into truly integrated weed management strategies otherwise these farmers will become very reliant on chemical weed control. CA acknowledges this by including continuous plant cover and a diverse crop rotation as the two additional pillars beside the no-tillage pillar. Nonetheless, farmers practicing no-tillage are often more reliant on herbicides, which has triggered a discussion about the sustainability of no-tillage cropping systems from a weed management point of view. Herbicide resistance has become a challenge in CA, as it has in tilled cropping systems, and has initiated an interest in re-thinking cropping practices while sticking to the overall principle of no-tillage. In the United States, the main focus has been on diversifying herbicide use and not relying solely on one mode of action by introducing more modes of action through genetic modification of the crop ('gene stacking') or by re-introducing selective herbicides. Many of the re-introduced herbicides are residual herbicides and their use may be problematic due to the reduced performance of many residual herbicides in no-tillage systems. Foliar-applied herbicides should be preferred but unfortunately, weeds are very prone to evolve resistance to some of the most widely used foliar-applied herbicides such as the ALS and ACCase inhibitors and these herbicides should therefore be used judiciously. In Europe, where the cultivation of genetically modified crops is not allowed and fewer herbicides are authorized than in other parts of the world, more focus has been on diversifying the crop rotation and optimizing the use of preventive weed control methods. In the foreseeable future, herbicides will continue to be an important tool for CA farmers, as well as for farmers who till the soil. It is pivotal that the use of preventive weed control methods is optimized to ensure the long-term sustainability of any cropping system but this is even more important in CA because of the absence of any primary tillage.

© Burleigh Dodds Science Publishing Limited, 2025. All rights reserved.

7 References

Adeux, G., et al. (2019). Mitigating crop yield losses through weed diversity. *Nature Sustainability* 2: 1018–1026.

Adeux, G., et al. (2021). Cover crops promote crop productivity but do not enhance weed management in tillage-based cropping systems. *European Journal of Agronomy* 123: 126221.

Adeux, G., et al. (2022). Multicriteria assessment of conservation agriculture systems. *Frontiers in Agronomy* 4: 999960.

Alletto, L., et al. (2010). Tillage management effects on pesticide fate in soils: a review. *Agronomy for Sustainable Development* 30: 367–400.

Andert, S., et al. (2016). The influence of crop sequence on fungicide and herbicide use intensities in North German arable farming. *European Journal of Agronomy* 77: 81–89.

Andert, S., et al. (2018). Patterns of pre-crop glyphosate use and in-crop selective herbicide intensities in Northern Germany. *European Journal of Agronomy* 97: 20–27.

Antier, C., et al. (2020). Glyphosate use in the European agricultural sector and a framework for its further monitoring. *Sustainability* 12: 5682.

Armengot, L., et al. (2016). Tillage as a driver of change in weed communities: a functional perspective. *Agriculture, Ecosystems & Environment* 222: 276–285.

Baek, Y., et al. (2021). Evolution of glyphosate-resistant weed. *Reviews of Environmental Contamination and Toxicology* 255: 93–128.

Banks, P. A. and Robinson, E. L. (1982). The influence of straw mulch on the soil reception and persistence of metribuzin. *Weed Science* 30: 164–168.

Banks, P. A. and Robinson, E. L. (1984). The fate of oryzalin applied to straw-mulched and nonmulched soils. *Weed Science* 32: 269–272.

Banks, P. A. and Robinson, E. L. (1986). Soil reception and activity of acetochlor, alachlor, and metolachlor as affected by wheat (*Triticum aestivum*) straw and irrigation. *Weed Science* 34: 607–611.

Basche, A. D., et al. (2016). Soil water improvements with long-term use of a winter rye cover crop. *Agricultural Water Management* 172: 40–50. https://doi.org/10.1016/j.agwat.2016.04.006

Benbrook, C. M. (2016). Trends in glyphosate herbicide use in the United States and globally. *Environmental Sciences Europe* 28: 1–15.

Blanco-Canqui, H., et al. (2015). Cover crops and ecosystem services: insights from studies in temperate soils. *Agronomy Journal* 107: 2449–2474. https://doi.org/10.2134/agronj15.0086

Blanco-Canqui, H. and Wortmann, C. S. (2020). Does occasional tillage undo the ecosystem services gained with no-till? A review. *Soil and Tillage Research* 198: 104534.

Blumhorst, M. R., et al. (1990). Efficacy of selected herbicides as influenced by soil properties. *Weed Technology* 4: 279–283.

Borger, C. P., et al. (2013). Increased carrier volume improves preemergence control of rigid ryegrass (Lolium rigidum) in zero-tillage seeding systems. *Weed Technology* 27: 649–655.

Boyd, S. A., et al. (1990). Sorption of nonionic organic compounds by corn residues from a no-tillage field. *Journal of Environmental Quality* 19: 734–738.

© Burleigh Dodds Science Publishing Limited, 2025. All rights reserved.

Buhler, D. D. and Daniel, T. C. (1988). Influence of tillage systems on giant foxtail, *Setaria faberi*, and velvetleaf, abutilon theophrasti, density and control in corn, *Zea mays*. *Weed Science* 36: 642–647.

Buhler, D. D. and Oplinger, E. S. (1990). Influence of tillage systems on annual weed densities and control in solid-seeded soybean (*Glycine max*). *Weed Science* 38: 158–165.

Carbonari, C. A., et al. (2016). Dynamics of sulfentrazone applied to sugarcane crop residues. *Weed Science* 64: 201–206.

Cardina, J., et al. (2002). Crop rotation and tillage system effects on weed seedbanks. *Weed Science* 50: 448–460.

Cassigneul, A., et al. (2015). Nature and decomposition degree of cover crops influence pesticide sorption: quantification and modelling. *Chemosphere* 119: 1007–1014.

Chauhan, B. S. and Abugho, S. B. (2012). Interaction of rice residue and PRE herbicides on emergence and biomass of four weed species. *Weed Technology* 26: 627–632.

Chauhan, B. S. and Opeña, J. (2012). Effect of tillage systems and herbicides on weed emergence, weed growth, and grain yield in dry-seeded rice systems. *Field Crops Research* 137: 56–69.

Colbach, N. and Cordeau, S. (2022). Are no-till herbicide-free systems possible? A simulation study. *Frontiers in Agronomy* 4, 823069.

Cordeau, S., et al. (2020). Is tillage a suitable option for weed management in conservation agriculture? *Agronomy* 10: 1746.

Crutchfield, D. A., et al. (1986). Effect of winter wheat (*Triticum aestivum*) straw mulch level on weed control. *Weed Science* 34: 110–114.

Dang, Y. P., et al. (2015). Strategic till in no-till farming systems in Australia's northern grains-growing regions. I. Drivers and implementation. *Soil and Tillage Research* 152: 104–114.

Dao, T. H. (1991). Field decay of wheat straw and its effects on metribuzin and S-ethyl metribuzin sorption and elution from crop residues. *Journal of Environmental Quality* 20: 203–208.

Dayan, F. E., et al. (2018). Origins and structure of chloroplastic and mitochondrial plant protoporphyrinogen oxidases: implications for the evolution of herbicide resistance. *Pest Management Science* 74: 2226–2234.

Derksen, D. A., et al. (1993). Impact of agronomic practices on weed communities: tillage systems. *Weed Science* 41: 409–417.

Derpsch, R., et al. (2010). Current status of adoption of no-till farming in the world and some of its main benefits. *International Journal of Agricultural and Biological Engineering* 3: 1–25.

Díaz-Zorita, M., et al. (2004). Soil structural disturbance effects on crop yields and soil properties in a no-till production system. *Agronomy Journal* 96: 1651–1659.

Erbach, D. C. and Lovely, W. G. (1975). Effect of plant residue on herbicide performance in no-tillage corn. *Weed Science* 23: 512–515.

FAO. (2011). *Save and grow: a policymaker's guide to the sustainable intensification of smallholder crop production*. 37pp. Rome, Italy: FAO.

Fernandez-Cornejo, J., et al. (2013). Conservation tillage, herbicide use, and genetically engineered crops in the United States: the case of soybeans. *AgBioForum* 15: 231–241.

Finger, R., et al. (2023). Glyphosate ban will have economic impacts on European agriculture but effects are heterogenous and uncertain. *Communications Earth & Environment* 4: 286.

© Burleigh Dodds Science Publishing Limited, 2025. All rights reserved.

Fonteyne, S., et al. (2020). Rotation, mulch and zero tillage reduce weeds in a long-term conservation agriculture trial. *Agronomy* 10: 962.

Friedrich, T. and Kassam, A. (2012). No-till farming and the environment: do no-till systems require more chemicals? *Outlooks on Pest Management* 23: 153–157.

Gaston, L. A., et al. (2001). Fluometuron wash-off from cover crop residues and fate in a loessial soil. *Soil Science* 166: 681–690.

Ghadiri, H., et al. (1984). Interception and retention of atrazine by wheat (*Triticum aestivum* L.) stubble. *Weed Science* 32: 24–27.

Gonese, J. U. and Weber, J. B. (1998). Herbicide rate recommendations: soil parameter equations vs. registered rate recommendations. *Weed Technology* 12: 235–242.

Grint, K. R., et al. (2022). Cereal rye cover crop terminated at crop planting reduces early-season weed dnsity and biomass in Wisconsin corn-soyban production. *Agrosystems, Geosciences & Environment* 5: e20245.

Guinet, M., et al. (2023). Fostering temporal crop diversification to reduce pesticide use. *Nature Communications* 14: 7416.

Haramoto, E. R. and Gallandt, E. R. (2004). Brassica cover cropping for weed management: a review. *Renewable Agriculture and Food Systems* 19: 187–198.

Harrison, G. W., et al. (1976). Herbicide phytotoxicity as affected by selected properties of North Carolina soils. *Weed Science* 24: 120–126.

Heap, I. (2024). The International Herbicide-Resistant Weed Database. Online. Thursday, May 2, 2024. Available www.weedscience.org

Hobbs, P. R. (2007). Conservation agriculture: what is it and why is it important for future sustainable food production? *The Journal of Agricultural Science* 145: 127–137.

Jørgensen, L. N., et al. (2019). Links between pesticide use pattern and crop production in Denmark with special reference to winter wheat. *Crop Protection* 119: 147–157.

Johnson, M. D., et al. (1989). The influence of herbicide formulation on weed control in four tillage systems. *Weed Science* 37: 239–249.

Karlen, D. L., et al. (1994). Crop residue effects on soil quality following 10-years of no-till corn. *Soil Tillage Research* 31: 149–167. https://doi.org/10.1016/0167-1987(94)90077-9

Kettler, T. A., et al. (2000). Soil quality assessment after weed-control tillage in a no-till wheat–fallow cropping system. *Soil Science Society of America Journal* 64: 339–346.

Khalil, Y., et al. (2018). Effect of crop residues on interception and activity of prosulfocarb, pyroxasulfone, and trifluralin. *PLoS One* 13: e0208274.

Khalil, Y., et al. (2019). Rainfall affects leaching of pre-emergent herbicide from wheat residue into the soil. *PLoS One* 14: e0210219.

Khalil, Y., et al. (2020). Pyroxasulfone efficacy for annual ryegrass control is affected by wheat residue height, amount and orientation. *Pest Management Science* 76: 861–867.

Kozak, J., et al. (1983). Adsorption of prometryn and metolachlor by selected soil organic matter fractions. *Soil Science* 136: 94–101.

Kudsk, P. and Mathiassen, S. K. (2020). Pesticide regulation in the European Union and the glyphosate controversy. *Weed Science* 68: 214–222.

Lamoreaux, R. J., et al. (1993). Efficacy of dimethenamid, metolachlor and encapsulated alachlor in soil covered with crop residue. In *Proceedings Brighton Crop Protection Conference-Weeds*, 1015–1020.

Légère, A., et al. (2013). Extreme grain-based cropping systems: when herbicide-free weed management meets conservation tillage in northern climates. *Weed Technology* 27: 204–211.

© Burleigh Dodds Science Publishing Limited, 2025. All rights reserved.

Locke, M. A. and Bryson, C. T. (1997). Herbicide-soil interactions in reduced tillage and plant residue management systems. *Weed Science* 45: 307–320.

Martínez, I., et al. (2016). Two decades of no-till in the Oberacker long-term field experiment: part I. Crop yield, soil organic carbon and nutrient distribution in the soil profile. *Soil and Tillage Research* 163: 141–151.

Melander, B., et al. (2013). European perspectives on the adoption of nonchemical weed management in reduced-tillage systems for arable crops. *Weed Technology* 27: 231–240.

Neve, P., et al. (2024). Current and future glyphosate use in European agriculture. *Weed Research*. https://doi.org/10.1111/wre.12624

Nichols, V., et al. (2015). Weed dynamics and conservation agriculture principles: a review. *Field Crops Research* 183: 56–68.

Nunes, J. J., et al. (2023a). Impact of cereal rye cover on the fate of preemergence herbicides flumioxazin and pyroxasulfone and control of *Amaranthus* spp. in soybean. *Weed Science* 71: 493–505.

Nunes, J. J., et al. (2023b). Planting into a living cover crop alters preemergence herbicide dynamics and can reduce soybean yield. *Weed Technology* 37, 226–235.

Pardo, G. and Martínez, Y. (2019). Conservation agriculture in trouble? Estimating the economic impact of an eventual glyphosate prohibition in Spain. *Planta Daninha* 37: e019197994.

Peigné, J., et al. (2016). How organic farmers practice conservation agriculture in Europe. *Renewable Agriculture and Food Systems* 31: 72–85.

Pedersen, H. J., et al. (1995). Adsorption and ED50 values of five soil-applied herbicides. *Pesticide Science* 44: 131–136.

Peter, C. J. and Weber, J. B. (1985a). Adsorption, mobility, and efficacy of alachlor and metolachlor as influenced by soil properties. *Weed Science* 33: 874–881.

Peter, C. J. and Weber, J. B. (1985b). Adsorption and efficacy of trifluralin and butralin as influenced by soil properties. *Weed Science* 33: 861–867.

Peter, C. J. and Weber, J. B. (1985c). Adsorption, mobility, and efficacy of metribuzin as influenced by soil properties. *Weed Science* 33: 868–873.

Rahman, A. and Matthews, L. J. (1979). Effect of Soil Organic Matter on the Phytotoxicity of Thirteen s-Triazine Herbicides. *Weed Science*, 27, 158–161.

Reddy, K. N., et al. (1997). Tillage and cover crop effects on cyanazine adsorption and desorption kinetics. *Soil Science* 162: 501–509.

Reddy, K. N., et al. (1995). Chlorimuron ethyl sorption and desorption kinetics in soils and herbicide-desiccated cover crop residues. *Journal of Agricultural and Food Chemistry* 43: 2752–2757.

Reed, H. K., et al. (2019) Planting green effects on corn and soybean production. *Agronomy Journal* 111: 2314–2325.

Reicosky, D. C. (2015). Conservation tillage is not conservation agriculture. *Journal of Soil and Water Conservation* 70: 103–108.

Schmitz, G. L., et al. (2001). The effect of wheat (*Triticum aestivum*) straw levels on chlorimuron, imazaquin, and imazethapyr dissipation and interception. *Weed Technology* 15: 129–136.

Soltani, N., et al. (2020). Off-target movement assessment of dicamba in North America. *Weed Technology* 34: 318–330.

Takeshita, V., et al. (2019). Effect of organic matter on the behavior and control effectiveness of herbicides in soil. *Planta Daninha* 37: e019214401.

© Burleigh Dodds Science Publishing Limited, 2025. All rights reserved.

Teasdale, J. R., et al. (2007). Cover crops and weed management. In *Non-chemical weed management: principles, concepts and technology*, eds Mahesh, K.U. and Blackshaw, R.E., pp. 49–64. Cabi Publishing.

Teasdale, J. R., et al. (2003). Influence of hairy vetch residue on atrazine and metolachlor soil solution concentration and weed emergence. *Weed Science* 51: 628–634.

Thomas, A. G., et al. (2004). A multistudy approach to understanding weed population shifts in medium- to long-term tillage systems. *Weed Science* 52: 874–880.

Travlou, E., et al. (2024). Chemical weed control and crop injuries due to spray drift: the case of dicamba. *Agrochemicals* 3: 22–28.

Trichard, A., et al. (2013). Identification of weed community traits response to conservation agriculture. *Agricultural Ecosystem Environment* 179: 179–186.

Trigo, E. J. and Cap, E. J. (2003). Impact of the introduction of genetically modified crops in Argentinean agriculture. *AgBioForum*6. 87-94

Unger, P. W. (1991). Organic matter, nutrient, and pH distribution in no-and conventional-tillage semiarid soils. *Agronomy Journal* 83: 186–189.

Van den Putte, A., et al. (2010). Assessing the effect of soil tillage on crop growth: a meta-regression analysis on European crop yields under conservation agriculture. *European Journal of Agronomy* 33: 231–241.

Van Deynze, B., et al. (2022). Are glyphosate-resistant weeds a threat to conservation agriculture? Evidence from tillage practices in soybeans. *American Journal of Agricultural Economics* 104: 645–672.

Vullioud, P., et al. (2006). Results of a 35-year ploughless tillage experiment at Changins (1970–2004). Part III: Weeds, diseases, pests and slugs. *Revue Suisse d'Agriculture* 38: 1–16.

Whalen, D. M., et al. (2019). Integration of residual herbicides with cover crop termination in soybean. *Weed Technology* 34: 11–18.

Wienhold, B. J. and Gish, T. J. (1994). Effect of formulation and tillage practice on volatilization of atrazine and alachlor. *Journal of Environmental Quality* 23: 292–298.

Wolf, T. M., et al. (2000). Optimizing postemergence herbicide deposition and efficacy through application variables in no-till systems. *Weed Science* 48: 761–768.

Young, B. G. (2006). Changes in herbicide use patterns and production practices resulting from glyphosate-resistant crops. *Weed Technology* 20: 301–307.

© Burleigh Dodds Science Publishing Limited, 2025. All rights reserved.

Chapter 7

Advances in the reduction of herbicide use in Conservation Agriculture: French case studies

Stéphane Cordeau[1], Maé Guinet, Nicolas Munier-Jolain, and Guillaume Adeux, Université de Bourgogne Europe, Institut Agro, INRAE, Agroécologie, France; and Violaine Deytieux, INRAE, U2E, Unité Expérimentale du domaine d'Epoisses, France

1 Introduction

According to the FAO, Conservation Agriculture (CA) is defined as a farming system or production strategy which implements three principles simultaneously and permanently over time (FAO, 2021): (i) absence of soil disturbance (exception made of the disturbance generated by the seeder); (ii) crop diversity in time (crop sequence, rotation) and space (intercropping, companion plants, cover crops); and (iii) maximum soil coverage by crop residues or sown cover crops (annual or permanent cover). This definition highlights that CA is currently characterized based on the technical means implemented, which have been described in greater depth and geographically contextualized in numerous studies (Kassam et al., 2009; Kassam and Friedrich,

1 Correspondence: Stéphane Cordeau, Agroécologie – INRAE, 17 rue Sully, BP 86510, F-21065 Dijon Cedex, France stephane.cordeau@inrae.fr

http://dx.doi.org/10.19103/AS.2024.0145.10
© Burleigh Dodds Science Publishing Limited, 2025. All rights reserved.

2011; Friedrich et al., 2012; Kassam et al., 2019). More specifically, CA systems are defined as those in which no mechanical disturbance of the soil is carried out except by the seed drill (less than 15 cm wide or 25% of the cultivated area), a practice known as no-till, no-tillage, or zero-tillage, which grow more than three different crop species in pure stands or mixtures (of varieties or species), and which cover at least 70% of the soil surface with stubble, crop residues or dead or living cover crops at the time of sowing (Kassam et al., 2019). CA systems combine a diverse set of agricultural management practices (rotation, cover, sowing, fertilization, weeding, etc.) which may change as the system evolves (Knowler and Bradshaw, 2007; Vankeerberghen and Stassart, 2016; Derrouch et al., 2020).

Managing weeds has always been a challenge in no-till systems. The development of non-selective, effective and inexpensive foliar herbicides for burn-down weed control prior to crop sowing such as aminotriazole (1958), paraquat (1963), glyphosate (1975) and glufosinate (1986) helped to control the development of perennial weeds, which are a major threat in reduced-tillage systems (Evans, 1972; Watson, 1974; Hoss et al., 2003). These herbicide applications at crop sowing (or just before) help managing all the weeds that are established during the summer fallow period, which may compete with crops in its early stage and become difficult to control late in the season, even with high dose of herbicide. The development of no-till and the increased use of glyphosate was then encouraged in some countries by the development of herbicide-tolerant GM crops in 1996 (Green, 2018).

The number of herbicide active ingredients on the market is decreasing (Chauvel et al., 2022), thereby limiting the number of chemical solutions for farmers. Irrational use of a single herbicide over a long period of time may result in the development of resistant weed biotypes (Délye et al., 2013), shifts in weed flora and negative effects on the succeeding crops (Owen, 2008; Grundy et al., 2011). Worldwide, the first glyphosate-resistant population of blackgrass (*Alopecurus myosuroides*) was identified in France in 2023 (R4P, 2023).

Today, managing weeds remains challenging during and after the transition to CA (Derrouch et al., 2020) because banning tillage and favouring the accumulation of organic matter on the soil surface limit the number of weed management techniques or their efficacy (Cordeau, 2022). In CA, there are fewer weed control options because tillage is not allowed, and in-crop mechanical weeding is not possible due to the presence of living or dead mulch (Chauhan et al., 2012). This impacts the weed community's taxonomic or functional composition (Derrouch et al., 2021a) and thus weeding tactics over time (Derrouch et al., 2020). Weed (including crop volunteer) management, as well as cover crop termination, mainly relies on herbicides such as glyphosate (Antier et al., 2020).

© Burleigh Dodds Science Publishing Limited, 2025. All rights reserved.

Reducing herbicide use in CA is a challenging priority. This chapter gathers French data and experiences collected since 2007 by INRAE on long-term cropping systems experiments, factorial experiments and farmers' network with the objective of quantifying the performance of CA systems aiming to reduce pesticide use, especially in terms of weed management, crop performance and other technical and economic indicators.

2 Transitioning from tillage-based systems to conservation agriculture while using herbicide to manage weeds

Integrated weed management (IWM) promotes the use of multiple techniques to achieve sustainable weed control while reducing the reliance on chemical control. However, IWM strategies that reduce both herbicide usage and tillage intensity remain difficult to implement, because tillage is a pivotal weed management tool (Nazarko et al., 2005; Lechenet et al., 2016). Combining several weed management tactics such as crop diversification, cover cropping and increased seeding rates is encouraged to ensure that weed pressure does not transcribe into yield loss. CA proposes to diversify crop rotations and use cover crops to increase ecosystem services. However, the effects of diverse crop rotations, including cover crops on weeds remain to be studied in the no-till system context (Chauhan et al., 2012; Adeux et al., 2021).

2.1 Long-term integrated weed management cropping system experiment

A unique technique rarely enables ambitious objectives to be achieved (Cordeau et al., 2020a). An in-depth redesign of the system is often necessary to achieve enhanced multiperformance (Hill and MacRae, 1995). In 2000, INRAE set up a long-term cropping system experiment to compare the efficacy of four IWM systems along a gradient of herbicide use, in comparison to a reference system.

The field experiment was conducted from harvest 2000 to harvest 2017 at the INRAE experimental farm in Bretenière (47° 14′ 11.2″ N, 5° 05′ 56.1″ E), 15 km southeast of Dijon, France. The site is subject to an oceanic climate (but with a greater temperature range than the Atlantic coast), characterized by cold wet winters (average daily temperature of 4°C and average monthly precipitation of 43 mm) and hot summers (average daily temperature of 18°C and average monthly precipitation of 69 mm). The experiment was set up as a completely randomized block design. The set of decision rules characterizing each of the CS was replicated on two blocks. To avoid complete overlap between crop:year

© Burleigh Dodds Science Publishing Limited, 2025. All rights reserved.

and CS effects, the two plots (1.7 ha) of each CS did not start with the same entry point (i.e. crop).

Over the 17 years of the experiment, the cropping system called S2 hereafter, implemented IWM principles during its transition toward CA. Indeed, no-plough, but quite frequent minimum tillage was applied for the first 10 years. Then, no-till was applied in S2 over the last 7 years of the experiment, soil was covered with highly diversified grass/legume-based cover crops during the summer fallow period and crops were directly seeded into glyphosate-terminated cover crops. Recent literature showed that fields conducted with CA principles are effectively CA after 5 years of permanent implementation of CA (Derrouch et al., 2020), particularly due to the shift in weed flora (Derrouch et al., 2021a; Derrouch et al., 2021b). That means that even the 7 first years of implementing CA principles in our experiment cannot be assessed as the effects of CA, but rather the effects of recently implemented CA principles, i.e. the initial and transitional phase to permanent CA. We thus considered here that the 17 years was a system simulating a farmer transitioning from minimum tillage-based to CA over the 17 years. Weed community before and after in-crop weed management, weed/crop biomass at crop flowering and crop yield at harvest were assessed over the 17 years.

We present here the conclusions for three out of five cropping systems (Adeux et al., 2019; Adeux et al., 2022b; Cordeau et al., 2022): the reference standard system (S1), the IWM system transitioning to CA (S2) and the tillage-based system mixing mechanical/herbicide weed management (S4). S1 was the reference cropping system, typical of the Burgundy region, designed to maximize financial return. It was characterized by a triennial oilseed rape–winter wheat–winter barley rotation, systematic moldboard ploughing in summer-autumn and herbicides as sole curative weed management tool. Nitrogen fertilization aimed to ensure the full needs of the crop. S4 was designed to mimic farmers aiming to reduce herbicide reliance and resulted in a more complex 6-year rotation which included winter sown crops (winter wheat (*Triticum aestivum* L.), winter barley (*Hordeum vulgare* L.), triticale (× *Triticosecale* Wittm. ex *A. Camus*) or faba bean (*Vicia faba* L.), autumn sown oilseed rape (*Brassica napus* L.), spring crop (oat (*Avena sativa* L.), sugarbeet (*Beta vulgaris subsp. vulgaris* L.), faba bean, lupin (*Lupinus albus* L.), spring barley or mustard (*Brassica juncea* (L.) Czern) and summer-sown crops (maize (*Zea mays* L.)), sorghum (*Sorghum bicolor* (L.) Moench), soybean (*Glycine max* (L.) Merr.), or sunflower (*Helianthus annuus* L.). S4 differed from S1 and S2 by the tillage intensity and weed management strategies. S4 implemented moldboard ploughing once every two or three years and a wide array of preventive and cultural weed management tools such as false seedbed technique, delayed sowing of winter cerealsand higher seeding rates. Herbicide was the only in-crop weed management in S1 at the recommended dose, while S4 resorted preferentially to preventive measures

© Burleigh Dodds Science Publishing Limited, 2025. All rights reserved.

and mechanical weeding. However, applications of specialized herbicides on target species remained possible in S4 when weather conditions were not suitable for mechanical weeding or to control weeds with a low sensibility to mechanical weeding. More detailed information concerning crop sequence, pesticide application, soil tillage, sowing and harvest dates can be found in Adeux et al. (2019), Deytieux et al. (2012) and Chikowo et al. (2009).

2.2 Similar herbicide use in CA compared to the reference

The S1 reference system resulted in an average annual herbicide treatment frequency index (HTFI) of 1.9. Weed management relied mainly on broad-spectrum herbicides typically used in winter cereal-based rotations (Figure 1). On average over the 17 years, the S2 system reduced herbicide use by 9% compared with S1. Half of herbicide use was represented by glyphosate-based herbicides applied to terminate crop volunteers, weeds and cover crops before sowing (Figure 1). The proportion of the HTFI used to control grass weeds was higher than in the other systems, which clearly indicates high grass weed pressure (e.g. *Alopecurus myosuroides*). The use of glyphosate increased during the transition from superficial tillage to no-till, even though the doses of individual treatments declined (about 2 kg/ha of active ingredient before 2010, about 0.9 kg/ha after 2010). The S4 IWM system enabled a 64% reduction in HTFI compared with S1. This was made possible by the implementation of a number of weed management tools, such as crop diversification, occasional ploughing (approximatively every two years), mechanical weeding including hoeing of certain crops (maize), delayed sowing dates for winter cereals (wheat, barley), the use of smothering species (triticale) and false seedbed operations during the fallow period.

2.3 Containing weed pressure and yield loss

Weed control was considered satisfactory on all the plots. No increased weed pressure was observed on the plots over time (Figure 2A). Weed species often considered as highly problematic in cereal-based cropping systems of the Burgundy region (e.g. *Alopecurus myosuroides*) were contained. Weed abundance before weeding averaged (over 17 years) 25 plants/m^2 in S1, 54/m^2 in S2 and 45/m^2 in S4. After any weeding operation, weed abundance averaged (over 17 years) 2 plants/m^2 in S1, 12/m^2 in S2 and 20/m^2 in S4. Weed control efficiency (computed based on weed densities before and after curative weed control operations) averaged 78% in S1, 69% in S2 and 63% in S4, which we considered acceptable.

The implementation of 17 years of contrasted tillage and herbicide use shaped weed assemblages and dominant traits: short-life cycle, short-cycle

© Burleigh Dodds Science Publishing Limited, 2025. All rights reserved.

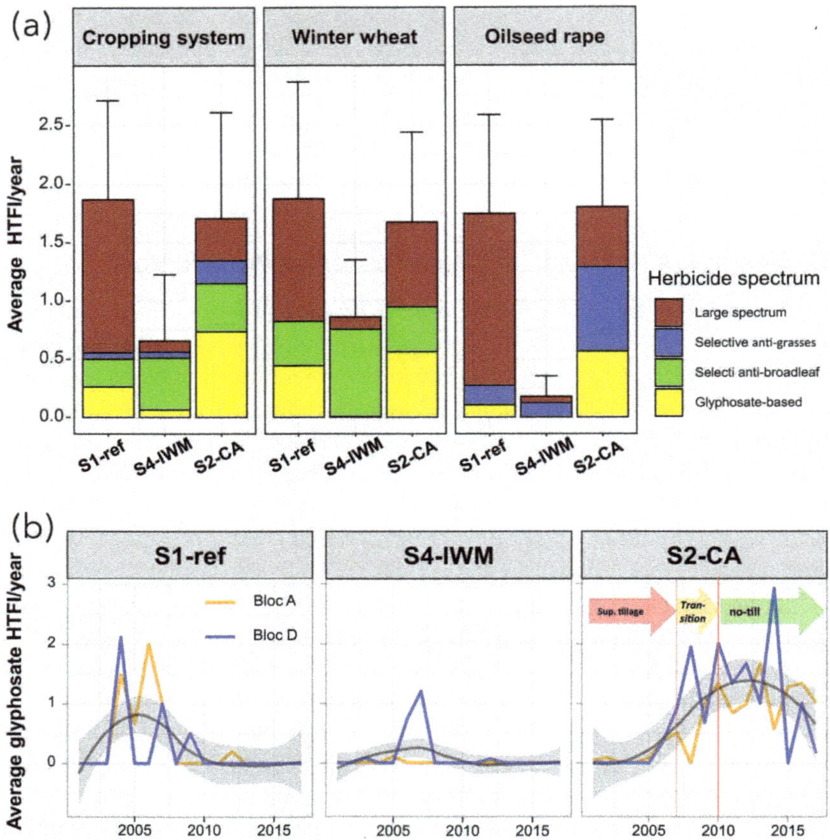

Figure 1 (A) Herbicide treatment frequency index (HTFI) per year, average across the whole 2001–2017 period of the experiment in the three cropping system, or in the winter wheat and oilseed rapes only, colored by the spectrum of the herbicide; (B) Dynamics of glyphosate TFI across the three cropping systems, repeated twice (blocs A and D). The trends (black line) are highlighted with a loess smoother and 95% confidence intervals (grey bands). According to Adeux et al. (2019)

winter annuals with a creeping growth form in S1, wind-dispersed Asteraceae and perennial species in S2 and spring and summer species in S4. *Asteraceae* (*Sonchus asper, Senecio vulgaris*), grasses (*Lolium* spp., *Echinocloa crus-galli*) and perennials (*Convolvulus arvense*, *Taraxacum officinale*) dominated the weed community in S2 during the last 7 years of strict CA.

Weed-induced yield losses (estimated from weed biomasses and relationships between crop and weed biomasses, see Adeux et al. (2019)) were negligible in all systems (Figure 2B). The percentage of weed biomass was less than 5% in 100% of plot:years (i.e. case situations) in S1, in 66% in S2 and

© Burleigh Dodds Science Publishing Limited, 2025. All rights reserved.

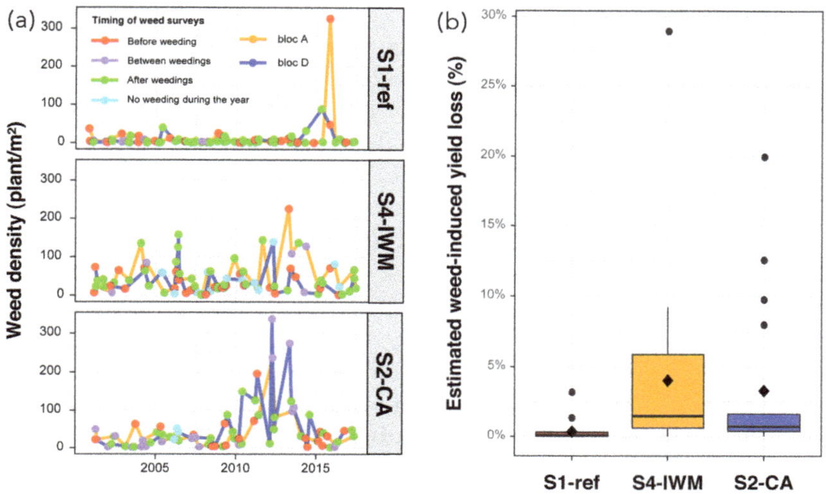

Figure 2 (A) Dynamic of total weed density (plant/m²) over 2001–2017 period in the two blocs of the three cropping systems. Each dot corresponds to a weed survey colored according to its relative timing to weedings. (B). Weed-induced yield loss (%) estimated by the reduction of crop biomass due to increased weed biomass sampled at crop flowering each year. According to Adeux et al. (2019)

in 79% in S4. When S2 transitioned to CA, the percentage of weed biomass was sometimes higher, particularly when companion plants intercropped with oilseed rape (considered as "weeds") were not winter-killed.

2.4 Lessons from the transition to CA applying IWM principles

S2 achieved satisfactory weed management. The weed densities observed remained controllable. Herbicide use was reduced by –21% during the superficial tillage phase (2000–2010) and by –9% across all 17 years of the experiment, compared to the standard reference system. During the no-till phase, social pressure on glyphosate use was not as pronounced as today in Europe (Antier et al., 2020), hence, the strategy adopted was to use glyphosate before crop sowing in order to limit herbicide use during the crop cycle. As a consequence, total herbicide use was similar to the reference but composed of 50% of glyphosate-based herbicides. In certain crops, no herbicides were required for weed control due to important crop residues on the soil surface, as was the case for winter faba bean after sorghum. The success of this technique resides in obtaining a sufficiently thick mulch to suppress weed emergence and establishment but not the crop (Teasdale and Mohler, 2000; Menalled et al., 2022).

© Burleigh Dodds Science Publishing Limited, 2025. All rights reserved.

However, reliance on glyphosate in S2 was high, even though the doses applied were reduced over time. It was often difficult to ensure maximum growth of cover crops due to high slug pressure, delayed sowing after harvest of the previous crop due to lack of manpower and non-optimized sowing density. Optimizing cover crop mixture richness and composition (Smith et al., 2020), seeding rates of each species (Bybee-Finley et al., 2022) and sowing date in light of unpredictable weather conditions (Mirsky et al., 2017) remains a challenge to optimize weed suppression in cover crops (Osipitan et al., 2019). We observed the rapid evolution of weed communities towards *Asteraceae* species, which easily establish in no-till (Cordeau et al., 2015b), thanks to wind dispersal. This flora has limited the possibilities of further reducing the frequency of herbicide treatments. The profitability of the S2 system was lower than the profitability in the S1 reference system, with a semi-net margin difference of −180€/ha/year, mainly due to the introduction of minor crops with a lower gross product.

3 Effect of strategic tillage on weed, crop yield and economy

3.1 Aims of strategic tillage after long-term no-till sequence

No-till ensures that weed seeds remain on the soil surface, a condition deemed to be unfavourable to weed seed germination because of poor seed:soil contact (Cordeau et al., 2015b) and increased weed seed mortality (Chauhan et al., 2006; Nichols et al., 2015). However, many studies have reported higher weed pressure under no-till than under ploughing (Cardina et al., 2002; Adeux et al., 2019). Occasional tillage could help to diversify selection pressures and cope with certain challenges encountered in no-till (Crawford et al., 2015), such as the management of herbicide resistant-weeds (Dang et al., 2015) or other pests (ex. slugs, voles) (Douglas and Tooker, 2012), soil compaction (Peixoto et al., 2019) or reduced crop productivity (Díaz-Zorita et al., 2004; Van den Putte et al., 2010; Çelik et al., 2019; Peixoto et al., 2019). Extensive research has focused on how tillage intensity drives weed dynamics and/or crop yields, but few studies have investigated tillage effects in fields previously managed under CA principles for a long time. Strategic tillage as used in this chapter describes the use of occasional tillage implemented after a long sequence of no-till (Conyers et al., 2019; Peixoto et al., 2020), needed to manage an emerging agronomical issue due to the no-till phase (e.g. increased weed/slugs pressure).

We investigated weed community (density, species richness, composition) and winter wheat (yield, yield components, grain quality) response to different levels of tillage intensity (no-till, superficial tillage, ploughing) in fields previously managed with two out of three CA principles during 17 years (i.e. diversified crop rotation and soil cover) and with all three principles (i.e. adding

© Burleigh Dodds Science Publishing Limited, 2025. All rights reserved.

no-till) over the last 7 years, i.e. fields of the S2 system previously presented in the long-term cropping system experiment (Adeux et al., 2019; Cordeau et al., 2020b; Cordeau et al., 2022).

3.2 Effect of tillage type

Within each S2 field, three strips were defined after main crop harvest in 2017: one strip was dedicated to conventional tillage (i.e. ploughing, CT), another to reduced tillage (i.e. superficial and non-inversion tillage, RT) and the last one was maintained under no-till (NT). Soil preparation prior to winter wheat sowing consisted in stubble cultivation in RT (8cm deep), ploughing with skim coulters in CT (25 cm deep) and rotary harrowing in both RT and CT (8cm deep). Glyphosate was applied at 1080 g/ha during the fallow period in NT. All field strips were then sown with winter wheat at the same weeding date and rate, and all management practices (fertilization, weeding, etc.) were similar. Weed community composition was assessed before and after weeding. Weed biomass was sampled at crop maturity.

Following the methodology and reference prices of Lechenet et al. (2014, 2017), production costs included mechanization, input and fuel costs. Gross products were computed as the mathematical product between the observed mean yield in each treatment and the selling price of winter wheat. The gross margin was computed as the difference between the gross product and the sum of the production costs (mechanization, fuel, inputs).

Tillage treatments significantly affected weed communities before weeding: total weed density and species richness before weeding were greatest in RT, intermediate in CT and lowest in NT (Table 1). Indeed, in the S2 system, 77.1% of weed seeds were concentrated in the top 0–10 cm soil layer (Cordeau et al., 2022), the only soil layer stimulated by stubble cultivation and rotary harrowing in RT. Higher weed emergence in CT than NT was however, unexpected. Reasons which could explain this result include the upwelling of persistent weed seeds (e.g. *Fumaria officinalis* or *Sinapis* spp.) (Chancellor, 1986; Thompson et al., 1993; Mulugeta and Stoltenberg, 1997; Soltani et al., 2016) and/or the incomplete burial of weed seeds initially present in the top horizon (*e.g. Alopecurus myosuroides*) (Roger-Estrade et al., 2001). Differences in species richness and abundance transcribed into a significant tillage effect on weed community composition before weeding (12.17% of partial variation explained, $P = 0.009$). Of the tenmost abundant species observed before weeding, four were associated to RT (*Sinapis* spp., *Sonchus asper*, *Veronica hederifolia* and *Viola arvensis*) and two to both NT and RT (*A. myosuroides* and *Lapsana communis*). Weed density (4 to 13 plants/m²), weed biomass (1 to 15 g of dry matter/m²) and species richness after weeding were similar across tillage treatments Higher crop yields under CT

© Burleigh Dodds Science Publishing Limited, 2025. All rights reserved.

Table 1 Effect of tillage treatments (CT: conventional tillage, RT: reduced tillage, NT: no-till) implemented before wheat sowing in fields conducted under CA principles during 17 years, on weeds and wheat yield and yield components. Tillage systems sharing identical letters are not significantly different at $P<0.05$. DM: dry matter. Credits: plowing by © INRAE/MAITRE Christophe, superficial tillage and direct seeding © INRAE/FARCY Pascal

Response variable	P	CT	RT	NT
Weed density before weeding (plants/m²)	<**0.0001**	12 ± 4 b	33 ± 10 c	5 ± 2 a
Weed density after weeding (plants/m²)	<**0.0001**	4 ± 1 a	13 ± 2 b	8 ± 2 b
Weed biomass at crop maturity (g DM/m²)	<**0.0001**	1 ± 0 a	15 ± 10 b	3 ± 2 ab
Number of wheat grains per ear	<**0.0001**	40.1 ± 1.1 c	34.1 ± 0.9 b	30.7 ± 0.8 a
Wheat grain yield (t DM/ha)	**0.0002**	7.1 ± 0.4 b	6.1 ± 0.3 a	5.5 ± 0.3 a

© Burleigh Dodds Science Publishing Limited, 2025. All rights reserved.

probably resulted from increased mineralisation of soil organic matter (Alvarez et al., 1998) or enhanced soil structure (Peixoto et al., 2019), rather than lower weed:crop competition, as reflected by an overall low weed biomass after weeding, across all tillage treatments. Gross margin was significantly higher in RT than in CT and NT ($P < 0.001$). Thus, higher winter wheat productivity in RT than in NT overcompensated for higher production costs, whereas it was not the case for CT.

These results shed light on the short-term effect of different types of tillage on weed communities, weed:crop interference and crop performance in fields conducted under CA principles for years (Blanco-Canqui and Wortmann, 2020; Peixoto et al., 2020).

4 Multicriteria assessment of conservation agriculture

Decades after its development by pioneer farmers, CA acreage is increasing exponentially worldwide (Kassam et al., 2019), although very few studies have assessed its performance on a multicriteria basis, including environmental, economic and societal aspects (Craheix et al., 2016). Here, we mobilize data from the French AGROSYST database, which gathers all farming practices and performances of the 3000 farms involved in the French DEPHY farmers' network (Adeux et al., 2022a). Performances (13 indicators) of CA (CA, $N = 36$) and pseudo-CA systems (pseudo-CA, $N = 19$, allowing one occasional superficial tillage) were compared to those of conventional tillage (plowed, CT, $N = 135$) and superficial tillage (ST, $N = 90$) based systems sharing similar production situations (climate, soil type, presence of livestock or irrigation, etc.).

4.1 Pesticide use in CA systems of French farmers

On average, the conclusions of this study are that CA systems require as much fungicide as 'neighbouring' CT or ST systems in similar production situations, slightly less insecticides (although not significant), but more herbicides (2.1 HTFI points on average, with around +1 and +0.4 HTFI points compared to 'neighbouring' CT or ST systems). Overall, CA systems are more reliant on total pesticide use than their 'neighbouring' CT systems, with +1.1 total TFI points on average, but no difference in total TFI was observed between CA systems and their ST neighbours (Figure 3).

4.2 Performance of CA on technical and economic indicators

We showed that CA (compared to STs and CTs, respectively) decreased time of traction/ha/year (–25 and –32%), fuel consumption (–21 and –39%), as well as mechanization costs (–20 and –26%), tended to slightly decrease profitability/

© Burleigh Dodds Science Publishing Limited, 2025. All rights reserved.

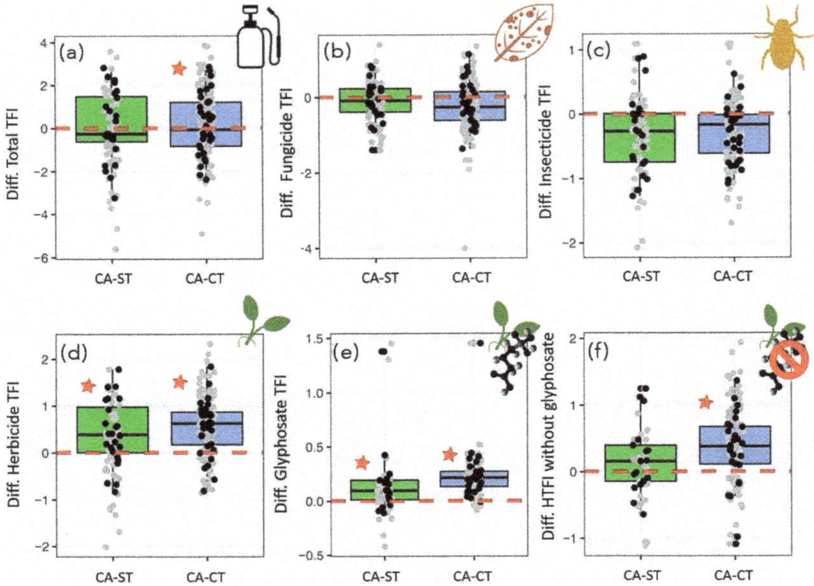

Figure 3 Difference in pesticide use between CA, conventional tillage (plowed, CT; blue) and minimum tillage (ST; green) based systems in the Dephy Farm French network, for the total treatment frequency index (Total TFI (A), fungicide TFI (B), insecticide TFI (C), herbicide TFI (D), glyphosate-based herbicide TFI (E) and herbicide TFI without glyphosate (F) (according to Adeux et al. (2022a)). Black dots represent average differences per similar production situation. Red stars indicate significant differences ($P < 0.05$).

ha (–7 and –19%, non-significant) due to slightly lower productivity (–19% and –25%) but resulted in better profitability per hour of field traction (+23% and +18%). Pseudo-CA did not implement the three CA principles since crop rotation was as diverse and cover crop as frequent as in ST and CT, and tillage occurred, albeit rarely. However, pseudo-CA decreased fuel consumption (–25% compared to CTs), and resulted in similar productivity and economic profitability (per hectare and per hour of field traction).

Further analysis is needed to identify the determinants of multi-performance in given production situations and track down innovative systems optimizing multiple performances and solving apparent trade-offs, such as tillage intensity and herbicide use and/or limited productivity.

5 Transitioning to pesticide-free conservation agriculture: the CA-SYS platform

Agriculture faces the complex challenge of transforming the way in which we produce food to ensure global food security whilst minimizing or

© Burleigh Dodds Science Publishing Limited, 2025. All rights reserved.

avoiding environmental harms (Foley et al., 2005; Vanbergen et al., 2020) and considering all facets of sustainability. Arguably, the most profound transformation would involve shifting the conventional paradigm of production towards an ecological intensification of agriculture. The implementation of ecological intensification therefore requires designing productive, sustainable agricultural systems that fundamentally integrate ecosystem services delivered by biodiversity into crop production by the manipulation of biotic interactions that assure profitable yields (i.e. high yield and/or low production cost) and minimize disservices (e.g. pests damage, soil degradation, greenhouse gas emission) (Dore et al., 2011; Garnett et al., 2013; Gaba et al., 2014).

5.1 Biodiversity-based systems at the core of the CA-SYS platform

Since summer 2018, different agroecological systems mobilizing diverse rotations of arable crops are experimented on the CA-SYS platform (Cordeau et al., 2015a), such as wheat, barley, rapeseed, buckwheat, soybean, faba bean, etc. The CA-SYS platform covers 125 ha of the INRAE 'U2E AgroEcology Experimental and Innovation Facility' experimental unit (47°14′11.2″N, 5°05′56.1″E, located close to Dijon, Eastern France (Figure 4)) and is divided into 42 fields of 2.5 ha, which are surrounded by 10 ha of ecological infrastructures.

The overarching objective of CA-SYS is to design pesticide-free agricultural systems that resort to both wild and cultivated biodiversity as a means of sustaining agricultural production, i.e. biodiversity-based systems (Petit et al., 2018; Petit et al., 2021). All pesticides are excluded, even bio-products and those authorized in organic farming. The CA-SYS platform is an agroecological system experiment testing nature-based solutions, defined by Maes and Jacobs (2017) as any transition that uses ecosystem services with decreased input of non-renewable natural capital and increased investment in renewable natural processes.

5.2 Conservation agriculture systems tested on the CA-SYS platform

A crucial aspect in the design of the CA-SYS platform has been the active co-design with farmers and other agricultural stakeholders to guide the implementation of ambitious but practicable biodiversity-based management systems. Four pesticide-free cropping systems are being tested on the CA-SYS platform, according to two agricultural pathways: (1) a plowing/tillage-based system (TS) inspired by organic agriculture (Crowder et al., 2010; Saffeullah

© Burleigh Dodds Science Publishing Limited, 2025. All rights reserved.

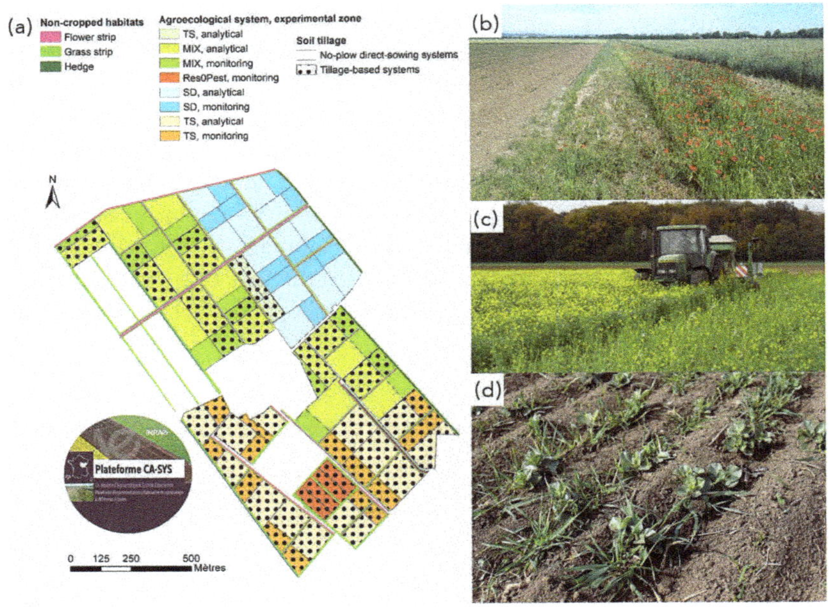

Figure 4 (A) Map of the CA-SYS platform (Hugard R. and Deytieux V. © INRAE 2018); (B) 3-m wide flower strip bordered by two 3-m wide grass strips (Cordeau S. © INRAE 2019); (C) Direct drilling in field conducted in Conservation Agriculture (Farcy P. © INRAE 2018); (D) Winter wheat intercropped with winter fababean (Cordeau S. © INRAE 2019), according to Vanbergen et al. (2020)

et al., 2021); (2) a no-plow/no-till direct-sowing system (SD) maximizing soil cover and inspired by CA (Kassam and Friedrich, 2011; Reicosky, 2015).

We will focus here on the two SD cropping systems (Figure 5) since they are defined as pesticide-free forms of CA (SD1 system) or occasional-tillage (SD2 system). Both systems implement a 6-year crop rotation, including summer and spring crops in order to diversify the typical winter crop-based rotation of the region. Crops are not fertilized with phosphorus and potassium but fertilized with sulphur and nitrogen according to their need and targeted yields (e.g. 7t/ha in winter wheat, i.e. about 10% less than the potential yield in conventional systems in the production situation). SD1 resorted to permanent cover and no-till, whereas superficial tillage was allowed in SD2 when necessary, but no more than once a year per plot, essentially before crop sowing to terminate weeds, crop volunteers or cover crops. Irrigation is allowed to ensure the establishment of summer crops in both systems (e.g. soyabean), to limit hydric stress at crop flowering or to ensure maximum soil cover provided by summer-sown cover crops in SD1 only. Cover crop termination is ensured by no-till methods in SD1 while superficial tillage is allowed in SD2 if necessary. Indeed, cover crop mixtures were designed to be terminated without tillage, e.g. by avoiding species such

© Burleigh Dodds Science Publishing Limited, 2025. All rights reserved.

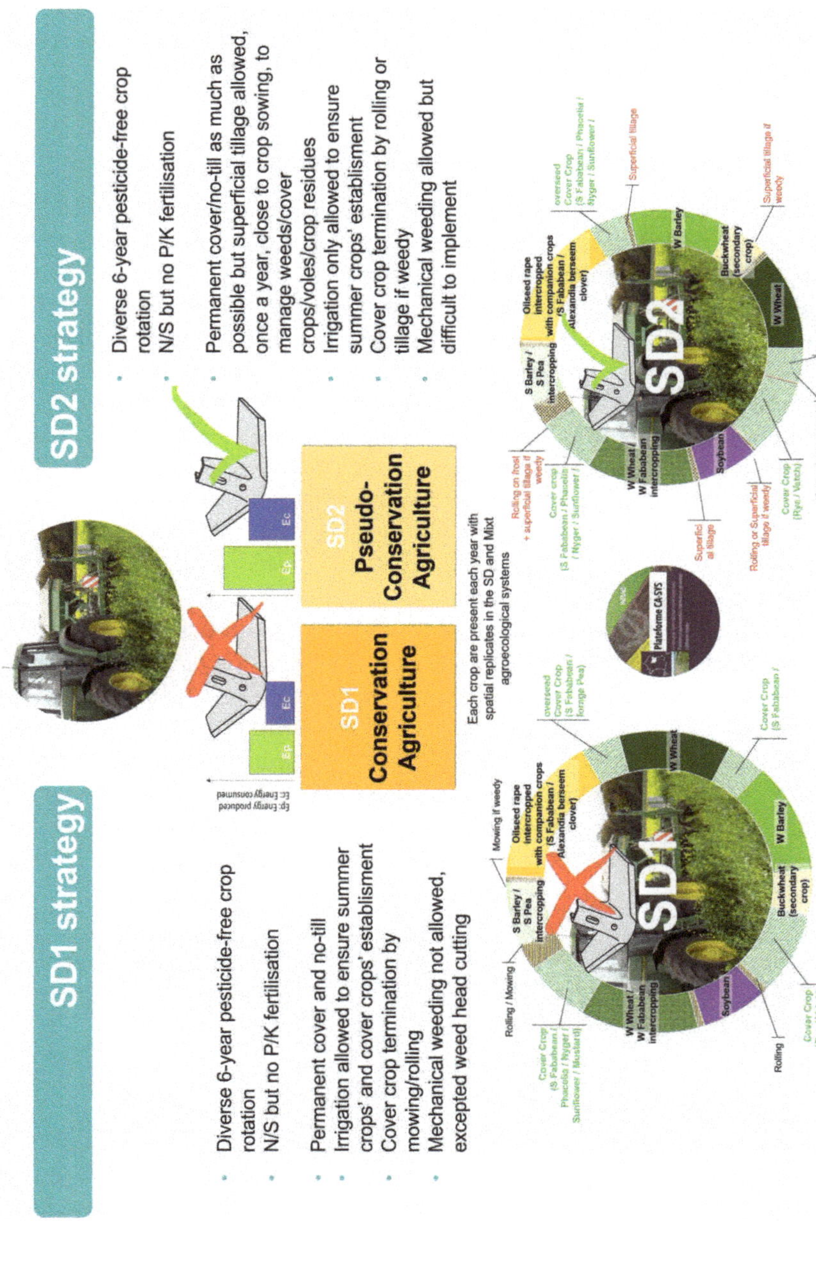

Figure 5 Farming strategies deployed in the SD1 and SD2systems in the CA-SYS platform over the 2018–2023 period.

© Burleigh Dodds Science Publishing Limited, 2025. All rights reserved.

as grasses that are difficult to terminate without soil disturbance in pesticide-free systems (Wortman et al., 2012; Keene et al., 2017). Mechanical weeding is allowed only in SD2 system since it disturbs the soil surface. However, it remains difficult to implement mechanical weeding in reduced tillage systems since crop and cover crop residues remain on the soil surface (Bates et al., 2012).

5.3 Lessons learned from the 2018–2023 phase

Managing pest and particularly weeds in reduced tillage (SD2) and no-till (SD1) pesticide-free systems was particularly challenging, especially during the first years (2018–2023). Our experience on reducing herbicide use in no-till systems presented above was useful (Adeux et al., 2019; Cordeau et al., 2022), but implementing pesticide-free systems was far more challenging as technical difficulties (e.g. direct sowing of a crop in perennial weeds) interacted with multiple pest pressure (e.g. slugs) and damages to accentuate yield loss or crop failure. Weed pressure increased because other pest. Even though alternative farming practices were combined across and within cropping seasons, pest pressure reduced crop growth and stunted crops were not able to suppress weeds. Despite their chaotic population dynamics over the years, voles and slugs were a major issue. In SD1, the initial crop rotation was redesigned to facilitate management of crop volunteers and grass weeds (Figure 6). Indeed, the prohibition of tillage and herbicides greatly complicated the management of crop volunteers during the fallow period, especially for winter cereals and oilseed rape (volunteers of summer crops being killed by frost in our region). We observed that weed communities rapidly evolved towards *Asteraceae*, grasses and perennials. Cover crop composition was adapted to ensure maximum efficacy of the termination method (see next section). We noted that climate change and dry summers challenged our ability to achieve/obtain well-developed and suppressive cover crops, considering irrigation is frequently forbidden during

Figure 6 Main weed issues in SD1 plots (A, Cordeau S. © INRAE 2022) with annual and perennial grasses such as *Lolium perenne*, *Lolium multiflorum*, *Alopecurus myosuroides*, *Vulpia myuros*, *Anisantha sterilis* and *Asteraceae* such as *Sonchus asper*, *Taraxacum officinale*, *Picris hieracoides*, *Picris echioides*; and in SD2 plots (B, Rodolphe Hugard and Guillaume Poussou © INRAE 2021) such as *Cirsium arvense*.

© Burleigh Dodds Science Publishing Limited, 2025. All rights reserved.

dry summers in our region. Furthermore, as I see one of our previous factorial experiments, irrigation tended to increase weed biomass rather than enhancing, possibly due to synchronized weed/cover crop seed germination (Rouge et al., 2022; Rouge et al., 2023). Our previous factorial experiment also showed that nitrogen fertilization did not enhance the weed suppressive ability of cover crops (Rouge et al., 2022; Rouge et al., 2023). Increasing crop and cover crop seeding rate appears primordial in pesticide-free reduced (SD2) and no-till (SD1) systems to ensure satisfactory emerged density.

In SD2, we learnt that a single superficial tillage operation may limit the impact on soil organisms (Rincon-Florez et al., 2016) but does not ensure complete cover termination. The selection of the best tool for plant cover termination depends on weather conditions and targeted plant community (e.g. dominated by annual vs. perennial weeds, crop volunteers, or grasses etc.). The humid weather conditions during autumn in clay soils limit the number of days to implement termination and frequently decrease the termination rate because of plant regrowth. The beneficial effects of superficial tillage are questioned since the development of *Cirsium arvense* appeared even quicker in this system (Figure 6). Indeed, alfalfa was established for 3 years in 7 out of 10 SD2 fields over the 2019–2023 phase to manage *Cirsium arvense*. The implementation of mechanical weeding remained challenging due to the presence of crop residues on soil surface. Only weed head cutting is feasible and allowed to limit weed seed shedding.

5.4 Redesign of pesticide-free cropping system transitioning to CA through a step-by-step process

An in-depth agronomic diagnosis of the SD1 and SD2 cropping systems over the first 5 years of the experiment led to the conclusion that the systems were in an agronomic deadlock, exhibiting different issues for different reasons. A redesign of the cropping system could not be circumvented and we opted for a new way to do so. All the CA-SYS systems were prototyped in 2017–2018 with the help of many stakeholders following the *de novo* approach (Reau et al., 2015; Jeuffroy et al., 2022). It consisted in designing, during workshops involving experts with diverse and complementary knowledge, a few virtual cropping system prototypes (SD1 and SD2), tailored to farmers' aims and resources (Vereijken, 1997). Then, the designed cropping systems had been were implemented in the field without a smooth and step-by-step transition, operating an abrupt contrast with the past farming practices. The knowledge required to design highly innovative agroecological cropping systems is often lacking and technical options ought to be reconsidered after witnessing unplanned effects. The process of cropping system design is not linear and ought to be considered as a series of decisions and actions forming a pathway

© Burleigh Dodds Science Publishing Limited, 2025. All rights reserved.

towards a desired target. As shown by Darnhofer et al. (2010), the evolution of cropping systems is a continuous learning process, not only opening new perspectives for action, but also sometimes renewing or specifying targets.

For these reasons, a step-by-step process (Meynard et al., 2023) was adopted to design pesticide-free cropping systems transitioning to CA. The SD1 and SD2 systems were redesigned in workshops with experts. Particularly, we defined a set of common and specific targets/objectives for both cropping systems. We decided to target pesticide-free, productive and resilient cropping systems that face climate change with enhanced soil health, that is:

- in SD1, maximize the use of biological regulation of pests, limit to its maximum the reliance on tillage (but if necessary, implement deep/ inversion tillage such as plowing), and for which the economic profitability could rely on the payment of a diverse set of ecosystem services (i.e. carbon storage, biodiversity preservation, etc. amongst crop production); and
- in SD2, make complementary use of biological and mechanical regulation of pests, limit to its maximum the depth of tillage when implemented (even if the frequency of tillage can be higher than in SD1), in order to achieve acceptable economic profitability mainly through the sale of agricultural products.

We now propose to explore whether avoiding tillage in synthetic pesticide-free cropping systems is feasible through two different pathways, which differ in their definition of minimum soil disturbance: deep tillage (potentially with soil inversion) but not frequent over time in SD1 as used in rotational no-till strategies (Mirsky et al., 2012), and superficial tillage in SD2 although possibly more frequent than in SD1 as used in organic reduced tillage strategies (Newton et al., 2020). We advocate here that it is necessary to assess the way to transition to CA with a flexible and open-minded way of thinking since it is more important to see CA for the ecosystem services it can deliver than through the agronomic practices implemented to achieve it (Giller et al., 2015; Cordeau, 2024).

Since we observed that the crop establishment was crucial in synthetic pesticide-free no-till system, we now implemented strip-till in SD1 to clean the row before sowing the crop and allow the use of biocontrol to manage slugs, only on sensible crops (mustard, rye, oilseed rape) at their sensible growth stage (seedlings) and after field examination of high pest pressure with adapted protocols. The use of biocontrol was rethought to ensure crop establishment and valorization of fertilization at crop sowing, to ensure initial

© Burleigh Dodds Science Publishing Limited, 2025. All rights reserved.

crop density and crop growth, in order to outcompete emerging weeds that may cause yield loss later.

6 Conclusions

CA relies on the implementation of no-till, permanent soil cover and diversification of crops in order to achieve multiple goals that can differ depending on farmers and production situations. However, reducing the use of herbicides in CA systems remains challenging since the development of these farming strategies was made possible by the development of non-selective wide wide-spectrum herbicides and the optimization of herbicide-based weed management strategies, coupled with farm equipment being able to minimize soil disturbance. Assessing the potential to reduce herbicide use in the context of CA requires an analysis of farmers' motivations for implementing CA, their farming equipment and recent evolution and finally their weed floristic context. We identified two pathways through which herbicide-free (or pesticide-free) no-till systems could be achieved, which are designed either in experimental stations or on commercial farms: (i) if conventional CA has been implemented in the past years and tillage is not an option, continuing towards the reduction of herbicide use in no-till systems by optimizing crop rotation, cover cropping, farm equipment to manage weeds with various tactics; (ii) if organic or pesticide-free agriculture has been implemented in the past years and strategic and occasional tillage is an option, looking for the minimal soil disturbance (e.g. strip-tillage) that allows quick crop establishment and growth, to compete with weeds later on and ensure the delivery of targeted ecosystem services (including crop production). Both pathways need to be explored and assessed to provide farmers with operational and scientific knowledge on how to achieve their agroecological transition towards pesticide-free CA with limited risk.

7 Acknowledgments

The authors thank the CA-SYS team and all the staff from the experimental farm U2E that strive every day for making these experiments exist, such as CA-SYS. We thank all the people (farmers and crop advisors) who took part in the workshops to design cropping systems and semi-natural habitats. We thank the farmers and farm advisors from the DEPHY network who shared and collected the data, respectively, as well as the AGROSYST team who extracted the data. The authors thank INRAE for their financial support. The authors acknowledge financial support from the French "Priority Research Program Alternative Crop Production and Protection", the ANR SPECIFICS (ANR-20-PCPA-0008), the ANR BeCreative (ANR-20-PCPA-0001), the European Horizon 2020 Adopt-IPM

© Burleigh Dodds Science Publishing Limited, 2025. All rights reserved.

and IWM PRAISE, the French ministries for agriculture and environment and the Office Français de la Biodiversité (OFB, DEPHY Expe II project ABC "Agroécologie en Bourgogne et Région Centre"), the Casdar projects RAID (Biological control of weeds) and Vancouver (Weeds management with cover crop), the French "Investissements d'Avenir" programme and the project ISITE-BFC (contract ANR-15-IDEX-03), the Ecophyto COSTRAA project funded by OFB (Office Français de la Biodiversité), the INRAE SumCrop ASPECT project and the COPRAA project which was supported by OFB as part of the Ecophyto call on "Global approaches to limit the use of phytopharmaceuticals: Combining preventive and curative products in sectors, from farmers to consumers" launched by the French Ministries in charge of Ecology, Agriculture, Health and Research.

8 References

Adeux, G., et al. (2019). Diversified grain-based cropping systems provide long-term weed control while limiting herbicide use and yield losses. *Agronomy for Sustainable Development* 39, 42. https://doi.org/10.1007/s13593-019-0587-x

Adeux, G., et al. (2021). Cover crops promote crop productivity but do not enhance weed management in tillage-based cropping systems. *European Journal of Agronomy* 123, https://doi.org/10.1016/j.eja.2020.126221.

Adeux, G., et al. (2022a). Multicriteria assessment of conservation agriculture systems. *Frontiers in Agronomy* 4, 999960. https://doi.org/10.3389/fagro.2022.999960

Adeux, G., et al. (2022b). Cropping system diversification does not always beget weed diversity. *European Journal of Agronomy* 133. https://doi.org/10.1016/j.eja.2021.126438.

Alvarez, R., et al. (1998). Nitrogen distribution in soil density fractions and its relation to nitrogen mineralisation under different tillage systems. *Soil Research* 36, 247–256.

Antier, C., et al. (2020). A survey on the uses of glyphosate in European countries. INRAE. https://doi.org/10.15454/A30K-D531.

Bates, R.T., et al. (2012). Integrating mechanical and reduced chemical weed control in conservation tillage corn. *Agronomy Journal* 104, 507–517.

Blanco-Canqui, H. and Wortmann, C.S. (2020). Does occasional tillage undo the ecosystem services gained with no-till? A review. *Soil and Tillage Research* 198, 104534.

Bybee-Finley, A.K., et alR. (2022). Finding the right mix: a framework for selecting seeding rates for cover crop mixtures. *Ecological applications* 32, e02484. https://doi.org/10.1002/eap.2484

Cardina, J., et al. (2002). Crop rotation and tillage system effects on weed seedbanks. *Weed Science* 50, 448–460.

Çelik, İ., et al. (2019). Strategic tillage may sustain the benefits of long-term no-till in a Vertisol under Mediterranean climate. *Soil and Tillage Research* 185, 17–28.

Chancellor, R.J. (1986). Decline of arable weed seeds during 20 years in soil under grass and the periodicity of seedling emergence after cultivation. *Journal of Applied Ecology* 23, 631–637.

© Burleigh Dodds Science Publishing Limited, 2025. All rights reserved.

Chauhan, B.S., et al. (2006). Influence of tillage systems on vertical distribution, seedling recruitment and persistence of rigid ryegrass (Lolium rigidum) seed bank. *Weed Science* 54, 669–676.

Chauhan, B.S., et al. (2012). Ecology and management of weeds under conservation agriculture: a review. *Crop Protection* 38, 57–65.

Chauvel, B., et al. (2022). From sea salt to glyphosate salt: a history of herbicide use in France. *Advances in Weed Science* 40. https://doi.org/10.51694/AdvWeedSci/2022;40:seventy-five008.

Chikowo, R., et al. (2009). Integrated Weed Management systems allow reduced reliance on herbicides and long term weed control. *Agriculture, Ecosystems and Environment* 132, 237–242.

Conyers, M., et al. (2019). The strategic use of minimum tillage within conservation agriculture in southern New South Wales, Australia. *Soil and Tillage Research* 193, 17–26.

Cordeau, S. (2022). Conservation agriculture and agroecological weed management. *Agronomy* 12, 867. https://doi.org/10.3390/agronomy12040867

Cordeau, S. (2024). Moving conservation agriculture from principles to a performance-based production system. *Renewable Agriculture and Food Systems* 39, 1–7. https://doi.org/10.1017/S1742170524000048

Cordeau, S., et al. (2015a). Towards the establishment of an experimental research unit on Agroecology in France. In Stockdale, E.A (Ed.). *Aspects of applied biology* 128: valuing long-term sites and experiments for agriculture and ecology. Association of Applied Biologists, (pp. 271–273).

Cordeau, S., et al. (2015b). Weed species differ in their ability to emerge in no-till systems that include cover crops. *Annals of Applied Biology* 166, 444–455. https://doi.org/10.1111/aab.12195

Cordeau, S., et al. (2020a). Diversity is the key for successful agroecological weed management. *Indian Journal of Weed Science* 52, 204–210.

Cordeau, S., et al. (2020b). Is tillage a suitable option for weed management in conservation agriculture? *Agronomy* 10, 1746, https://doi.org/10.3390/agronomy10111746.

Cordeau, S., et al. (2022). Legacy effects of contrasting long-term integrated weed management systems. *Frontiers in Agronomy* 3, 769992. https://doi.org/10.3389/fagro.2021.769992

Craheix, D., et al. (2016). Using a multicriteria assessment model to evaluate the sustainability of conservation agriculture at the cropping system level in France. *European Journal of Agronomy* 76, 75–86.

Crawford, M.H., et al. (2015). Changes in the soil quality attributes of continuous no-till farming systems following a strategic tillage. *Soil Research* 53, 263–273.

Crowder, D.W., et al. (2010). Organic agriculture promotes evenness and natural pest control. *Nature* 466, 109–112.

Dang, Y.P., et al. (2015). Strategic tillage in no-till farming systems in Australia's northern grains-growing regions: I. Drivers and implementation. *Soil and Tillage Research* 152, 104–114.

Darnhofer, I., et al. (2010). Adaptiveness to enhance the sustainability of farming systems: a review. *Agronomy for Sustainable Development* 30, 545–555.

Délye, C., et al. (2013). Deciphering the evolution of herbicide resistance in weeds. *Trends in Genetics* 29, 649–658.

© Burleigh Dodds Science Publishing Limited, 2025. All rights reserved.

Derrouch, D., et al. (2020). Weed management in the transition to conservation agriculture: farmers' response. *Agronomy* 10, 843.

Derrouch, D., et al. (2021a). Functional shifts in weed community composition following adoption of conservation agriculture. *Weed Research* 62, 103–112. https://doi.org/10.1111/wre.12517

Derrouch, D., et al. (2021b). Weed community diversity in conservation agriculture: post-adoption changes. *Agriculture, Ecosystems & Environment* 312, 107351.

Deytieux, V., et al. (2012). Is integrated weed management efficient for reducing environmental impacts of cropping systems? A case study based on life cycle assessment. *European Journal of Agronomy* 36, 55–65.

Díaz-Zorita, M., et al. (2004). Soil structural disturbance effects on crop yields and soil properties in a no-till production system. *Agronomy Journal* 96, 1651–1659.

Dore, T., et al. (2011). Facing up to the paradigm of ecological intensification in agronomy: revisiting methods, concepts and knowledge. *European Journal of Agronomy* 34, 197–210.

Douglas, M.R. and Tooker, J.F. (2012). Slug (Mollusca: Agriolimacidae, Arionidae) ecology and management in no-till field crops, with an emphasis on the Mid-Atlantic region. *Journal of Integrated Pest Management* 3, C1–C9.

Evans, D. (1972). "Filed performance of glyphosate derivatives in the control of Agropyron repens and other perennial weeds", in: *Brit Weed Contr Conf Proc*.

Fao. (2021). *Conservation Agriculture principles* https://www.fao.org/conservation-agriculture/overview/principles-of-ca/en/, accessed on 13th December 2021 [Online]. [Accessed 2021].

Foley, J.A., et al. (2005). Global consequences of land use. *Science* 309, 570–574.

Friedrich, T., et al. (2012). Overview of the global spread of conservation agriculture. *Field Actions Science Reports* 6, 1–7.

Gaba, S., et al. (2014). Managing biotic interactions for ecological intensification of agroecosystems. *Frontiers in Ecology and Evolution* 2, 29.

Garnett, T., et al. (2013). Sustainable intensification in agriculture: premises and policies. *Science* 341, 33–34.

Giller, K.E., et al. (2015). Beyond conservation agriculture. *Frontiers in Plant Science* 6, 870.

Green, J.M. (2018). The rise and future of glyphosate and glyphosate-resistant crops. *Pest Management Science* 74, 1035–1039.

Grundy, A., et al. (2011). The impact of herbicide management on long-term changes in the diversity and species composition of weed populations. *Weed Research* 51, 187–200.

Hill, S.B. and Macrae, R.J. (1995). Conceptual framework for the transition from conventional to sustainable agriculture. *Journal of Sustainable Agriculture* 7, 81–87.

Hoss, N.E., et al. (2003). Efficacy of glyphosate, glufosinate, and imazethapyr on selected weed species. *Weed Science* 51, 110–117.

Jeuffroy, M.-H., et al. (2022). Design workshops for innovative cropping systems and decision-support tools: learning from 12 case studies. *European Journal of Agronomy* 139, 126573.

Kassam, A. and Friedrich, T. (2011). "Conservation agriculture: principles, sustainable land management and ecosystem services", in: *Proceedings of the 40th National Convention of the Italian Agronomy Society*, 7–9.

© Burleigh Dodds Science Publishing Limited, 2025. All rights reserved.

Kassam, A., et al. (2019). Global spread of conservation agriculture. *International Journal of Environmental Studies* 76, 29–51.

Kassam, A., et al. (2009). The spread of conservation agriculture: justification, sustainability and uptake. *International Journal of Agricultural Sustainability* 7, 292–320.

Keene, C., et al. (2017). Cover crop termination timing is critical in organic rotational no-till systems. *Agronomy Journal* 109, 272–282.

Knowler, D. and Bradshaw, B. (2007). Farmers' adoption of conservation agriculture: a review and synthesis of recent research. *Food Policy* 32, 25–48.

Lechenet, M., et al. (2014). Reconciling pesticide reduction with economic and environmental sustainability in arable farming. *PLoS One* 9, e97922.

Lechenet, M., et al. (2016). Profiling farming management strategies with contrasting pesticide use in France. *Agricultural Systems* 149, 40–53 https://doi.org/10.1016/j.agsy.2016.1008.1005.

Lechenet, M., et al. (2017). Reducing pesticide use while preserving crop productivity and profitability on arable farms. *Nature Plants* 3, 17008 https://www.nature.com/articles/nplants20178.

Maes, J. and Jacobs, S. (2017). Nature-based solutions for Europe's sustainable development. *Conservation letters* 10, 121–124.

Menalled, U., et al. (2022). Cereal rye mulch biomass and crop density affect weed suppression and community assembly in no-till planted soybean. *Ecosphere* 13, 6. https://doi.org/10.1002/ecs2.4147

Meynard, J.-M., et al. (2023). Unravelling the step-by-step process for farming system design to support agroecological transition. *European Journal of Agronomy* 150, 126948.

Mirsky, S.B., et al. (2012). Conservation tillage issues: cover crop-based organic rotational no-till grain production in the mid-Atlantic region, USA. *Renewable Agriculture and Food Systems* 27, 31–40.

Mirsky, S.B., et al. (2017). Hairy vetch biomass across the Eastern United States: effects of latitude, seeding rate and date, and termination timing. *Agronomy Journal* 109, 1510–1519. https://doi.org/10.2134/agronj2016.09.0556

Mulugeta, D. and Stoltenberg, D.E. (1997). Increased weed emergence and seed bank depletion by soil disturbance in a no-tillage system. *Weed Science* 45, 234–241.

Nazarko, O.M., et al. (2005). Strategies and tactics for herbicide use reduction in field crops in Canada: a review. *Canadian Journal of Plant Science* 85, 457–479.

Newton, P., et al. (2020). What is regenerative agriculture? A review of scholar and practitioner definitions based on processes and outcomes. *Frontiers in Sustainable Food Systems* 4, 194.

Nichols, V., et al. (2015). Weed dynamics and conservation agriculture principles: a review. *Field Crops Research* 183, 56–68.

Osipitan, O.A., et al. (2019). Impact of cover crop management on level of weed suppression: a meta-analysis. *Crop Science* 59, 833–842.

Owen, M.D. (2008). Weed species shifts in glyphosate-resistant crops. *Pest Management Science: formerly Pesticide Science* 64, 377–387.

Peixoto, D.S., et al. (2019). A soil compaction diagnosis method for occasional tillage recommendation under continuous no tillage system in Brazil. *Soil and Tillage Research* 194, 104307.

Peixoto, D.S., et al. (2020). Occasional tillage in no-tillage systems: a global meta-analysis. *Science of the Total Environment* 745, 140887.

© Burleigh Dodds Science Publishing Limited, 2025. All rights reserved.

Petit, S., et al. (2018). Biodiversity-based options for arable weed management: a review. *Agronomy for Sustainable Development* 38, 48. https://doi.org/10.1007/s13593-018-0525-3

Petit, S., et al. (2021). Landscape-scale approaches for designing and assessing biodiversity-based agricultural systems enhancing biological pest control. *Environmental Monitoring and Assessment* 193, 75. https://doi.org/10.1007/s10661-020-08812-2

R4p (2023). *Update of the Herbicide Resistance maps 2023.* https://www.r4p-inra.fr/en/update-of-the-herbicide-resistance-maps-2023/. [Accessed 12/02/2024 2024].

Reau, R., et al. (2015). "De novo design workshop: a method for co-designing innovative cropping systems", in: *5th International Symposium for Farming Systems Design, 7–10 September 201.*

Reicosky, D.C. (2015). Conservation tillage is not conservation agriculture. *Journal of Soil and Water Conservation* 70, 103A–108A.

Rincon-Florez, V.A., et al. (2016). Short-term impact of an occasional tillage on microbial communities in a Vertosol after 43 years of no-tillage or conventional tillage. *European Journal of Soil Biology* 74, 32–38.

Roger-Estrade, J., et al. (2001). Modelling vertical and lateral weed seed movements during mouldboard ploughing with a skim-coulter. *Soil and Tillage Research* 63, 35–49.

Rouge, A., et al. (2022). Weed suppression in cover crop mixtures under contrasted levels of resource availability. *European Journal of Agronomy* 136, 126499. https://doi.org/10.1016/j.eja.2022.126499

Rouge, A., et al. (2023). Carry-over effects of cover crops on weeds and crop productivity in no-till systems. *Field Crops Research* 295, 108899. https://doi.org/10.1016/j.fcr.2023.108899

Saffeullah, P., et al. (2021). "Organic agriculture: principles, current status, and significance," in *Microbiota and Biofertilizers*. Springer, 17–37.

Smith, R.G., et al. (2020). Are cover crop mixtures better at suppressing weeds than cover crop monocultures? *Weed Science* 68, 186–194. https://doi.org/10.1017/wsc.2020.12

Soltani, E., et al. (2016). A quantitative analysis of seed dormancy and germination in the winter annual weed Sinapis arvensis (Brassicaceae). *Botany* 94, 289–300.

Teasdale, J.R. and Mohler, C.L. (2000). The quantitative relationship between weed emergence and the physical properties of mulches. *Weed Science* 48, 385–392.

Thompson, K., et al. (1993). Seed size and shape predict persistence in soil. *Functional Ecology* 7, 236–241.

Vanbergen, A.J., et al. (2020). Transformation of agricultural landscapes in the Anthropocene: Nature's contributions to people, agriculture and food security. *Advances in Ecological Research* 63, 193–253. https://doi.org/10.1016/bs.aecr.2020.08.002

Van Den Putte, A., et al. (2010). Assessing the effect of soil tillage on crop growth: ā meta-regression analysis on European crop yields under conservation agriculture. *European Journal of Agronomy* 33, 231–241.

Vankeerberghen, A. and Stassart, P.M. (2016). The transition to conservation agriculture: an insularization process towards sustainability. *International Journal of Agricultural Sustainability* 14, 392–407.

© Burleigh Dodds Science Publishing Limited, 2025. All rights reserved.

Vereijken, P. (1997). A methodical way of prototyping integrated and ecological arable farming systems (I/EAFS) in interaction with pilot farms. *European Journal of Agronomy* 7, 235–250.

Watson, G. (1974). "Motivation and response in the development of minimal tillage techniques outside the United Kingdom", in: *Proceedings*.

Wortman, S.E., et al. (2012). Optimizing cover crop benefits with diverse mixtures and an alternative termination method. *Agronomy Journal* 104, 1425–1435.

© Burleigh Dodds Science Publishing Limited, 2025. All rights reserved.

Part 2

Weed management in specific production systems

Chapter 8

Weed management in perennial Conservation Agriculture systems: optimising use of groundcover

Emilio J. Gonzalez-Sanchez, Universidad de Cordoba, Spain; Julio Roman-Vazquez, European Conservation Agriculture Federation (ECAF), Belgium; Oscar Veroz-Gonzalez, Asociación Española Agricultura de Conservación Suelos Vivos, Spain; Rosa Carbonell-Bojollo, IFAPA Centro Alameda del Obispo, Spain; Francisco Marquez-Garcia, Antonio M. Conde-Lopez and Miguel A. Repullo-Ruiberriz de Torres, Universidad de Cordoba, Spain; Manuel Moreno-Garcia and Rafaela Ordoñez-Fernandez, IFAPA Centro Alameda del Obispo, Spain; and Jesus A. Gil-Ribes, Universidad de Cordoba, Spain

1 Introduction

Perennial crops are normally considered as those that are permanent, requiring a number of growth cycles before fruit is produced. One of the characteristics of these plants is their requirement for winter chilling before flowering the following spring (Hatfield & Walthall, 2014). The importance of perennial crops is obvious. In addition to being traditional crops, always rooted in local cultures, in recent years, perennial crops have gained considerable popularity among farmers across the world for their economic benefit. The profitability of perennial crops is based on reduced expenses for seed and fertiliser (since the crop is seeded once and harvested many times), and reduced costs for weed control, compared to other cultural practices associated with annual crops (Cui et al., 2018).

http://dx.doi.org/10.19103/AS.2024.0145.14
© Burleigh Dodds Science Publishing Limited, 2025. All rights reserved.

In the last decade, awareness of diverse environmental situations, and also those related to agriculture, has grown significantly throughout the world. For instance, according to the European Environment Agency, the main environmental problems arising from agriculture are soil erosion, water pollution by sediment, fertilisers and pesticides, the decrease in biodiversity, greenhouse gas (GHG) emissions, low biodiversity and landscape degradation (EEA, 2024).

The first factor affecting tree productivity is the climate, and more specifically rainfall, which is usually irregular and often scarce in some areas. Water is the factor that most influences production, and competition from weeds greatly limits productivity. For this reason, farmers have fought against the presence of weeds in crops. In perennial crops, weed control systems based on soil tillage have caused serious problems with high erosion rates, water availability, a decrease in biodiversity and associated ecosystem services. Indeed, soil erosion and the consequent contamination of water by sediment are one of the most important environmental problems caused by conventional agriculture. These problems are even more pronounced in perennial crops than in annual crops, since, in some regions, trees are grown on steep slopes. Therefore, applicable solutions to soil management in perennial crops should be based on Conservation Agriculture (CA) principles. CA, as defined by the FAO, is a farming system that promotes minimum soil disturbance (i.e. no-tillage), maintenance of a permanent soil cover and diversification of plant species (FAO, 2024). Even though CA is generally more linked to annual crops, it is applicable to both annual crops and perennial crops (Saavedra-Saavedra et al., 2015).

Weed control is essential in perennial CA systems to ensure optimal resource use and sustainable long-term productivity. Weeds compete with perennial crops for water, nutrients and sunlight, leading to reduced yields and weaker plants, which can undermine the sustainability of these systems. Effective weed management is key to maintaining soil health, as it reduces the need for herbicides, preserves beneficial soil organisms and enhances biodiversity. According to Smith et al. (2011), proper weed control in perennial systems can improve crop resilience and reduce the spread of pests and diseases, further contributing to the stability of the ecosystem. Additionally, studies by Jones and Dehaan (2017) emphasise that integrated weed management strategies are vital for balancing the ecological and economic goals of CA, ensuring that these systems remain productive and sustainable over time. In this chapter, CA practices for perennial crops and weed management under those conditions will be developed.

© Burleigh Dodds Science Publishing Limited, 2025. All rights reserved.

2 Conservation Agriculture in perennial crops: growing green strips

The case of CA for perennial crops is somehow special since spontaneous plants are sometimes considered as part of the solution offered to farmers. This can be perceived as a paradox, but the protective groundcovers for woody crops, in some cases, spontaneous groundcovers, are composed of what conventional farmers identify as 'weeds'. In the late 1970s, trials were already being carried out with groundcovers in olive groves in Spain. Almost from the beginning, it was observed that the need to maintain the cover was limited to the space in between tree rows, where the capacity for water infiltration in the soil was limited by compaction that occurs due to trampling and the transit of machinery (Saavedra-Saavedra, 2017). Indeed, keeping the soil protected by groundcovers is a CA agronomic principle in perennial crops, whereby the soil surface between rows of trees remains protected against erosion. With this technique, at least 30% of the soil is protected either by sown vegetative covers, spontaneous vegetation or inert covers, such as pruning residues or tree leaves. For the establishment of sown groundcovers and the spread of inert covers, techniques must be in agreement with the CA principle of minimum soil disturbance (Gonzalez-Sanchez et al., 2015). As a matter of fact, those spontaneous groundcovers are the native species of plants that normally grow in the woody crops. Conventional farmers either perform tillage or apply herbicides to control them, whereas CA farmers use them for the purpose of protecting the soil. In any case, those sown or 'weed covers' must also be managed in CA.

The need to control living groundcovers to avoid competition for water with the perennial crop is needed. The competition for nutrients could be corrected with foliar fertilisation, but in rainfed conditions, since it is not possible to provide trees with water, crops normally suffer and loose yields. Besides, in the area under the canopy, soils normally have a much greater capacity to infiltrate water, since there, organic matter content is higher than in between tree rows. Therefore, if a groundcover is left under the trees, it would be much more competitive and also more difficult to control. The conclusion reached was that the living groundcover should be reduced to space between tree lines (Saavedra-Saavedra, 2017) (Fig. 1). The same approach was reached for different perennial crops, such as citrus (Arenas-Arenas et al., 2015) and almond trees (Arquero-Quilez et al., 2015).

As stated in the literature, the most effective way to combat erosion is to cover the soil. In perennial crops, various types of covers could be used, since well-managed groundcovers between the tree rows of woody plantations do not hinder the performance of other agricultural practices such as phytosanitary treatments, pruning, fertilisation; on the contrary, it improves accessibility to the

© Burleigh Dodds Science Publishing Limited, 2025. All rights reserved.

Figure 1 Groundcover in between olive trees in Spain.

plots since the soil cover facilitates tractor traffic. This fact is vital because it allows important operations to be carried out in time, such as harvesting in wet conditions.

There are different types of groundcovers that can be used, such as living plants or pruning residues. With the dead materials, there is no competition with the trees, but the living vegetation cover in dryland plantations can compete for nutrients or water. Groundcovers normally compete for water with the main crop, in periods of scarcity, when rainfall is low and water demand is high. Competition for water is greatest from late winter or early spring until autumn when rain falls again with intensity and the needs of the trees are lower. This competition forces the growth of the groundcover to be limited or to be removed during this period since otherwise the harvests would be seriously reduced (Saavedra-Saavedra, 2017).

Plants that form the vegetation cover develop their roots in the soil, loosening it and forming a strong support framework, and at the same time, they provide infiltration channels for water when they die. Both factors have a very positive influence on soil structure and on increased infiltration rate. On the other hand, the above-ground part protects the soil and buffers the direct impact of raindrops, preventing particle disaggregation, and also forms a trap that reduces the speed of the water in the event of runoff, so that the erosive capacity of the water is significantly reduced.

© Burleigh Dodds Science Publishing Limited, 2025. All rights reserved.

For all these reasons, and because of the organic matter they provide, the use of living groundcovers is, at minimum cost, the best and most natural way to improve the soil and fight erosion. Although it will be necessary to use certain management techniques to reduce competition and maintain the quantity and quality of crop production.

In dry conditions, or when there are limitations on water use, an 'ideal groundcover' (Table 1) should not consume water during the period of scarcity, or in other words, it would have to complete its cycle of active vegetation at

Table 1 Characteristics of an 'ideal' groundcover.

Underdeveloped height	A groundcover with tall vegetation species hinders operations and has a greater amount of plant material, contributing to high water transpiration, which results in lower water availability when the crop needs it.
Rapid autumn establishment	A groundcover that develops with the first autumn rains will quickly protect the soil from possible erosion events that may occur with subsequent precipitation.
Shallow root development	Generally, the roots of woody crops tend to explore deeper soil horizons than herbaceous species. A groundcover with deep-rooted species will compete more with the crop for water and nutrients than those with shallow roots.
Minimally competitive	The ideal groundcover is one that completes its cycle during the period of abundant rainfall and finishes before the crop enters its phase of higher water and nutrient demand.
Not a host for insects/pests	The groundcover should not consist of species that host pests posing a risk to the crop. On the contrary, it is considered beneficial if it comprises species that have hardly any pest that could affect the crop.
Low combustibility	One of the main concerns for farmers who plant groundcover is the risk of fire, especially during the period when the cover remains on the ground as dry plant material, primarily in the summer. Therefore, the less flammable the dry plant material, the better.
Capable of mobilising nutrients	If the groundcover, in addition to protecting the soil from erosion and runoff, is capable of mobilising nutrients, it will improve the natural fertility of the soil. In this regard, legumes, with their ability to fix nitrogen, are the best at fulfilling this function.
Slow decomposition	Groundcovers with a high C/N ratio are recommended, as this implies a slow decomposition rate and therefore a longer time covering the soil. In this regard, grasses are the most advantageous species, while legumes tend to have a higher decomposition rate.
Self-sowing capability	If the groundcover can produce viable seeds, we can take advantage of this and avoid having to reseed it season after season.

Source: Adapted from AEACSV, 2024a.

© Burleigh Dodds Science Publishing Limited, 2025. All rights reserved.

the beginning of spring. On the other hand, it would have to have formed enough dry matter by that time, and these dry remains would not degrade easily. Finding species that meet these qualities while eliminating the rest of the spontaneous vegetation is not easy. Therefore, vegetative groundcovers, whether cultivated or spontaneous species, are the best means to combat soil erosion.

3 Types of groundcover in perennial CA systems

Groundcovers can be classified according to the type of plant material used as a protective cover. Each type of groundcover is briefly described below (Fig. 2 and Table 2).

3.1 Non-selected spontaneous cover

This type of cover allows the spontaneous growth of vegetation in the spaces between the tree rows (Fig. 3). The advantage of this groundcover is the cost savings on items like seed, planting operations and selective treatment. However, a disadvantage is that the plant species involved are often very diverse, making management more difficult. This means that if mechanical mowing is chosen, there may be a shift towards low-growing, perennial or easily regenerating species. In the case of chemical mowing, high doses of herbicide might be needed. Additionally, these species decompose quickly, offering little soil protection.

This type of groundcover is normally composed of what farmers describe as weeds. But not all weeds are harmful to the crop all the time, so managing groundcover will bring benefits for both the farmer and the agri-environment.

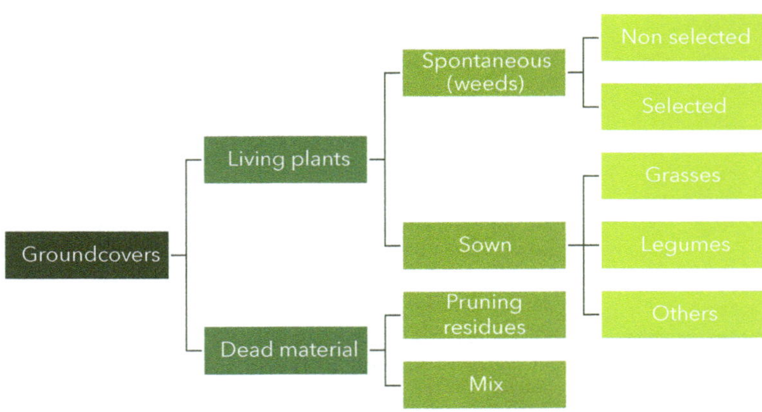

Figure 2 Types of groundcovers for perennial crops.

© Burleigh Dodds Science Publishing Limited, 2025. All rights reserved.

Table 2 Characteristics of different living groundcovers and pruning residues with respect to various agricultural and environmental factors.

Species	Underdeveloped Height	Rapid autumn establishment	Shallow root development	Minimally competitive	Not a host for insects/pests	Low combustibility	Slow decomposition	Self-sowing capability
Bromus rubens	✓	✓	✓	↔	✓	✗	✓	✓
Brachypodium distachyon	✓	✓	✓	↔	✓	✗	✓	✓
Hordeum vulgare	↔	✓	✓	↔	✓	✗	✓	✓
Vicia sativa	✓	✗	↔	↔	✓	✗	✗	✗
Lupinus albus	✓	✗	↔	↔	✓	✗	✗	✗
Trifolium pratense	✓	✗	↔	↔	✓	✗	✗	✗
Sinapis alba	✗	✓	✗	↔	✓	✗	✗	✗
Eruca vesicaria	✗	✓	✗	↔	✓	✗	✗	✗
Pruning residues	*Does not apply*	*Does not apply*	*Does not apply*	✓	↔	↔	✓	*Does not apply*

Legend: ✓ : Yes ; ✗: No ; ↔: Intermediate
Source: AEACSV, 2024b.

© Burleigh Dodds Science Publishing Limited, 2025. All rights reserved.

Figure 3 Unselected spontaneous cover.

3.2 Selected spontaneous cover

This type of groundcover is recommended for soils that have been cultivated for years, resulting in a seed bank with a wide variety of species and a high seed density. In this case, the native flora is managed so that a single species, or a mixture of species from a single family, dominates the cover. This is achieved by gradually encouraging a natural vegetation cover of grasses (ryegrass, barley, among others), which are very common and effective in controlling erosion, while progressively eliminating the more problematic and difficult-to-control species, like broadleaves, such as *Malva spp., Ecballium elaterium* (Fig. 4).

3.3 Sown grass cover

This is one of the most widespread types of groundcover and serves as an alternative to natural or spontaneous vegetation. It is recommended for highly eroded soils or those with a very poor or low-density seed bank. This is often the case in woody crops that have been managed with tillage and bare soil. Since the objective of the cover is not to achieve a high yield from the species sown, but rather to protect the soil, non-certified seeds or natural grass seeds can be used for sowing. The trend with this type of groundcover is to use species that have early emergence, so that with the first autumn rains, they cover the soil

© Burleigh Dodds Science Publishing Limited, 2025. All rights reserved.

Figure 4 Selected spontaneous grass cover in orange trees.

and protect it. A short life cycle species, which would allow them to complete their cycle without needing to be controlled, would produce seeds that would eliminate the need to sow the cover every season (Fig. 5).

3.4 Sown legume cover

This is an interesting alternative due to its ability to fix atmospheric nitrogen, thereby saving on nitrogen fertiliser. However, in terms of soil protection, they are considered less effective, mainly because of their rapid decomposition. Additionally, herbicide treatments for their control are expensive, making mechanical mowing the best management option, especially with species with little regrowth capability, like vetch. These species, along with clover, are the most suitable for use as a groundcover (Fig. 6).

3.5 Sown cruciferous cover

This type of groundcover is a good option to introduce into a rotation with grass covers, as after several years of using the same species as a groundcover, its effectiveness deteriorates, reducing soil protection. Advantages of using cruciferous covers include being popular among farmers, having rapid growth and abundant biomass production, quickly covering the soil. Being winter crops, they coexist with the crop during the rainy season and low evapotranspiration,

© Burleigh Dodds Science Publishing Limited, 2025. All rights reserved.

Figure 5 Sown grass cover.

Figure 6 Sown legume cover in vineyard.

© Burleigh Dodds Science Publishing Limited, 2025. All rights reserved.

Figure 7 Sown cruciferous cover in pistachio.

resulting in minimal competition. Moreover, their deep taproot can break up the plough pan and improve poor water infiltration in the soil (Fig. 7).

3.6 Pruning residue cover

This type of cover involves spreading shredded pruning residues, leaving them on the soil surface in the center of the row, so that they decompose slowly and provide prolonged soil protection. This is an alternative use for residue that traditionally has had little value to farmers. Pruning residues should be chopped into small pieces to avoid becoming a habitat for certain pests, like the bark beetle, and to avoid hindering other farming operations, such as harvesting (Fig. 8). Table 2 (see earlier) summarizes the characteristics of different living groundcovers and pruning residues with respect to various agricultural and environmental factors.

4 Weed and groundcover control in perennial crops

Weeds and groundcover can be controlled in two ways:

- **Chemical control:** This is a method based on the use of herbicides. Table 3 provides the main recommendations for managing weeds that are difficult to control. This table was prepared by the Spanish Association for Conservation Agriculture and Living Soils guidance for field technicians.

© Burleigh Dodds Science Publishing Limited, 2025. All rights reserved.

Figure 8 Pruning residue cover.

- **Mechanical control:** In this case, control is performed by mowers. Even if it can be applied to any kind of living groundcovers, this type of control is more effective for plant species that have a low regrowth capacity. To be effective, it is recommended to carry out several control operations at the time when the height is sufficient for this operation before drying out the cover in the spring.

The choice between chemical and mechanical mowing will depend on the type of groundcover, first of all, and the need to carry out highly effective control, as well as the cost of the operation on each type of farm. Two different periods must be distinguished in terms of water availability and its influence on weed management. The first, from mid-autumn and during winter, and even in some places until spring, when rainfall is usually abundant and the needs for trees are minimal, which in many cases allows the presence of weeds without causing crop losses. The second period, from late winter or early spring, when the trees begin vegetative activity, evapotranspiration is high and as do the demands of the crop. From that moment until the end of summer, the presence of grass in the plantations is counterproductive and can greatly reduce the harvest. For

© Burleigh Dodds Science Publishing Limited, 2025. All rights reserved.

Table 3 Active ingredients and treatment doses against some difficult-to-control weeds and recommendations about its control.

Weed	Active ingredient	Dose	Comments
Malva and *Lavatera* spp.	Fluroxypyr	0.2–0.3 kg/ha treated	With a certain degree of development.
	Glyphosate + oxyfluorfen	1.4–2.2 + 1.10–1.14 kg/ha	Low developed grass.
	Glyphosate + MCPA	0.36 + 0.36 kg active ingredient (ai)/ha	Very small.
Ecballium ellaterium	Diuron (80%)	0.8–3.2 kg ai/ha	Perform the application in pre-emergency.
	Isoxaben (50%)	1–2 kg ai/ha	Perform the application in pre-emergency.
	Fluroxypyr	0.40 kg/ha	Apply in early post-emergence, repeating on regrowth.
Chenopodium	Oxyfluorfen	0.6 l/ha	They emerge in early winter. Application in very early post-emergence (height not exceeding 4–5 cm).
Solanum	Oxyfluorfen	0.6 l/ha	
Amaranthus	Oxyfluorfen	0.6 l/ha	For young plantations less than 4 years old. Apply in early post-emergence (1–2 leaves).
Conyza spp.	Glyphosate +diflufenican	4 kg/ha de producto comercial con 16.8 + 4.12% pv	Difficult to control, with species that have shown resistance. Application is early post-emergence. In well-established covers that cover the ground well, the conyza population usually decreases.
Ononis	Glyphosate + ammonium thiocyanate	0.72–1.08 kg/ha de glyphosate +0.75–1 kg/ha ammonium thiocyanate	Weed that is very difficult to control. Apply during the vegetative stage and always before the plant shows signs of water stress or forms seeds.
Sinapis alba, *Sinapis arvensis*	Glyphosate + MCPA	0.4 + 0.4 and 0. 8+ 0.8 kg/ha	When the plants are somewhat larger. Do not delay the application too much at the end of winter or the beginning of spring.
	Fluroxypyr	0.2–0.25 kg/ha	
Convolvulus arvensis, *Convolvulus althaeoides*	Fluroxypyr	0.3 kg/ha	Application during flowering.
	Glyphosate	2 and 2.25 kg/ha	

(*Continued*)

© Burleigh Dodds Science Publishing Limited, 2025. All rights reserved.

Table 3 (Continued)

Weed	Active ingredient	Dose	Comments
Cynodon dactylon	Glyphosate	2 kg/ha	Application at the beginning of flowering.In drought conditions, it is recommended to apply the treatment earlier (no more than 100 l/ha of broth). In drought conditions, the effectiveness is improved by adding ammonium sulphate at a rate of 2 to 3% of the broth volume.

Source: AECSV, 2024b.

example, in Córdoba, located in the heart of the Andalusian in Spain, olive groves with average rainfall between 500 and 600 mm per year, it is possible to maintain grass in the plantation aisles until mid-March, but once this date has passed, it is necessary to remove the grass if a good harvest is to be obtained.

Spontaneous species have the advantage of not requiring sowing every year, but they do need to be managed properly with herbicides or mechanical means. Annual grasses are very suitable. From a sparse population of *Hordeum murinum L.*, *Lolium rigidum* Gaudin, Bromus spp. and Vulpia spp., it has been possible to install a vegetative cover between tree lines, by applying selective herbicides for grasses, allowing them to grow and bear fruit. Once an adequate density has been reached (500 to 1000 plants per m^2) in one or two years, the cover will be controlled by applying glyphosate at the end of winter or beginning of spring, in order to reduce competition for water (which is important in dry lands), but a narrow strip is left untreated to allow fruiting and seed production that will allow the establishment of the cover the following year. Chemical mowing, if done with herbicides of foliar absorption and translocation, such as glyphosate, allows effective control of the plants and limits transpiration shortly after application. If the dose of herbicide is adequate, regrowth does not occur. Mechanical mowing, with a brush cutter or a mower, is also an option. However, farmers must be vigilant, since mechanical mowing rarely achieves the effect of total control of the vegetative activity of the cover, because the plants sprout, and even some species, with a creeping habit, sometimes escape from control.

5 Equipment for controlling weeds/groundcover

5.1 Chemical control

For applying phytosanitary products, a hydraulic sprayer with a projected jet, also known as a boom sprayer, is normally used. Application volumes can vary between 75 l/ha and 500 l/ha. It is necessary that the boom sprayer is well

© Burleigh Dodds Science Publishing Limited, 2025. All rights reserved.

calibrated. To do this, a periodic check must be carried out on the condition of the filters and nozzles, although other sprayer parts must also be examined. As a general rule, the following recommendations are suggested when carrying out an application:

- The sprayer pump must work at low pressure (no more than 4 bar).
- The pressure gauge must measure the pressure at an appropriate scale.
- It is recommended that the equipment has at least two filters, one at the pump outlet and another before each nozzle. This way, the nozzles are prevented from becoming clogged by solid elements that may be dragged by the liquid. The filters must be cleaned at least every time an application is completed.
- It is recommended to use symmetrical flat fan nozzles at all application points, except at the ends of the boom, in which case, asymmetrical flat fan nozzles would be used. As with the filters, it is essential to clean the nozzles after application.
- The nozzles must be arranged on the application boom in such a way that they distribute the product as evenly as possible. To do this, the nozzles must be rotated between 3° and 4° to prevent the jets applied by adjacent nozzles from colliding with each other. In addition, to achieve an ideal overlap between the jets, it is recommended that, in the case of 110° flat fan nozzles, they be placed 50 cm apart from each other, and the application bar should be placed 50 cm from the target canopy.

Using sensors to detect weeds is an innovative approach that plays a crucial role in reducing the reliance on herbicides in agriculture. This technology offers significant benefits for both environmental sustainability and economic efficiency. Sensors equipped with advanced imaging technologies, such as multispectral and hyperspectral cameras, can accurately identify and differentiate weeds (Fig. 9). This precision allows for targeted application of herbicides, where chemicals are only applied to the weeds detected. By avoiding blanket spraying, farmers can significantly reduce the overall amount of herbicides used, leading to more sustainable agricultural practices.

The overuse of herbicides has long been associated with negative environmental impacts, including soil degradation, water contamination and harm to non-target plant and animal species. By using sensors to detect and selectively treat weeds, the volume of herbicides released into the environment is minimised. This approach not only protects ecosystems but also promotes biodiversity by reducing the impact on beneficial plants and insects. On top of that, saving herbicide use will also have an impact on costs, since less inputs would be required.

© Burleigh Dodds Science Publishing Limited, 2025. All rights reserved.

Figure 9 Boom sprayer equipped with sensors to detect weeds.

5.2 Mechanical control

Mowers are a key tool in the mechanical control of groundcovers in woody crops. Various types of mowers are used, depending on the specific needs of the crop and the groundcover. Flail mowers (horizontal shaft hammers), e.g., are commonly used in orchards and vineyards because they effectively chop up and mulch groundcovers, leaving the organic matter on the soil surface to decompose and enhance soil fertility (Fig. 10). Rotary mowers with vertical shafts are also widely used, particularly for managing taller or denser groundcovers, as they provide a clean cut (Fig. 11).

There are different types of mowers that can be classified depending on the mobile element used for controlling the weeds or living groundcovers (Table 4).

As mentioned before, in order to prevent excessive growth of the cover, several passes with a mower should be carried out throughout the year. Managing the weed/groundcover with a mower may have some drawbacks:

- Possible regrowth after mowing increases water consumption, which can lead to a decrease in the yield of the olive grove.

© Burleigh Dodds Science Publishing Limited, 2025. All rights reserved.

Figure 10 Mechanical groundcover control with a horizontal shaft mower in almond trees.

- The remains of the groundcover are finely chopped up after clearing, which accelerates their degradation, reducing the protection of the soil.
- Clearing can cause a reversal of flora towards perennial weeds with a great capacity for regrowth or creeping species that may escape mechanical clearing.

Using the appropriate mower for the specific conditions of the woody crop system allows for precise control of groundcover growth. This ensures that the groundcover benefits, such as soil protection and weed suppression, are retained without compromising the health and productivity of the woody crops.

6 Conclusion

CA favours the presence of groundcovers in perennial crops, that include spontaneous weeds. Sown or spontaneous groundcovers control is key to maintaining sustainable agricultural practices, ensuring long-term crop vitality and enhancing overall farm profitability.

© Burleigh Dodds Science Publishing Limited, 2025. All rights reserved.

Figure 11 Vertical shaft chain mower.

While groundcovers can offer benefits like erosion control and soil moisture retention, they can also compete with perennial crops for essential resources such as water, nutrients and light. Uncontrolled groundcovers may hinder crop growth and reduce yields. By managing groundcovers effectively, farmers can strike a balance that maximises the benefits of groundcovers while minimising their potential negative impacts on perennial crops. Weeds compete with

Table 4 Type of mowers for mechanical weed or groundcover control.

Type of mower	Comment
Vertical shaft chain	Widely used, especially on soils with a large number of stones. They have limitations on very uneven terrain, so it's necessary for them to have multiple working bodies to ensure their length is not excessive.
Vertical shaft blade	Provides better cutting than the chain brush cutter by leaving less chopped residue, which improves its longevity. However, they do not work well with stones or on uneven terrain.
Horizontal shaft hammer	These offer the best cutting performance of all and are, therefore, highly recommended. They have limitations such as reduced width and require more power compared to other brush cutters.

© Burleigh Dodds Science Publishing Limited, 2025. All rights reserved.

crops for vital resources, which can significantly reduce yields and the overall health of perennial plants. Minimising weed competition allows perennial crops to establish and thrive, ensuring long-term productivity and economic stability for farmers. Overall, a well-executed weed control strategy is crucial for maximising crop performance, maintaining soil health, and supporting the ecological balance of agricultural systems.

7 References

AEACSV. 2024a. Manual about Best Management. Guide on practical application of Conservation Agriculture techniques Practices. (https://lifeagromitiga.eu/wp-content/uploads/2019/12/BMPs-Life-Agromitiga-Manual.pdf); accessed on August 26th 2024.

AEACSV. 2024b. Technical Guides (In Spanish). (https://agriculturadeconservacion.org/fichas-tecnicas); accessed on August 18th 2024.

Arenas Arenas, F.J., et al. (2015). Guía de Cubiertas Vegetales en cítricos. Sevilla (España): Junta de Andalucía. Consejería de Agricultura, Pesca y Desarrollo Rural Instituto de Investigación y Formación Agraria y Pesquera. (https://www.juntadeandalucia.es/agriculturaypesca/ifapa/servifapa/registro-servifapa/aa0532aa-eede-450d-b799-6ff1648c391b); accessed on July 23rd 2024.

Arquero Quilez, O., et al. 2015. Guía de Cubiertas Vegetales en almendro. Sevilla (España): Junta de Andalucía. Consejería de Agricultura, Pesca y Desarrollo Rural Instituto de Investigación y Formación Agraria y Pesquera. (https://www.juntadeandalucia.es/agriculturaypesca/ifapa/servifapa/registro-servifapa/78cbd014-6939-452d-b996-56478b48210f); accessed on July 23rd 2024.

EEA. 2024. State of Europe's Environment. (https://www.eea.europa.eu/en/topics/at-a-glance/state-of-europes-environment); accessed on August 21st 2024.

FAO. 2024. Conservation Agriculture. (https://www.fao.org/conservation-agriculture/en); accessed on August 1st 2024.

Jones, M. and Dehaan, L. 2017. Integrated Weed Management -for Sustainable Perennial Cropping Systems. *Agriculture, Ecosystems & Environment*, 240, 145–156.

Hatfield, L. and Walthall, C.L. 2014. Climate Change: Cropping System Changes and Adaptations. In Neal K. Van Alfen (Ed.), *Encyclopedia of Agriculture and Food Systems*, Academic Press, pp. 256–265.

Cui, L., et al. 2018. Development of Perennial Wheat Through Hybridization Between Wheat and Wheatgrasses: A Review. *Engineering*, 4(4), pp. 507–513.

Saavedra-Saavedra, M. 2017. 35 Years Researching and Experimenting on Groundcovers in Olive Groves (In Spanish). Agricultura de conservación: AC. (http://www.agriculturadeconservacion.org/images/descargas/Revistas_AC/REVISTA-AC35-web.pdf); accessed on July 16th 2024.

Smith, R., et al. 2011. Weed Management in Perennial Agriculture. *Journal of Sustainable Agriculture*, 35(2), 150–165.

© Burleigh Dodds Science Publishing Limited, 2025. All rights reserved.

Chapter 9

Weed management in no-tillage vegetable cultivation: using the Sistema de Plantio Direto de Hortaliças to achieve an agroecological transition

Marcelo Zanella and Darlan Rodrigo Marchesi, Agricultural Research and Rural Extension Company of Santa Catarina (EPAGRI), Brazil; Arcângelo Loss and Guilherme Wilbert Ferreira, Federal University of Santa Catarina State (UFSC), Brazil; and Jamil Abdalla Fayad, Agricultural Research and Rural Extension Senior, Brazil

1 Introduction

From its origins in 1950s, the 'green revolution' has become dominant in Brazil, backed by heavy public and private subsidies. Agriculture is now increasingly

http://dx.doi.org/10.19103/AS.2024.0145.15
© Burleigh Dodds Science Publishing Limited, 2025. All rights reserved.

controlled by large corporations that dominate the transnational agri-food market. In this model, plants other than those sown or planted in the cultivated area are considered undesirable and called weeds or invasive species. This means they must be eliminated because they compete with crops for resources (water, nutrients and light) and host insects and pathogens.

Green revolution agriculture is characterized by monocultures, where the elimination of spontaneously occurring plants (typically characterized as 'weeds') contributes to a decrease in plant and wider biodiversity. This lack of diversity affects the dynamic equilibrium of the agroecosystem, increasing the imbalances that manifest themselves in the form of diseases, pests and nutritional disorders. This logic of this form of cultivation presupposes and necessitates the intensive use of agrochemicals that both further disrupts agroecosystems and increases the economic and environmental costs of production, increasing the dependency and vulnerability of small family farms, in particular.

The transition from this conventional farming system to an agroecological model represents a profound change in the way we deal with food production. The cultivation of vegetables through the no-tillage cultivation system developed in Brazil (SPDH+: Sistema de Plantio Direto de Hortaliças) provides a promising alternative with a range of benefits for the environment, farmers and consumers (Fayad *et al*., 2019). SPDH+ cultivation has been developed since the 1990s as a method for agroecological transition with family farmers as producers of 'real food',[1] increasing the productivity of crops and livestock, and decreasing the economic and environmental cost of production. Based on a rich and growing body of knowledge, SPDH+ has established itself as a proven method for achieving an agroecological transition path in vegetable cultivation. It is now also being applied to the production of grains, fruits, milk, and meat. The development of SPDH+ is on two levels: political-pedagogical and technical-scientific. The first axis is based on the 'dialectical methodological conception' relating to social change, while the second axis is based on scientific and extension research relating to the 'promotion of plant health' (Fayad *et al*., 2019).

Plant are living beings. However, this can be lost in conventional agricultural and wider language. Statements such as 'that subject is in a vegetative state'; 'the soil provides food for the plant', 'the health of the plant depends on the health of the soil', 'weed plant' and 'cover plant' among others risk reducing the importance of the plant as a passive 'subject'. This reductionist view renders invisible the plant's ability to perceive its environment, generate information about it, and deal proactively with abiotic and biotic. challenges present in its environment. All these abilities of plants form the 'praxis' which we have

1 For SPDH+, real food means food that are produced without the use of agrochemicals.

© Burleigh Dodds Science Publishing Limited, 2025. All rights reserved.

termed 'sessile culture'. The denial of this reality has blurred the understanding that at the heart of an agroecosystem is the way plants actively form and transform their environment, including soil and the atmosphere. In this sense, the relationships between plants and other living things are those of cooperation and facilitation. This understanding overrides the utilitarian and reductionist view of competition between cultivated and spontaneous plants which underpins conventional monocultures. This misunderstanding reduces the dynamic balance and health of an agroecosystem, in favour of a production model dependent on herbicides, insecticides, fungicides and high loads of a few highly soluble mineral nutrients.

2 Promoting plant health: fundamentals

Understanding and promoting plant health, which lies at the center of an agroecosystem, requires following two lines of enquiry. The first is to deepen knowledge about the plant, its distinctive characteristics and the strategies it uses to grow healthily when facing the challenges of the agroecosystem environment. The second is to understand and improve the environment it inhabits in order to support healthy plant development. In practice, these two areas go together because plant evolution is the story of co-evolution with other organisms and with the environment in which it lives.

When they evolved from an original aquatic environment to a terrestrial one, more than 430 million years ago, plants began colonization based on a few species and relationships with a small number of other organisms, mainly fungi. Subsequent development and co-evolution has been marked by a huge increase in plant species diversity which, in turn, has promoted the increase in and diversity of other organisms. These increasingly complex relationships have allowed negative imbalances in ecosystems (in the form of competition, domination, and exploitation by one organism over another) to be minimized to maintain ecosystem stability. The evolutionary strategy of plants is based in part on strategies of cooperation, exchange, and mutual protection. A key feature of an ecosytem is the quality of relationships between species (rather than the number of species) in achieving ecosystem stability and resilience. This requires going beyond more narrow productivist and economic perspectives to understand the complex social relationships and environmental conditions that promote the health of agroecosystems and the plants that live within them.

Once plants left the aquatic to inhabit the terrestrial environment, they played a key role in helping to change the composition of the atmosphere, including the formation of the ozone layer, an increase of O_2 concentration up to 21%, reduction in the atmospheric concentration of CO_2 and other gases harmful to life, as well as production of high energy carbon chains to serve as food and oxidizing sites in the terrestrial environment. These new

© Burleigh Dodds Science Publishing Limited, 2025. All rights reserved.

conditions made the emergence of other life forms possible, as well as starting a revolutionary step in plant evolution by establishing a supportive habitat organically connecting the soil and atmosphere. These conditions relied on fundamental relationships of cooperation between species in order to create self-sustaining ecosystems (Fayad *et al.*, 2019).

As they evolved, plants built an increasingly complex root system to mine nutrients and absorb water from the soil. Roots also evolved to exude organic and mineral compounds. These contributed to increase the biodiversity of organisms in the rhizosphere, the area immediately around the roots. This has resulted in a developing symbiotic association with diverse soil microorganisms which have enhanced the ability of plants to access nutrients and respond to abiotic and biotic stresses. The rhizosphere can be called 'the home of multifunctional microorganisms', because it is an extremely complex region of the soil that hosts a diverse community of microorganisms, including bacteria, fungi, actinomycetes, and other microorganisms. The rhizosphere is a resource-rich environment where microorganisms find a source of carbon, nutrients, and moisture. The roots secrete organic compounds, such as root exudates, which attract and feed the microorganisms present in this environment. This close symbiotic relationship between plants and microorganisms benefits both, contributing to plant health and growth (Chamkhi *et al.*, 2022; Loss *et al.*, 2019).

The diversity of the soil microbiome has been found to be critical for the sustainable functioning of agroecosystems (Tian *et al.*, 2015). Rhizosphere microorganisms perform a number of vital functions. Microorganisms play an important role in nutrient cycling (Li *et al.*, 2021). For example, some nitrogen-fixing bacteria are able to convert atmospheric nitrogen into plant-usable forms, thus reducing the need for nitrogen fertilization. Other decomposing microorganisms, such as fungi and bacteria, play a key role in the decomposition of organic matter, releasing essential plant nutrients such as phosphorus. In addition, microorganisms present in the rhizosphere can help plants defend themselves against pathogens. Some species of beneficial microorganisms have the ability to produce antimicrobial substances or compete directly with pathogens, helping plants to resist disease. This beneficial interaction between plants and microorganisms in the rhizosphere is an important example of natural biological control (Lourenzi *et al.*, 2022; Igiehon *et al.*, 2018).

Plant root systems also work together with the aerial part of the plant which has the primary function of collecting solar energy, releasing O_2 and capturing CO_2 to build highly energetic carbon chains, as well as transpiring water vapour. The plant's vascular system has the capacity to transport, via water, nutrients to plant tissues. The combination of these two systems (linked by the vascular system), one based on soil and the other on the atmosphere, has transformed both soil and atmosphere, and places plants at the center of ecosystem function. Part of these two systems are the social relations built with

© Burleigh Dodds Science Publishing Limited, 2025. All rights reserved.

the community of fungi, bacteria, actinomycetes, nematodes, and animals both below and above the ground.

In these ways, plants have evolved the capacity to adapt to the challenges presented by their environment, from exploiting optimal growth conditions to dealing with stresses such as deficiency or excess of water, mineral nutrients, temperature, light; the physical condition of the soil, CO_2 and O_2 concentrations, salinity; biotic predators; and even cultivation practices introduced by humans such as pruning. Understanding these mechanisms of action, which we call 'sessile culture', is fundamental to building the SPDH+ cultivation system in order to facilitate these processes and relationships, promoting plant and broader ecosystem health and productivity.

A fundamental example of this capacity is to understand nutrition in ecophysiological terms, considering the following factors:

- The plant self-regulates its own absorption capacity as a function of its growth. This correlation between nutrient uptake and crop growth is expressed by the 'daily uptake rate' (DUR) and the 'absolute growth rate' (AGR) (Fayad *et al.*, 2019; Bouain *et al.*, 2019).
- Throughout the life cycle, the plant maintains a balance between daily biomass production and the amounts of absorbed nutrients (Fayad *et al.*, 2019).
- Managing plant nutrition is related to climatic conditions, mainly solar radiation and temperature, and plant signals in response to these conditions.
- Plant interactions and associations with fungi, bacteria, and other living soil organisms, which play a role in nutrient availability.
- Plant communities in the agroecosystem are able to modify the physicochemical properties of the soil matrix directly by root organic and inorganic substances and by total aerial and root dry mass (Wt) production or indirectly through the microbiome, leading to changes in nutrient availability.
- The plant's ability to utilize a nutrient is dependent on and proportionate to the availability of other nutrients (Fayad *et al.*, 2019; Lemaire *et al.*, 2019; Medici *et al.*, 2019).

Biodiversity is key to building resilient and productive agroecosystems. This includes promoting the range of organisms that have symbiotic relationships with plants. Amongst other things, this requires developing a plant diversification and rotation plan able to produce more than 10 tons of phytomass (dry mass) per hectare per year on the soil surface. This provides a suitable habitat and source of nutrition for a wide range of organisms as well as a providing a source of nutrition for plants themselves. Broad and diverse root systems promoted by

© Burleigh Dodds Science Publishing Limited, 2025. All rights reserved.

phytomass also increase root exudation of organic and mineral compounds, increasing the population and diversity of the soil microbiome, 'drilling and folding the mineral matrix' and ultimately helping to form and transform soil physical and chemical structure. Building phytomass thus contributes to a healthy, self-regulating and resilient agroecosystem.

3 Planning for a transition to Sistema de Plantio Direto de Hortaliças cultivation

Family farmers who historically manage areas in simplified systems, such as conventional monocultures, need to make fundamental changes if they wish to transition to more complex agroecosystems such as the SPDH+. To enable proper decision-making in a change of system, it is recommended to start by performing ground and aerial diagnostics of the cultivation area. At ground level, stratified soil sampling at 0–10 cm and 10–20 cm makes it possible to identify and correct mismatches in chemical fertility indicators. This may include measuring pH levels to correct to values between 5.5 and 6.0, measuring and improving the amount of nutrients such as P, Ca, Mg, and K to appropriate levels, depending on crop requirement. Another need is to identify compacted layers in the soil profile, which might impede root growth, requiring mechanical decompaction using a subsoiler (Comin *et al.*, 2024). Above ground analysis needs to take account of the spatial and temporal arrangement of crops in the landscape to optimize the health of both individual species and overall plant biodiversity. An example is sowing and establishment of green manures with due care in the number of species, number of seeds, and sowing time according to the season, appropriately integrated with planting and healthy growth of the next crop.

Based on the underpinning science, SPDH+ cultivation that successfully promotes plant health is based on the following factors (Fayad *et al.*, 2019):

- Minimization of nutritional, salinity, water availability, temperature, light, pH, and oxygen diffusion velocity stresses around the root system.
- Use of spatial arrangements associated with the architecture of the root system, the size of the plant, and the quantity of fruit, according to the needs of each crop. For example, pruning in Cucurbitaceae and the verticalization of conduction systems in tomatoes which balance root and aboveground growth.
- Plant nutrition based on growth rates and nutrient uptake, adjusted to environmental conditions (mainly light and temperature), the dynamics of the nutrients in the soil and to signals given by the plant itself relating, e.g., to nutrient deficiency.
- Addition of more than 10 tons of phytomass (dry mass) of cover plants per hectare and per year in rotation plans.

© Burleigh Dodds Science Publishing Limited, 2025. All rights reserved.

- Crop and green manure rotation (cultivated and spontaneous), with high diversity, which can evolve to include rotation with animals managed according to the Rational Voisin Grazing (RVG) system (developed by Andre Voisin), aiming to balance animal productivity and overall ecosystem sustainability (e.g. by allowing pasture sufficient time to recover between grazing periods).
- Soil disturbance restricted to the planting lines or supporting plants at the nursery stage.
- Cultivation of sown and spontaneous green manures, with selection of plants better adapted to local conditions and without harming production of the food-producing cash crop.
- Management of green manures and spontaneous plant growth in a synchronized operation, e.g., by rolling using a knife roller in the phase after sowing or planting of seedlings. This facilitates direct drill planting in the green cover, complemented, e.g., by mechanical mowing or cutting to manage the cover (so it co-exists with its companion cash crop) and remove the need for herbicides (Fig. 1).
- Decreasing the use of highly soluble, synthetic fertilizers and pesticides to the point that they can be eliminated altogether.
- Selection and production of healthy seedlings and seeds well adapted to abiotic and biotic stress conditions in the field.

Figure 1 Cultivation of bell pepper (*Capsicum annuum*, L.) where the spontaneous ones are in the phase of being managed with a backpack manual cutter or with a mini tractor (Credits: Municipality of Anitápolis, 2021).

© Burleigh Dodds Science Publishing Limited, 2025. All rights reserved.

- Modifying the composition of the wider landscape, by developing and linking woodland and other ecological corridors (e.g. field margins or riparian buffers) and integrating them with crops, livestock, roads, and buildings with the objective of promoting biodiversity and achieving more stable and milder microclimates (e.g. areas shaded by trees).
- (Re)establishment of the soil microbiome through plant diversity, the addition of green and organic manures, no-tillage, and continuous plant cultivation.

An example of applying these principles in practice is the cultivation of parsnip cassava (*Arracacia xanthorrhiza*) plants in mountainous areas in Brazil. This includes assessing nutritional requirements based on daily nutrient absorption rates, soil nutritional reserves, and reading plant signals related to nutritional status (e.g. leaf size and colour); use of crop rotation and cover crops to promote addition of dry matter exceeding 10 tons per hectare and per year; and minimal soil disturbance restricted to planting lines with maintenance of stable ridges over time (Fayad *et al.* 2019).

Effective monitoring of the success of implementing these principles can, e.g., be seen in the cultivation of cauliflower (*Brassica oleracea* var. botrytis) where the uniformity of the plants, the erect position of the leaves, the difference in green colour between new, mature leaves and lower leaves indicate that fertilization is carried out according to the daily uptake rate of nutrient absorption (DUR) and its appropriate adjustment. Mass retranslocation of nutrients thus does not occur (absence of yellowish lower leaves) (Fig. 2).

Figure 2 Cauliflower cultivation in mountainous areas under black oat (*Avena strigosa*) vegetation cover managed with a knife roller without the use of herbicide. (Credits: Municipality of Angelina, 2021).

© Burleigh Dodds Science Publishing Limited, 2025. All rights reserved.

Figure 3 Eggplant cultivation with spontaneous plants between the rows, close to the moment of mechanical management with a brush cutter (Credits: Municipallity of Águas Mornas, 2021).

Another example is growing eggplant (*Solanum melongena*) in monitoring the difference in green colour between mature and new leaves during fruiting. The dark green lower leaves indicate that fertilization matches DUR without signs of mass reallocation of nutrients from the old leaves to the fruits and new shoots (Fig. 3).

4 The role of green cover in Sistema de Plantio Direto de Hortaliças cultivation: using spontaneous plants ('weeds')

All plants, including so-called 'spontaneous plants', produce food for other organisms in the environment. Using photosynthesis, plants produce organic compounds exuded by the roots. The magnitude of this exudation can exceed the equivalent of 30% of the plant's photoassimilates (Marschner, 2012). These exudate compounds produced by a wide range of plants are fundamental to the biodiversity and functioning of the rhizosphere ecosystem (Lourenzi *et al.*, 2022; Loss *et al.*, 2019; Marschner, 2012).

To promote rhizosphere health, it is essential to adopt sustainable management practices in the cropping system. The type of agroecological management recommended by SPDH+ cultivation, such as the addition of organic matter to the soil and the use of crop rotations, contributes to the promotion of microbial diversity in the rhizosphere. In addition, reducing the use of pesticides and excessive chemical fertilizers can prevent the destruction or reduction of beneficial microorganisms present in the rhizosphere (Khmelevtsova *et al.*, 2022). Continuous soil cover using cover crops or

© Burleigh Dodds Science Publishing Limited, 2025. All rights reserved.

spontaneous plants also protects and supports the rhizosphere environment (Gao *et al*., 2022). This then promotes soil health which then boosts growth and yields of cash crops in soils cultivated using SPDH+ principles (Fayad *et al*., 2020; Loss *et al*., 2019).

In SPDH+ cultivation, the presence of permanent vegetation covering the soil contributes to the development of a healthy rhizosphere. This vegetation cover provides organic matter to the soil, feeding the beneficial microorganisms present in this region. A diversity of species also contributes to the promotion of rhizosphere health. Crop rotation prevents the excessive accumulation of species-specific pathogens and promotes microbial diversity in the soil. This microbial diversity benefits vegetables by helping to protect against disease and stimulate healthy root growth (Du *et al*., 2021; Berg *et al*., 2017). Maintaining a healthy rhizosphere in SPDH+ cultivation requires management that combines agroecological practices, such as the use of mulch, crop rotation, and control of soil compaction. In addition, the adoption of soil and rhizosphere analysis techniques, such as nutrient analysis and microbiological analysis, can assist farmers in making decisions about proper rhizosphere management in SPDH+ cultivation.

Biodiversity has been shown to be very sensitive to agricultural practices (Bedano *et al*., 2021). Structural and functional changes in the soil microbiome can serve as an indicator of the impact of different agricultural methods on soil quality (Wang *et al*., 2021).

Agricultural practices used in conventional tillage systems (ploughing, harrowing and synthetic fertilizers) can negatively affect the richness and diversity of the soil microbiome. Conventional tillage (ploughing, harrowing and without use of cover plants) has been found to significantly reduce the richness and diversity of microbial species compared to minimum cultivation and promotion of soil cover (Legrand *et al*., 2018). For example, the continued use of inorganic fertilizers (NPK) for 34 years has been found to reduce the biodiversity and abundance of bacteria in Chernozems (a class of soil), with this reduction being more significant for the diversity and total number of bacterial species (Zhou *et al*., 2015).

To circumvent this problem, we need to use soil management systems that promote and preserve the structural and functional richness of the soil, such as SPDH+ cultivation. Practices such as cover crops and crop rotation promote the soil microbial community (Khmelevtsova *et al*., 2022). The use of crop rotation and cover crops favours soil microorganisms, which are often antagonists of pathogens and, therefore, are capable of suppressing diseases of agricultural plants (Peralta *et al*., 2018). Özbolat *et al*. (2023) found that, after 10 years of reduced soil preparation with management of spontaneous vegetation in orchard areas, the bacterial community increased significantly, with emphasis on bacteria beneficial to crop growth.

© Burleigh Dodds Science Publishing Limited, 2025. All rights reserved.

Green manures make it possible to keep soil covered all year round (as seen in Figures 2 and 3) but need proper management. This involves identifying species that can add physical (protection of the soil surface, reduction of temperature extremes, reduction of erosion, structuring of the soil profile), chemical (cycling and accumulation of nutrients and soil organic matter (SOM)), and biological (increase in activity and multiplication of soil microbiota) benefits during the development of the commercial crop, whilst promoting a favourable environment for the elimination of herbicides. The vegetable crops grown in SPDH+ system are one element in an overall plant composition (including plants providing soil cover) that promotes biodiversity and overall agroecosystem function. The establishment and management of green manures should therefore be carried out with the same care as commercial crops.

A key requirement is to understand and categorize vegetable species according to their development cycle, growth characteristics and how this relates to the wider growing environment, including how best they will fit in with green manures and spontaneous plant cover in a way that is mutually beneficial. An example is vertical cultivation of crops of the Solanaceae and Cucurbitaceae families (e.g. using trellises) which facilitates management of green cover with a knife roller and mower, enabling a 'planting-green' no-till system without herbicides. On the other hand, Brassicas require greater care related to weed management than the previous ones, particularly Liliaceous and leafy crops, such as Asteraceous (lettuce) and brassicaceous (aragula, cauliflower etc) (Fayad *et al.*, 2019). This requirement for differentiated management is partly related not only to the crops grown but also to the rotation scheme adopted relating to the timing of crop and green manure cultivation. In addition, it is essential to adjust plant arrangements and densities, how and when to manage cover crops and green manures, plant nutrition strategies and the amount of the phytomass retained on the soil surface (Fig. 4).

The management of plants providing soil cover depends on three fundamental parameters: quantity, intensity and functionality of the phytomass. Quantity is related to establishing the minimum threshold for the system's emergent properties[2] to manifest themselves. Intensity relates to achieving the greatest possible photosynthesizing plant activity over the year necessary for functional biodiversity. Functionality relates the other parameters to the contribution of the phytomass to optimizing agroecosystem function which benefits vegetable cultivation as one component in the system. This requires smart choices in selection and management of cover plants (including spontaneous plants) to achieve adequate amounts of phytomass throughout

2 Emergent properties are a consequence of the relationship between the components of the system. They can therefore only be accessed and measured once the components are integrated into the system. They appear when all parts of a system work together to achieve some objective.

© Burleigh Dodds Science Publishing Limited, 2025. All rights reserved.

Figure 4 Planting of bell pepper (*Capsicum annuum*, L.) after lodging of green manures without the use of herbicides (Credits: Municipality of Anitápolis, 2023).

the year to promote the health of both the commercial crop and the wider environment. Table 1 suggests potential combinations of main and cover crops to achieve functional rotations within SPDH+ cultivation.

In tropical and subtropical environments, the rotation plan is designed to produce at least 10 tons of dry phytomass per hectare and per year (Fayad *et al.*, 2019). This amount can reach more than 20 tons per hectare and per year when, in the summer season, we associate cultivation of cover plants with the occurrence of spontaneous plants, such as papuan (*Brachiaria plantaginea*), mucuna (*Strizolobium aterrimum*) or crotalaria (*Crotalaria juncea*), millet (*Penissetum Americanum*) or maize (*Zea mays*) and sunflower (*Helianthus annuus*). This scheme is even more enriched when in winter we include cultivation of black oats (*Avena strigosa*), rye (*Secale cereale*), vetch (*Vicia sativa*), ryegrass (*Loliun multiflorum*) and forage radish (*Raphanus sativus*).

Cutting green manures from the milky grain stage onwards (at the early grain-filling stage after flowering) using a knife roller allows better uniformity of coverage which also delays the appearance of spontaneous plants, favouring the initial development of the commercial crop. Rolling occurs either simultaneously or just before opening of the furrow and planting of vegetables. Long-cycle vegetables, such as Solanaceae, Fabaceae and cucurbits that use, e.g., trellises (cucumber and chayote) allow establishment of spontaneous plants which need to be managed with a brush cutter or mechanized mower (Fayad *et al.*, 2019).

© Burleigh Dodds Science Publishing Limited, 2025. All rights reserved.

Table 1 Indications in the use of diversifier plants in production systems by the vegetable group

Main crop	Indication in the use of diversifier plants		
	Cover crops species	Management approach	Notes
Tomatoes, bell string, beans	Fall/winter: black oat (*Avena strigosa*), white oat (*Avena sativa* L.), rye (*Secale cereale*), ryegrass (*Loliun multiflorum*), radish forage (*Raphanus sativus*). Spring/summer: Millet (*Penissetum Americanum*), papuan (*Brachiaria plantaginea)*, maize (*Zea mays*), sunflower (*Helianthus annuus*).	Roll the plants in the milky grain stage. Immediately afterwards, open up the rows for planting/seeding with green plants.	If necessary, in the initial phase, carry out one or two mowings between the rows of vegetables.
Cauliflower, broccoli, cabbage, moganga, pumpkin and watermelon	Fall/winter: black oat (*A strigosa*), white oat (*A. sativa* L.), rye (*S. cereale*), fodder pea (*Pisum sativum* subsp. *Arvense*), radishforage (*R. sativus*) Spring/summer: millet (*P. Americanum*), papuan (*B.plantaginea*), maize (*Z. mays*), sunflower (*H. annuus*).	Roll the plants in the milky grain stage. Immediately after rolling, open the rows for planting/seeding with green plants.	Generally, it is sufficient to carry out only one mowing in between the crop rows.
Lettuce	Spring/summer: Millet (*P. Americanum*), papuan (*B. plantaginea*), buckwheat (*Fagopyrum esculentum*). Fall/winter: black oat (*A. strigosa)*, white oat (*A. sativa* L.), rye (*S. cereale*).	In the case of using cultivation beds, sow the diversifier plants on the bed. Roll the plants in the dough grain stage and transplant them immediately afterwards.	It is important to obtain an adequate density of plants (green manures) so that the vigorous rooting helps the Soil structure.
Onion	Spring/summer: millet (*P. Americanum*), papuan (*B. plantaginea*), velvet bean (*M. pruriens*). Fall/winter: black oat (*A. strigosa*), white oat (*A. sativa* L.), rye (*S. cereale*).	Rolling of the plants in the milky grain stage. Immediately after rolling, open the rows for planting- green.	In the initial phase, carry out one or more cuts in between the rows, if necessary.
Cassava and arracacha or Peruvian carrot	Spring/summer: millet (*P. Americanum*), papuan (*B. plantaginea*). Fall/winter: black oat (*A. strigosa)*, white oat (*A. sativa* L.), rye (*S. cereale*).	Roll the plants at the dough grain stage and plant on the same day.	Generally, one or two inter-row cuts are sufficient.

© Burleigh Dodds Science Publishing Limited, 2025. All rights reserved.

5 Case studies: management of spontaneous plants in the cultivation of Solanaceae and Fabaceae

Plants from the botanical families Solanaceae and Fabaceae are cultivated in summer using practices such as trellising. In the specific case of tomatoes, peppers and beans, vertical cultivation promotes plant health and promotes coexistence with diversifying plants. However, these crops require more rigor in planning the rotation with green manures so that they last as long as possible, i.e. until the end of the harvest. The composition of the winter combination or cocktail of plants (with a higher percentage of grasses) and their termination should occur as late as possible, when they are more fibrous (at the dough grain stage).[3]

Throughout the growing cycle of these crops, spontaneous vegetation and other long cycle cultivated plants such as vetch (*V. sativa*) and ryegrass (*Lolium multiflorum*) emerge, develop and must be managed with a mower. Management that prioritizes coexistence of certain spontaneous and cultivated green manures with this group of food-producing crops is a central part of the planting-green no-tillage vegetable (SPDH+) cultivation system (Fig. 5).

Examples of common green manures used in southern Brazil are black and white oats (A. *strigosa* and *A. sativa*), rye (*S. cereale*), vetch (*V. sativa*), ryegrass (*L. multiflorum*), and forage radish (*R.sativus*). In the intercropping system, there is a preference for a mixture of black oats + vetch + forage radish sown over the straw of summer crops such as papuan (*B. plantaginea*), mucuna (*S. aterrimum*) or crotalaria (*C. juncea*), millet (*Penissetum americanum*) or corn (*Zea mays*) and sunflower (*Helianthus annuus*).

6 Case studies: management of spontaneous plants in the cultivation of Brassicae and Cucurbits

Brassicas are grown in summer, mid-season and winter; have a short-to-medium growth cycle and are harvested before the emergence of the floral stem (cabbage) or in the immature flower stage (cauliflower and broccoli). In contrast, cucurbits are grown in the spring/summer and are grown along the ground like strawberries, pumpkins and watermelon. These crops are more demanding to manage, especially in pollination, nutrition and pruning.

More rigorous planning of the rotation with green manures is required and results in a thicker mat of plants covering the soil. This mat should be kept until harvest which means avoiding long-cycle species such as vetch and ryegrass. A winter cocktail made up of grasses, forage radish and forage peas can be used,

3 Managing plant comfort: Conceptually, it refers to the set of practices for preparing and managing the growing environment that result in health for the plants. In commercial crops, plant health translates into increased yield, reduced diseases and pest attacks, reduced production costs (economic, environmental and social) and better quality of the food produced.

© Burleigh Dodds Science Publishing Limited, 2025. All rights reserved.

Figure 5 Growing chayote (*Sechium edule*) (a) and tomato (*Solanum lycopersicum*, L.) in planting-green SPDH+ cultivation (b) (Credits: Municipality of Anitápolis, 2019).

Figure 6 Cauliflower (*Brassica oleracea*, var. botrytis) and tetsukabuto hybrid moganga (*Curcubita maximus*) crops with sufficient straw accumulation for spontaneous plants management without herbicide use (Credits: Municipality of Angelina, 2021).

but preference should be given to grasses such as black oats + rye, which are rolled as late as possible, when they are more fibrous (at the floury grain stage) (Fig. 6). In both families, brassicae and cucurbits, it is possible to roll the cover crops only once during the crop cycle.

7 Case studies: management of spontaneous plants in the cultivation of leafy vegetables and Liliaceae (*Allium cepa* L.)

Short-cycle leafy vegetables, such as lettuce (*Lactuca sativa*), are grown in winter, mid-season and summer. A green manure cycle can be combined with one or more commercial crop cycles interspersed throughout the year. This is the case for lettuce, which has a small size and a shorter cycle. Its cultivation on grass litter, e.g., makes it possible to carry out up to two cycles. Liliaceous plants, such as onions, are grown in winter, with a small size and long growing cycle. This requires careful management of crop rotation and green manures to manage emergence of spontaneous plants in the Spring (Fayad *et al.*, 2018a). Green manures need to produce greater matt thickness on the soil surface that

© Burleigh Dodds Science Publishing Limited, 2025. All rights reserved.

Figure 7 A second planting of lettuce on the same green manure straw grown in flowerbeds (a) and onions on summer plus winter green manure straw with rye (b) (Credits: Municipality of Antonio Carlos and Ituporanga, 2023).

lasts until harvest. Summer green manure crops, such as millet (*P. Americanum*), papuan (*B. plantaginea*), maize (*Z. mays*), and sunflower (*H. annuus*), should be planted first. These crops should be managed with a knife roller and then the rye should be sown; the rye should then be rolled when the onion seedlings are planted (Fig. 7).

8 Case studies: management of spontaneous plants in cassava (*Manihot esculenta*) and arracacha or Peruvian carrot (*Arracacia xanthorrhiza*, Bancroft) cultivation

Manihot esculenta (Cassava) and *Arracacia xanthorrhiza* Bancroft are plants with a long cycle (more than 240 days). Cassava is planted in the spring/summer and harvested in an annual or biannual cycle (Fayad *et al.*, 2018b), while *Arracacia* spp is planted in winter and harvested in the following autumn/winter (Fig 8).

Figure 8 Long-cycle plants need more fibrous green manures in quantities greater than 10 tons per hectare, and spontaneous ones can be managed with one and, at most, two cuts (Credits: São João Batista and Major Gercino, 2018).

© Burleigh Dodds Science Publishing Limited, 2025. All rights reserved.

More lignified green manures, such as those derived from grasses, are recommended because they last longer, covering the soil and preventing early appearance of spontaneous plants. It is recommended to cultivate green manures (Poaceas) in the summer, such as *P. americanum*, and reseed grasses in the winter, such as *A. strigosa* (Schreb), using around 80 kg per hectare and 120 kg per hectare, respectively, via broadcast sowing. Green manures can be rolled with a knife roller on the same day as commercial crops are planted. This favours the planting operation and favours maintaining phytomass on the soil. This favours weed control, reducing the need for herbicides.

9 Conclusion

The implementation of sistema de plantio direto de hortaliças (SPDH+) offers a sustainable alternative for weed management without relying on herbicides. By maintaining permanent soil cover with cover crops, no-till suppresses weed germination and growth through physical and allelopathic effects. This approach reduces the weed seed bank over time, minimizing the need for mechanical or chemical interventions.

Despite these advantages, transitioning to SPDH+ requires strategic planning, particularly in selecting cover crops that maximize weed suppression while supporting soil fertility. Crop rotation, intercropping, and innovative mulching techniques play a crucial role in enhancing the effectiveness of no-till weed control.

Future research should further explore the interactions between cover crops and weed dynamics, optimize residue management for different vegetable species, and develop adaptive strategies for diverse climatic and soil conditions. As sustainable agriculture continues to evolve, SPDH+ stands as a viable pathway to reducing herbicide dependence while maintaining productivity and soil health.

10 Acknowledgements

The authors would like to thank the National Council for Scientific and Technological Development (CNPq) for the financial support received, referring to the research productivity grant for the fourth author (process no. 311474/2021-7) and the CNPq postdoctoral grant for the fourth author (process no. 150956/2024-0).

11 References

Bedano, J.C. Soil biodiversity for the sustainability of agroecosystems. *Acta Oecol.* 2021;110:103705.

© Burleigh Dodds Science Publishing Limited, 2025. All rights reserved.

Berg, M., et al. Plant microbial diversity is suggested as the key to future biocontrol and health trends. *FEMS Microbiol Ecol.* 2017;93(5):fix050. https://doi.org/10.1093/femsec/fix050.

Bouain, N., et al. Getting to the Root of Plant Mineral Nutrition: Combinatorial Nutrient Stresses Reveal Emergent Properties. *Trends in Plant Science*. 2019;24(6):542-552. https://doi.org/10.1016/j.tplants.2019.03.008

Chamkhi, I., et al. Is the rhizosphere a source of applicable multi-beneficial microorganisms for plant enhancement? *Saudi J Biol Sci.* 2022;29(2):1246–1259. 10.1016/j.sjbs.2021.09.032.

Comin, J., et al. Soil Health in No-tillage Vegetable Production Systems– SPDH. In: Mendes, I.C.; Cherubin, M.R. (Eds.). Soil Health Series: Volume 3 Soil Health and Sustainable Agriculture in Brazil. ASA, CSSA, and SSSA Books, 2020. https://doi.org/10.1002/9780891187448.ch7

Du, J.X., et al. Synergistically promoting plant health by harnessing synthetic microbial communities and prebiotics. *iScience*. 2021;24(8):102918. 10.1016/j.isci.2021.102918.

Epstein, E. & Bloom, A.J. *Nutrição mineral de plantas*. 2ª ed. Londrina: Editora Planta, 2004, p. 401.

Fageria, N.K. & Moreira, A. The role of mineral nutrition on root growth of crop plants. *Adv Agronom.* 2011;110:251–331.

Fayad, J.A., et al. *Sistema de Plantio Direto de Hortaliças: Método de transição para um novo modo de produção*. Epagri: Florianópolis, 2019, p. 428.

Fayad, J.A., et al. (coord). *Sistema de Plantio Direto de Hortaliças (SPDH): O cultivo do tomate*. Florianópolis: Epagri, 2016, p. 96. (Epagri. Boletim Didático, 132).

Fayad, J.A., et al. *Sistema de plantio direto de hortaliças (SPDH): o cultivo da cebola*. Florianópolis: Epagri, 2018a, p. 78. (Epagri. Boletim Didático, 146).

Fayad, J.A., et al. *Sistema de plantio direto de hortaliças (SPDH): o cultivo da mandioquinha-salsa*. Florianópolis: Epagri, 2018b, p. 68. (Epagri. Boletim Didático, 147).

Fayad, J.A., et al. *Management of Vegetable Conservation Agriculture Systems*. Burleigh Dodds Series in Agricultural Science. 01ed. Burleigh Dodds Science Publishing, 2020, v.1, pp. 279–326.

Gao, H., et al. Cover crop species composition alters the soil bacterial community in a continuous pepper cropping system. *Front Microbiol*. 2022;12:789034. https://doi.org/10.3389/fmicb.2021.789034.

Igiehon, N.O. & Babalola, O.O. Rhizosphere microbiome modulators: Contributions of nitrogen fixing bacteria towards sustainable agriculture. *Int J Environ Res.* 2018;15(4):574. https://doi.org/10.3390/ijerph15040574.

Kerbauy, G.B. *Fisiologia vegetal*. 2ª ed. Rio de Janeiro: Guanabara Koogan, 2013, p. 431.

Khmelevtsova, L.E., et al. Influence of agricultural practices on bacterial community of cultivated soils. *Agriculture*. 2022;12(3):371. https://doi.org/10.3390/agriculture12030371.

Lemaire, G., *et al.* Allometric approach to crop nutrition and implications for crop diagnosis and phenotyping. A review. *Agron. Sustain. Dev.* 2019;39:27. https://doi.org/10.1007/s13593-019-0570-6

Legrand, F., et al. Effect of tillage and static abiotic soil properties on microbial diversity. *Appl Soil Ecol*. 2018;132:135–145.

© Burleigh Dodds Science Publishing Limited, 2025. All rights reserved.

Li, M., et al. Fifteen-year no tillage of a Mollisol with residue retention indirectly affects topsoil bacterial community by altering soil properties. *Soil Tillage Res.* 2021;205:104804.

Loss, A., et al. Rizosfera e as reações que ocorrem no seu entorno. In: Fayad, J.A.;Arl, V.; Comin, J.J.; Mafra, A.L. Marchesi, D.R. (Org.). *Sistema de plantio direto de hortaliças : método de transição para um novo modo de produção*. 01ed. São Paulo: Expressão Popular, 2019, v. 01, pp. 175–212.

Lourenzi, C.R., et al. The role of PGPR secondary metabolites in alleviating allelopathic effects (biotic stress) and induced tolerance in plants. In: Sayyed, R.Z., Uarrota, V.G. (Org.). *Secondary Metabolites and Volatiles of PGPR in Plant-Growth Promotion*. 1ed. Cham: Springer, 2022, v. 1, pp. 133–152.

Marschner, H. *Marscher's Mineral Nutrition of Higher Plants*. 3rd ed. Amsterdam: Elsevier, Academic Press, 2012, p. 651.

Medici, A., et al. Identification of Molecular Integrators Shows that Nitrogen Actively Controls the Phosphate Starvation Response in Plants. The Plant Cell. 2019;31(5):1171–1184. https://doi.org/10.1105/tpc.18.00656

Özbolat, O., et al. Long-term adoption of reduced tillage and green manure improves soil physicochemical properties and increases the abundance of beneficial bacteria in a Mediterranean rainfed almond orchard. *Geoderma*. 2023;429. https://doi.org/10.1016/j.geoderma.2022.116218.

Pausch, J. & Kuzyakov, Y. Photoassimilate allocation and dynamics of hotspots in roots visualized by 14C phosphor imaging. *J Plant Nutr Soil Sci.* 2011;174:12–19.

Peralta, A.L., et al. Crop rotational diversity increases disease suppressive capacity of soil microbiomes. *Ecosphere*. 2018;9:9.

Taiz, L. & Zeiger, E. *Fisiologia vegetal*. 5. Ed. Porto Alegre: Artmed, 2013, p. 918.

Tian, W., et al. Responses of microbial activity, abundance, and community in wheat soil after three years of heavy fertilization with manure-based compost and inorganic nitrogen. *Agric Ecosyst Environ.* 2015;213:219–227.

Wang, H., et al. Eight years of manure fertilization favor copiotrophic traits in paddy soil microbiomes. *Eur J Soil Biol.* 2021;106:103352.

Yin, Y.G., et al. Visualising spatio-temporal distributions of assimilated carbon translocation and release in root systems of leguminous plants. Nat Sci Rep. 2020;10:8446.

Zanella, M., et al. Cultivo de brássicas: couve-flor, Brócolis e repolho. In: *SPDH: Método de transição para um novo modo de produção*. Florianópolis, 2019, pp. 327–357.

Zhou, J., et al. Influence of 34-years of fertilization on bacterial communities in an intensively cultivated black soil in northeast China. *Soil Biol Biochem*. 2015;90:42–51.

© Burleigh Dodds Science Publishing Limited, 2025. All rights reserved.

Chapter 10

Weed management in organic Conservation Agriculture systems

Paolo Bàrberi, Stefano Carlesi and Federico Leoni, Scuola Superiore Sant'Anna, Italy

1 Introduction

Awareness of the key function that soils and their biodiversity (accounting for 25% of the total biodiversity on Earth (European Commission, 2022)) play in agricultural productivity and the health of the entire planet is increasing. This has been translated into important regulatory initiatives and actions at global and continental levels to maintain or increase soil health. For example, goals #6 (drinking water and sanitation), #12 (responsible production and consumption), #13 (climate action) and #15 (life on earth) of the United Nations' 17 Sustainable Development Goals (SDGs) for 2030 (United Nations, 2015) are directly related to soil health; and others, including goal #2 (zero hunger), depend indirectly on soil quality and health.

The current European policy has been strongly inspired by the SDGs and is backed by the Farm to Fork and Biodiversity 2030 strategies. In these documents, on the one side, the importance of improving soil health emerges in all its centrality, and on the other side, organic agriculture is considered one of the most important approaches to reach this goal. The Farm to Fork strategy

http://dx.doi.org/10.19103/AS.2024.0145.16
© Burleigh Dodds Science Publishing Limited, 2025. All rights reserved.

aims for 25% of the total agricultural area in the EU managed under organic agriculture methods by 2030. The importance of this focus is further confirmed by the new European Soil Strategy (EU Soil Strategy for 2030), closely linked to two other strategies, Adaptation to Climate Change and the Zero Pollution action plan, in which the reduction of nutrient losses and pesticide use takes an important role (European Commission, 2021).

This massive effort on the regulatory side is finding adequate correspondence in programmatic research directions at the European scale. Over the past decade, the number of soil-focused research projects funded by the EU has increased considerably, and soil health topics have been among the prevailing ones in calls for proposals under the Horizon 2020 and Horizon Europe framework programmes dedicated to agriculture. As further evidence of this, soil is the theme of a specific Mission of Horizon Europe, a new funding instrument for the period 2021–2027 that aims to develop concrete solutions to major societal problems for 2030 through collaboration between researchers and different actors, including citizens. The Soil Mission, called "A Soil Deal for Europe", consists of the following objectives: to reduce desertification, conserve the stock of organic matter, reduce compaction and promote reuse of urban soils, reduce pollution and promote soil restoration, prevent erosion, improve the structure and increase biodiversity, reduce the global footprint on soil in the EU and improve knowledge about soil in society (European Commission, 2022). The Soil Mission is also a flagship initiative for a long-term vision of rural area development and aims to create 100 living labs and lighthouses. Living labs are territories in which innovative and virtuous soil management practices will be co-created and tested in different experimental sites under real-world conditions through collaboration among different actors such as researchers, farmers, foresters, land planners and managers, nongovernmental organizations and citizens. Lighthouses are individual sites (e.g. a farm or park) where best practices will be implemented and disseminated through collaboration between researchers and site managers.

Many of these initiatives consider conservation agriculture and regenerative agriculture, especially when framed in the context of organic agriculture, very important production approaches to meet these key soil-related objectives. To date, however, the take up of these methods by farmers is quite limited, mainly for two reasons:

- The difference between conservation and regenerative agriculture is quite blurred, and the latter method still does not even have a commonly agreed definition.
- Farmers fear that they may not adequately control weeds when shifting to no-till, mulch-based systems, especially where chemical herbicide use is forbidden, like in organic agriculture.

© Burleigh Dodds Science Publishing Limited, 2025. All rights reserved.

In this chapter, we first shed some light on the similarities and differences between organic conservation and regenerative agriculture. Then, we frame the issue of weed management in organic Conservation Agriculture systems by focusing on specific weed problems that may arise and on the toolbox available to farmers to minimize them. Then, we focus on the two most promising components of the weed management toolbox, *viz*. cover crops/dead mulches and intercropping/living mulches. Finally, we highlight some opportunities, challenges and future perspectives related to weed management in organic conservation and regenerative agriculture systems.

2 Organic conservation vs regenerative agriculture

Among the approaches that aim to improve the quality of agricultural soils, Conservation Agriculture and regenerative agriculture take a central place. Both aim to improve soil health through a combination of management practices that enhance the carbon storage (sink) capacity of the soil.

According to the FAO definition, Conservation Agriculture is an approach to economically sustainable agricultural production that aims to conserve natural resources while achieving high and sustainable yields and environmental conservation (Hobbs, 2007). The aim is to activate natural biological processes in and above the soil, minimizing mechanical tillage and using external inputs (synthetic or organic fertilizers and pesticides) in an optimal manner such that they do not interfere with the aforementioned biological processes. Conservation Agriculture is based on the integration of three basic principles: (i) posing minimal mechanical disturbance to the soil; (ii) achieving permanent vegetation cover; and (iii) employing diversified crop rotations (in annual crops) or plant associations (in perennial crops). It should be noted that although the FAO definition is very clear about the need to integrate these three principles, several researchers, farmers and technicians have often identified conservation agriculture with the exclusive application of reduced tillage or no-tillage techniques, often combined with the use of glyphosate in herbicide-resistant transgenic crop systems (Nalewaja, 2003). This interpretation has completely misrepresented the intentions of conservation agriculture, effectively leading to the simplification of cropping systems instead of their diversification, which has caused numerous technical and environmental problems, such as the uncontrolled increase in the use of glyphosate – by 15-fold between 1996 and 2016 (Benbrook, 2016) – and the resulting exponential increase in weed biotypes that have become resistant to this herbicide, especially in regions where glyphosate-resistant transgenic crops are grown (Heap, 2024). It is clear that in organic farming, since synthetic herbicides cannot be used, only the combination of reduced- or no-tillage and plant cover in diversified rotations (i.e. the application of true Conservation

© Burleigh Dodds Science Publishing Limited, 2025. All rights reserved.

Agriculture) can ensure satisfactory yields and incomes, an aspect highlighted in numerous works, such as the meta-analysis by Cooper et al. (2016).

Regenerative agriculture is an emerging production approach, for which the interest of both the scientific and operational worlds is increasing. The number of scientific articles published where the term is used explicitly has increased 8-fold from 2019 to 2022, although they still account for less than 0.4% of the total articles pertaining to agriculture (Fig. 1). To date, there is no universally accepted definition of regenerative agriculture. From the perspective of the principles underlying this approach, the overlaps with conservation agriculture are obvious, as the three core principles of the latter are also key in regenerative agriculture. Recently, a review paper found that there are numerous definitions and facets of the use of the term, within both the scientific literature and the companies and associations involved in promoting regenerative agriculture (Newton et al., 2020). However, in our opinion, there are three differences that differentiate regenerative agriculture from Conservation Agriculture: (i) a fourth basic principle: the importance of integration between crops and livestock; (ii) a greater emphasis on the importance of greatly reducing and, if possible, eliminating the use of external inputs such as fertilizers and pesticides, while maximizing biomass recycling within a farm or territory, in a circular economy perspective; and (iii) the explicit reference to the importance of creating a network of relationships between producers, citizens and other local actors,

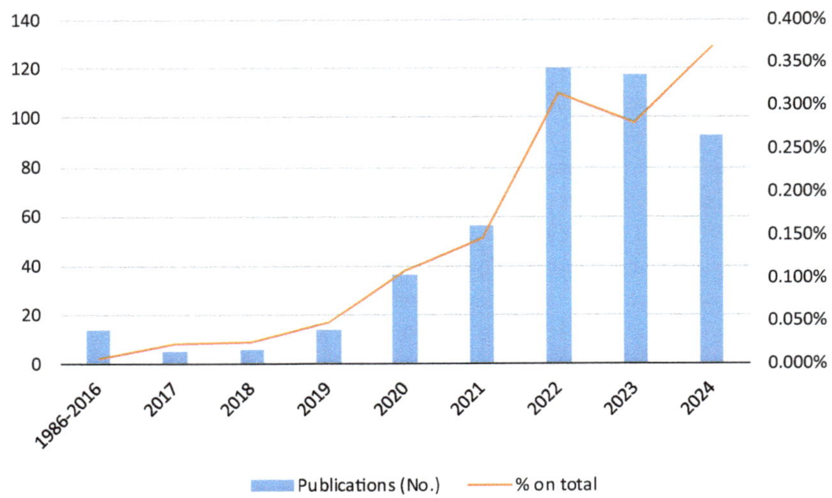

Figure 1 Number of articles published from 1986 to date that include the term 'regenerative agriculture' in the title, abstract or keywords (source: Scopus™, accessed July 29, 2024). The left and right vertical axes show, respectively, the total number of articles and the percentage of these out of the total number of articles reporting the term 'agriculture'.

© Burleigh Dodds Science Publishing Limited, 2025. All rights reserved.

within a short or very short supply chain perspective, which is absent in the definition of conservation agriculture.

Thus, regenerative agriculture appears to be an evolution of conservation agriculture towards a more agroecological approach; in fact, all basic principles of regenerative agriculture are perfectly aligned with the modern view of agroecology (see, e.g. Wezel et al., 2009) and, therefore, appear particularly suitable for application to those types of organic agriculture that are particularly sensitive to these issues.

In Europe, one of the main reasons given for supporting organic farming is precisely its perceived contribution to improving soil health and the environment in general (Reganold and Wachter, 2016). However, the application of conservation and regenerative farming methods in organic cropping systems, although increasing, is still at a sub-optimal level, despite a rising use of cover crops and more diverse rotations, because of the reluctance to abandon ploughing as the standard main tillage technique. Ploughing is considered, with good reason, to be a key technique for weed control, an aspect often highlighted by producers as among the major technical issues of the organic method (Bàrberi, 2002). In addition, it is considered important for the incorporation of soil amendments (e.g. manure and compost) and other biomasses, such as in green manuring. Although experimental evidence has shown that reduced tillage or no-tillage techniques combined with cover crops can be successfully applied in organic annual (e.g. Ranaldo et al., 2020) or perennial (e.g. Warren Raffa et al., 2021) cropping systems, many organic farmers are still reluctant to abandon the plough. By doing so, they fail to take advantage of the vast possibilities offered by the agroecological approach for improving the management of organic systems and expose themselves to justified criticism on the environmental impact of their practices, including increased greenhouse gas emissions and soil structure deterioration related to excessive reliance on tillage (Bàrberi, 2006).

In the recent past, there has been an increasing amount of research conducted on the possible combination of organic and conservation agriculture. As mentioned above, such a 'marriage' for years was considered impractical because of the dependence of organic systems on frequent and, often, intensive tillage required for weed control and the incorporation of soil amendments. The trait-filtering effect on weeds exerted by reduced- and no-tillage techniques, which results in the selection of geophyte (perennial) species (e.g. *Cynodon dactylon* (L.) Pers.) is well known, often necessitating the use of systemic herbicides for their control (Nichols et al., 2015).

Recently, however, mechanical innovations and the increasing focus on agroecological aspects of cropping systems management have prompted many research groups already active in organic farming to investigate these issues, with thrust often coming from organic practitioners. The first literature

© Burleigh Dodds Science Publishing Limited, 2025. All rights reserved.

review on the subject was carried out by Peigné et al. (2007), who made a clear picture of the technical and agro-environmental opportunities, limitations and barriers to the application of conservation techniques in organic farming. The TILMAN-ORG (2012–2014) project (www.tilman-org.net) addressed for the first time in a systematic way the study of the effects of conservation practices applied to herbaceous cropping systems on crop yields, chemical, physical and biological soil fertility and weed dynamics in a European network of field experiments, either short/medium term or long term. The main results of the project were compared to the available literature in Cooper et al.'s meta-analysis (2016), which showed how the application of minimum tillage techniques (shallow tillage with or without layer inversion and medium-deep tillage without layer inversion, e.g. chiselling) can be an excellent compromise to conserve soil fertility without jeopardizing organic crop yields. Compared to ploughing, the authors reported a yield reduction of only 7 percent resulting from the application of minimum tillage techniques. A very interesting aspect reported in the paper is that weeds, despite being clearly increased because of plough avoidance, did not limit crop yields, whereas the reduced availability of crop nutrients due to lack of inversion tillage may play a more important yield-limiting role depending on climate and crop conditions. Appropriate fertilization strategies and sufficiently long and well-designed diversified rotations involving the systematic use of cover crops and intercropping should be considered as key elements to promote high soil biological activity and ultimately the creation of a soil environment conducive to crop development and to better ecological suppression of pests, diseases and weeds (Deguine et al., 2023). These interactions do exist but are very difficult for organic farmers to predict and manage. However, systematic diversification of cropping systems, in particular the alternation of crops characterized by different cycles, root morphology, ecophysiology and cultivation techniques, altogether can contribute to the formation of a complex, synergistic and efficient edaphic community able to provide multiple ecosystem services, including improved weed suppression or containment (e.g. see Canali et al., 2015).

3 Weed management in organic conservation and regenerative systems: problems and toolbox

The main reason why some soil tillage is needed in organic systems is to facilitate weed management. Most studies that have analysed weed community dynamics following conversion from conventional to organic agriculture have shown increased weed problems (Belde et al., 2000), especially when reduced or no-till systems were adopted compared to organic plough-based systems (Gruber et al., 2000). In particular, no-till increases the abundance of grasses, perennial weeds and anomochorous (wind-dispersed) species (Zanin et al.,

© Burleigh Dodds Science Publishing Limited, 2025. All rights reserved.

1997). These species are particularly favoured by reduced soil disturbance because of a lack of primary dormancy, vegetative propagation and possession of ecological traits favourable to site recolonisation, respectively. The same effect is usually observed in reduced or no-till conventional systems, but direct (e.g. mechanical, thermal) weed control methods applied in organic systems are usually less effective than chemical herbicides. This reduces the selection pressure against weeds and hence the risk of undesired evolution in weed community composition (Bàrberi, 2002), but accelerates the build-up of weed seedbank densities that can more often turn into competitive weed stands. In addition, minimum tillage coupled with organic fertilisation can increase the abundance of troublesome arable weeds like *Galium aparine* (McCloskey et al., 1996). It can then be assumed that, at least in temperate regions, management of organic systems without returning to tillage from time to time would soon become unsustainable due to intolerable weed pressure, although this negative effect could be retarded by the adoption of an appropriately diversified crop rotation (Rasmussen et al., 1999).

On the other hand, the alternation between different tillage systems across time is an element of cropping system diversification, which may help keep a diversified weed community with positive consequences on crop/weed competition. Indeed, there is a growing body of scientific evidence suggesting that more diversified weed communities can indirectly preserve crop yield by reducing the presence of highly competitive species. For example, in arable cropping systems higher crop yield has been correlated with higher weed community evenness (Adeux et al., 2019b) or higher Shannon diversity values (Antichi et al., 2022). These effects can be explained by the lack of clear selection of weed traits occurring when management practices change continuously in time and space. For example, in a diversified crop rotation, the alternation between winter and spring-summer crops impairs the selection of weed species characterised by a given time pattern, thereby maintaining a more balanced weed community composition based on seasonality. Although, in temperate environments, there are some non-seasonal weed species that can potentially grow whenever minimum temperature and soil moisture requirements are met (e.g. *Capsella bursa-pastoris* (L.) Medik. and *Poa annua* L.), they usually do not have a high competitive ability with major crops. Furthermore, a diversified crop rotation often includes crops with different ecological and agronomic requirements, e.g. in terms of rooting depth, fertiliser rates, irrigation needs. Adapting agronomic practices to the requirements of a variety of crops has a strong filtering effect on weed communities, avoiding one or few species become adapted to a standard, recurrent management regime and hence dominant (Adeux et al., 2019a).

In this perspective, it is quite clear that to minimise the risk of undesired weed community dynamics, regenerative organic agriculture in rotational

© Burleigh Dodds Science Publishing Limited, 2025. All rights reserved.

Table 1 Classification of the different methods ascribed to EWM according to their main effect, application timing and category, with relevant examples.[1]

Method	Main effect	Application timing	Category	Examples
Diversified crop rotation	Reduction in weed emergence, (indirect) increase in crop competitive ability	–	Preventive	Alternation between winter and spring-summer crops
Cover crops	Reduction in weed emergence	Before crop cycle	Preventive	Single-species cover crops or cover crop mixtures
Photocontrol	Reduction in weed emergence	Before crop cycle	Preventive	Night-time cultivation
Mulching	Reduction in weed emergence/ increase in crop competitive ability	Before/during crop cycle	Preventive/ cultural	Dead mulching/plastic mulching
SOWING/planting pattern	Increase in crop competitive ability	During crop cycle	Cultural	Increased seeding rate, reduced inter-row distance
Transplanting	Increase in crop competitive ability	During crop cycle	Cultural	Use of transplants instead of seeds
Fertilization (localized)	Increase in crop competitive ability	During crop cycle	Cultural	Banded fertilization, seed dressing
Irrigation (localized)	Increase in crop competitive ability	During crop cycle	Cultural	Drip row irrigation
Competitive/resistant genotypes	Increase in crop competitive ability	During crop cycle	Cultural	Use of cultivars with higher tillering ratio and/or allelopathic potential
Polycultures	Increase in crop competitive ability	During crop cycle	Cultural	Intercropping, living mulches, agroforestry
Seed predation	Reduction in weed emergence	Before crop cycle	Preventive	Untilled field margin strips to attract seed predators
Seed decay	Reduction in weed emergence	Before crop cycle	Preventive	Incorporation of residues, green manures or composts
Increased germination	Reduction in weed emergence	Before crop cycle	Preventive	False/stale seedbed technique, use of germination stimulants
Preventing seed shedding	Reduction in weed emergence	Before/during crop cycle	Preventive/ direct	Stubble cultivation, (selective) spraying, selective cutting
Direct weed control methods	Elimination of emerged weeds	During crop cycle	Direct	Chemical, physical (e.g. mechanical, thermal), biological

[1]Modified after Bàrberi (2019).

© Burleigh Dodds Science Publishing Limited, 2025. All rights reserved.

systems should avoid the continuous use of no-till, aiming for example at alternating no-till crops with minimum-tilled crops. The case of perennial organic systems (e.g. orchards) is different because in these conditions a good, competitive and well-managed perennial vegetation stand (either a single-species or mixed perennial cover crop, or the natural vegetation) would usually exert good weed suppression without the need of periodical soil disturbance (Bradshaw and Lanini, 1995).

Application of the principles and methods of Integrated Weed Management (IWM) or – even better – Ecological Weed Management (EWM) (Bàrberi, 2019) should be the guiding light for the planning of an effective weed management system in regenerative organic agriculture. As such, methods pertaining to each of the three components of a good IWM/EWM strategy should be included in a weed management toolbox for organic conservation/regenerative agriculture. These components are (i) preventive methods, applied before the crop cycle with the main goal of reducing weed emergence during the crop cycle; (ii) cultural methods, with the main goal of increasing crop competitive ability against weeds; (iii) direct methods, with the goal of eliminating surviving weeds whenever their in-field density is deemed too high (Bàrberi, 2019). A list of EWM methods to be potentially included in such a toolbox is shown in Table 1. Among them, two appear particularly interesting for conservative and regenerative organic agriculture and are treated in detail in the next two sections: cover crops and dead mulches (among preventive methods), and intercropping and living mulches (among cultural methods).

Although some novel direct methods mainly aimed at killing perennial weeds are potentially applicable in organic conservation and regenerative systems with some success, they are of lesser importance in a dedicated toolbox compared to preventive and cultural methods. Indeed, a good EWM system should mainly rely on ecological interactions triggered by soil-improvement methods resulting in improved weed suppression, and not on direct methods of whatever type, whose use should be minimised. This is why these methods are not treated in detail but just recalled here: among them we could cite the Combcut™, a selective scissor-like cutter targeting tall and stiff weeds (e.g. *Cirsium arvense* L. and other thistles) that might have escaped control (https://lyckegard.com/en/products/combcut/on-what-weeds-does-combcut-work); use of hot water/steaming apparatuses for precision control of perennials (Latsch et al., 2017), and electrocuting, i.e. the use of electric current delivered by safe appliances to kill weeds (Schreier et al., 2022).

4 Cover crops and dead mulches

Cover crops are plants cultivated to provide a service to the cropping system they are part of. In some cases, to keep the definition as broad as possible,

© Burleigh Dodds Science Publishing Limited, 2025. All rights reserved.

they are also called 'subsidiary crops', highlighting the fact that they are grown to sustain the main crop. The primary direct function attributed to cover crops is indeed to cover the soil during the off-season when primary crops are not grown, thereby providing continuous soil cover and preventing bare soil conditions, contributing – among other effects – to reduced nutrient leaching ('catch crop' service).

As a consequence of this primary function, cover crops elicit other services, hereafter briefly reported. Firstly, by intercepting direct solar radiation and turning it into biomass, they create more stable conditions in terms of soil temperature and moisture and, in general, a more suitable environment for agroecosystem biota. A typical micro-climatic effect induced by cover crops is the reduction of soil temperature amplitude, keeping the soil warmer at night and cooler during the day.

Cover crop above- and below-ground biomass provides feed and shelter for different categories of organisms, from bacteria and fungi to nematodes, invertebrates and small mammals. Cover crops allow nitrogen fixation (legume species), host arbuscular mycorrhizal fungi (AMF), increase P mobilization, enhance carbon fixation, increase soil organic matter and feed the detritus chain.

Another crucial ecological function developed by cover crops while growing is niche occupation, a core role in reducing weed abundance. By reducing resource availability in the agroecosystem (such as light, water, space and soil nutrients), cover crops reduce the capacity of weeds to grow, photosynthesize, and thus produce new seeds and buds to colonize or survive in the agroecosystem. In this way, cover crops represent an interesting preventive weed management method.

Once the cover crop growing season is over or it is time to sow the subsequent cash crop, the cover crop must be terminated. The termination method should be chosen wisely according to the intended function of the cover crop residues. In organic conservation/regenerative agriculture, termination methods are restricted to a few mechanical methods, such as roller crimpers or the natural decay of the cover crop, favoured by higher organic matter mineralisation rates in warmer climates and by frost conditions in colder climates.

If cover crop residues are kept on the soil surface, this biomass may become a dead mulch. Dead mulch can be considered any plant biomass deliberately kept above the soil surface to exert any kind of planned function related to agroecosystem management. Conservation agriculture has as one of its pillars the preservation of soil health by reducing mechanical disturbance to soil biota, also reducing soil organic matter degradation and aiming to reach a new and more effective equilibrium to grow healthy and prosperous crops. On the other hand, reducing mechanical disturbance to soil implies a strong

© Burleigh Dodds Science Publishing Limited, 2025. All rights reserved.

limitation in direct methods to manage weeds or cover crops, thus requiring a deep rethinking of the weed management strategy, particularly in cropping systems where other effective direct methods to manage weeds are limited, as in organic farming. Due to these limitations, preventive and cultural weed management methods like cover crops and dead mulches assume a crucial role in organic conservation/regenerative agriculture. Despite this, the actual application of these methods is quite rare, and effective case studies are not as widespread as expected.

4.1 Mechanisms involved in weed suppression by cover crops and dead mulches

The main general mechanism involved in weed suppression by cover crops is niche occupation. Cover crop growth should ideally be spatially homogeneous and rapid (Nosratti et al., 2023), leaving as little space and light as possible for spontaneous vegetation. The window of opportunity for weeds to occupy their own niche should be minimal in both space and time. For the same reason, cover crops need to be efficient in using available agroecosystem resources. When limiting, water, soil nutrients and light should be intercepted more efficiently by cover crops than by weeds, which compete for occupying the same niche. Ideally, a ruderal competitive species (*sensu* Grime, 1979), with fast and broad (in hydrothermal needs) germination, should be considered a good cover crop ideotype.

An effect related to fast niche occupation is the microclimatic interference with weed seed germination. By affecting and reducing light availability and temperature amplitude at soil level, cover crops may significantly reduce weed germination rate, preventing opportunities for individuals to even begin competing for resources. Thus, while niche occupation reduces available resources, microclimate modification reduces the number of individual plants to compete with.

Another mechanism involved in cover crop-based weed suppression is the allelopathic effect. We consider an allelopathic effect as any interference on the growth, survival, development and reproduction of neighbouring plants created by another species through the release of allelochemicals in the nearby environment (Cheng and Cheng, 2015). The capacity to produce metabolites with allelopathic effects is an interesting characteristic for species that may be used as cover crops. Through direct negative influence on biomass development, reduced germination capacity or even ostracizing the creation of favourable interactions with other biota, cover crops may reduce the capacity of sensitive weed species to develop in the agroecosystem, thus decreasing their seed/bud bank density and their ability to damage subsequent cash crops.

© Burleigh Dodds Science Publishing Limited, 2025. All rights reserved.

Some cover crops exert their allelopathic effect while growing or germinating, while other species may exert it through biomass decay, with the donor plant either still alive or dead. However, in a conservation agriculture context, it should be noted that some cover crop species need to be finely ground and incorporated into the soil to exert their full allelopathic capacity, such as rye (Barnes and Putnam, 1983), *Brassica* spp. (Haramoto and Gallandt, 2004), Sudangrass (*Sorghum bicolor* L. Moench. ssp. *sudanense*; Pierce and Nattrass, 2023) and buckwheat (*Fagopyrum esculentum* Moench.; Sturm et al., 2018).

All these direct effects (except for allelopathic effects exerted by decaying biomass) are effective in reducing or interfering with the population dynamics of weed species germinating and relying on resources available during the same season as cover crop growth to endure in the agroecosystem. Consequently, the direct effects on the following cash crop are often reported as scarce or of low magnitude (Adeux et al., 2021). However, some indirect effects can reverberate over a longer period, pointing out the importance of having a long-term perspective in weed management. Cover crop residues can sustain the activity of several biota at ground- and below-ground levels, which can contribute to weed seed decay and seed predation. These effects are difficult to measure directly but have wide and general effects on reducing the weed seedbank and weed abundance (Sias et al., 2021).

In summary, cover crop growth during fallow periods may help manage weeds in a long-term perspective by reducing seed rain and bud bank accumulation, occupying weed niches, reducing weed germination and interfering with weed growth. Creating a microclimate at ground level and increasing soil health also increases weed seed predation and decay, while allelopathic effects may reduce seed germination during cover crop growth or after cover crop termination, directly affecting the following cash crop growth.

Another weed suppression effect during the following cash crop growth can be obtained by using the cover crop biomass as a dead mulch. When the biomass of the cover crop is left on the soil surface, different interactions can develop with the potential weed flora present in the agroecosystem. The first effect is related to light interception, since cover crop residues reduce both the light needed for smaller seeds to start the germination process when close to the soil surface and the available light for seedlings to develop biomass when seed energy is exhausted. The second mechanism is the physical obstruction posed by residue thickness, which impedes weed seedlings or new sprouts reaching the surface.

Dead mulches then have similar effects as living cover crops on microclimate alteration, reducing soil temperature and temperature amplitude and altering the environmental thermal signal needed for certain species to germinate. When the biomass C:N ratio is unbalanced towards carbon, biomass

© Burleigh Dodds Science Publishing Limited, 2025. All rights reserved.

degradation can cause nitrogen sequestration (Perry et al., 2010), modifying another environmental cue needed by weed seeds to start germination, i.e. nitrate concentration. Increased weed seed predation and decay may also occur, also during cash crop growth. This effect, often neglected in conventional tillage agriculture, is interesting in conservation/regenerative systems because minimal soil disturbance increases the potential for soil biota preservation and facilitates surface layering of weed seeds.

4.2 How to include cover crops proficiently in organic conservation and regenerative systems

Cover crop establishment must be done before the main crop seeding, and cover crops must be terminated before cash crop seedlings emerge. Thus, the first element to consider in cover crop management is timing relative to overall crop rotation organization. In both field and vegetable crops, the fallow period must be identified and arranged properly according to the cropping system needs, looking for a balanced compromise between cash crop and cover crop needs relative to the services that the latter must provide.

In cover crops, the ecological service of weed management is mainly biomass-dependent. This implies that biomass production should be considered a key aspect when cover crops are included in a crop sequence. In arable crop rotations, a fallow period between two subsequent cash crops can usually be found, although some temporal adjustments may need to be done. In vegetable crops, the introduction of cover crops in a crop sequence usually occurs at the expense of some cash crops, given the generally shorter duration of growth cycle in vegetables than in arable crops. Thus, the worth of the weed management service offered by a cover crop must compensate for the lack of income due to the sacrificed cash crop.

In temperate climates, cover crops are usually grown after an early harvested cash crop as autumn-winter crops, before a spring-summer crop. In dry areas or rigid northern climates, cover crops can find wider fallow periods but also face harsher growing conditions, thus the physiology of the cover crop must be selected considering the extreme conditions they will pass through. In Sicily, for example, cover crops are usually seeded in October and terminated in April before the main summer-spring crop (Restuccia et al., 2020), while in Sweden, cover crops are seeded in August and terminated in early spring. Genetic selection plays a special role in these limiting climates (Brandsæter et al., 2000).

Cover crops can adapt to the different needs of the main cash crop, thus they can potentially be grown in any period that leaves enough time to develop sufficient biomass to perform the desired service. It is quite common to grow Sudangrass (Creamer and Baldwin, 2000) or *Crotalaria juncea* L. (Bilenky et al.,

© Burleigh Dodds Science Publishing Limited, 2025. All rights reserved.

2022) as cover crops in spring-summer in warmer climates such as Arizona or North Carolina, not to mention the use of cover crops in tropical agricultural systems.

4.3 Developing an efficient dead mulch in organic conservation and regenerative systems

When cover crops are grown to serve as dead mulch, three main aspects must be considered:

- Sufficient biomass must be developed by the planned termination date, which is usually very close to the seeding date of the following cash crop.
- Homogeneity of biomass development, as uneven biomass distribution may cause a strong negative effect on the effectiveness of the dead mulch function.
- Proper devitalization, avoiding cover crops to become unwanted species (volunteer plants) competing with the cash crop.

Regarding biomass, higher amounts lead to more effective weed suppression. Different minimum amounts of biomass needed to obtain effective weed control have been reported by various authors, ranging from 10 t ha^{-1} (Mirsky et al., 2013) to 3 t ha^{-1} (Stagnari et al., 2014). Dead mulch degradation, and thus the effectiveness of weed suppression, is strongly context-dependent: the vegetal matrix, climate and weed composition may interact differently in determining weed suppression effects.

An excess of nitrogen in the dead mulch biomass (low C:N ratio) may favour mulch degradation (Vincent-Caboud et al., 2019) and weed germination, making nitrogen available when the cash crop is not yet ready to efficiently utilize it. An ideal dead mulch biomass should have a balanced nitrogen content (C:N ratio of ca. 10) to avoid subtracting soil nitrogen from cash crop use (high C:N ratio) or favouring weed germination and quickly losing soil cover capacity (low C:N ratio). If the main aim is to control weeds, a long-lasting dead mulch should be preferred over a nitrogen-rich cover crop.

Concerning stand homogeneity, cover crop sowing should be done with high-quality seeds with a high germination rate, and using machinery that allows complete adhesion of seeds to soil. Organic conservation/regenerative agriculture sometimes leads to reduced germination power due to less soil softness and reduced nitrogen availability, resulting in decreased seed germination. It is crucial that the seeding machinery creates a proper environment for seed development, favouring homogeneous contact between soil and seeds and a minimum seeding depth, as soil preparation in conservation tillage could be of poorer quality than in conventional tillage (Vincent-Caboud

© Burleigh Dodds Science Publishing Limited, 2025. All rights reserved.

et al., 2019). A careful investment in and maintenance of the seeding machinery is thus necessary to obtain evenly seeded and germinated cover crops. Row seeding and a slightly increased seeding rate to compensate for potentially reduced germination are advisable (Mirsky et al., 2012).

Regarding proper termination, two elements are of crucial importance: the devitalisation method and timing. Different methods of devitalization can be used, each with advantages and disadvantages. The most common methods to obtain dead mulch from a cover crop by active devitalization are:

- Mowing or use of any cutting means that separate the above-ground biomass from roots, causing the death of the above-ground biomass without touching the basal stem or soil. Some species may re-sprout from the basal stem.
- Undercutting, which usually consists of blades or other mechanical tools that separate roots from the above-ground biomass, cutting also the basal stem and minimally disturbing the soil.
- Roller-crimping, which uses a bladed cylinder with applied weight to devitalise (crimp) the above-ground tissues without cutting them. This causes a more effective termination of the entire canopy by creating stress on plant tissues that lose water if properly applied. However, if crimping is not effective (not enough pressure applied, or use on species tolerant to crimping), it may lead to improper, uneven plant killing and a consequent higher probability of competition between the survived cover crop and the cash crop.

For this reason, other methods that are easier to manage, such as roto-cutting, have been developed. On top of these methods, additional devitalisation techniques like flaming (Frasconi et al., 2019) can be used to increase the effectiveness of devitalization and reduce the risk of re-sprouting. Passive methods to obtain dead mulch from a cover crop include the use of predictable extreme seasonal conditions, such as frost for cold-sensitive species (Lorin et al., 2015) or lack of water for drought-sensitive species.

Another crucial point is devitalisation timing, particularly relative to the phenological phase when the cover crop must be terminated in relation to the cash crop sowing requirements. Each species used as a cover crop has a different sensitivity to the termination method used according to its phenological phase. Usually, mono- and dicotyledonous species are more sensitive to mechanical disturbance when in full flowering (Lawson et al., 2015; Mirsky, 2009). Often, a compromise between the timing of cover crop and cash crop growing cycles should be sought, avoiding excessive rigidity. Cash crop growing techniques can be efficiently adapted to cover crop optimization in a no-till dead mulch, as demonstrated by Antichi et al. (2022). In this experiment, postponing sowing

© Burleigh Dodds Science Publishing Limited, 2025. All rights reserved.

the cash crop (sunflower) allowed maximization of the benefits reaped from hairy vetch (*Vicia villosa* Roth.) dead mulch upon optimisation of cover crop termination time, achieving pesticide-free and no-till management of weeds without affecting cash crop yield and revenue.

4.4 How to manipulate planned agrobiodiversity to select an efficient cover crop and dead mulch

All the above-mentioned considerations are strictly dependent on extensive knowledge of the genetic characteristics of the plant material to be used as cover crop. Here we propose a three-step approach to find the needed functional cover crop characteristics from the available germplasm:

Step I: Select the service(s) that must be provided in the agroecosystem by the dead mulch. According to the composition of the weed assemblage, different services may be desired to control weeds. In the case of a prevalence of problematic annual ruderal species (*sensu* Grime, 1979), two main services could be desirable from the cover crop: prevent weed germination and sustain cash crop growth and coverage. In case of the prevalence of competitive annual species, we may want to prevent seed germination while avoid providing too much available nitrogen to weeds. In the case of the prevalence of competitive perennial species, we should aim at magnifying the physical barrier effect provided by the cover crop instead of contrasting germination or favouring seed predation. The core idea of Step I is that the specific services requested from cover crop application must be planned before choosing species.

Step II: Once the desired services have been selected, suitable species capable of providing such services must be found among the available genetic material, and the most adapted varieties to the local context should be chosen. Once a pool of species and varieties are selected, timing of the cover crop (e.g. sowing and termination dates) must be chosen in relation to the needs of the following cash crop, allowing cover crop termination to be proficiently performed with physical tools. Biomass production and the phenological phase of the cover crop at the planned termination time must be carefully considered. EU-funded projects like OSCAR (Crossland et al., 2015) or the Midwest Cover Crop Council in the USA (Baas et al., 2011) have produced interesting tools to help farmers select species according to desired services in different parts of the world. However, at the moment, a universal toolbox does not exist, and local sources must be consulted to find the most appropriate species.

Step III: Diversification of the cover crop (i.e. using mixtures instead of single species) and the techniques to establish it should then be considered. Following the 'insurance hypothesis', an increase in diversity (both of species and varieties) should ensure the effectiveness of the cover crop due to two main functional effects ascribed to increased diversity: (i) increasing the

© Burleigh Dodds Science Publishing Limited, 2025. All rights reserved.

breadth of conditions in which the cover crop mixture could prosper and develop (Wendling et al., 2019); (ii) niche complementarity, upon which some cover crop assemblages should lead to transgressive over-yielding, determining higher biomass production than the best sole crop grown under optimal conditions. Despite some clear examples of increasing productivity of biomass for increased weed and functional group diversity (Ranaldo et al., 2020), some recent studies have stated that in cover crop biomass production, species mixtures rarely over-yield the best sole crop (Smith et al., 2020). However, these studies were conducted with a substitutive seeding scheme, i.e. sowing a fraction of the suggested seeding rate according to the number of species present in the mixture, leading to incomplete occupation of the available ecological space. For this reason, a semi-additive scheme, even if it implies higher investment in seed resources, must be considered as the most conservative approach to obtain insurance on the adaptability of cover crop to the increasing variability of weather conditions.

Cover crops and dead mulches play a crucial role in fostering organic conservation and regenerative agriculture by enhancing soil health, supporting biodiversity and effectively managing weeds. Their ability to provide continuous soil cover, regulate microclimate and suppress weed growth through niche occupation and allelopathic effects makes them very interesting methods in a sustainable farming toolbox. Implementing these practices requires careful selection of cover crop species, seeding rates and timing of sowing and termination, as well as understanding the ecological interactions they promote in the agroecosystem.

5 Intercropping and living mulches

Organic conservation/regenerative agriculture advocates moving away from direct weed control methods and instead prioritize preventive agroecological approaches based on the maintenance of the soil cover to ensure constant weed competition and promote more diverse weed communities, thereby minimizing adverse effects on crops. Among preventive weed management methods, the use of permanent living mulches (PLMs), involving the intercropping of a service crop alongside the cash crop, merits a special mention because of their contribution to maintain sufficient weed control throughout the crop rotation and drive weed dynamics towards a more diversified and less competitive community. PLMs differ from intercropping in a strict sense in terms of their temporal persistence in the field. In intercropping, co-cultivation occurs between two cash crops, and it is often limited to the duration of the main crop growing cycle. Therefore, while intercropping may offer specific advantages in standard organic and low-input systems, its restricted temporal duration makes it less suited for achieving the long-term

© Burleigh Dodds Science Publishing Limited, 2025. All rights reserved.

and continuous soil cover required in organic conservation/regenerative systems.

5.1 Mechanisms regulating weed suppressive capacity of living mulches

The ideal living mulch system would suppress weeds through multiple mechanisms (reducing the likelihood of weed escapes) without significantly limiting resource availability to the main crop. One of the key principles of adopting PLMs is to suppress tillage or reduce it and maintain soil constantly covered through the establishment of a permanent living cover throughout the crop rotation, ensuring continuous competition with weeds for resources such as space, light, and nitrogen (N). Additional mechanisms may include competition for other nutrients, allelopathy, and changes to arthropod communities. In the case of PLMs, the principles of conservation/regenerative agriculture can be applied in two phases: (i) within the cash crop, through the establishment of service crops for filling the ecological niches otherwise occupied by weeds and (ii) between crops in the rotation, increasing the duration of the vegetation cover in a given field, eliminating the fallow period. Weed management without the use of tillage and chemical tools can be therefore better achieved at the crop rotation level and it is driven by three key factors:

- Provided that the best choices in terms of cultivar selection and row-spacing are done, living mulches can improve weed control in a main crop by filling the ecological niches otherwise occupied by weeds, through competition for resources.
- After the main crop is harvested, maintaining living mulches in the field ensures continuous soil cover during the fallow period. This strategic use of living mulch serves as a natural barrier, inhibiting weed germination and impeding the spread of weeds.
- By sowing a subsequent crop alongside the living mulch or following strip tillage, farmers can capitalize on the diminished seed bank, taking advantage of reduced weed pressure in the soil.

5.2 How to establish permanent living mulches in organic conservation and regenerative agriculture

The establishment of the living mulch can be implemented either before, simultaneously with the main crop (contemporary intercropping, CI), or with delayed sowing in the already developed crop (relay intercropping, RC). In arable cropping systems, contemporary sowing of the living mulch with the main crop is often considered by farmers as more technically feasible, whereas RC is often preferred to mitigate the risk of competition between the main crop

© Burleigh Dodds Science Publishing Limited, 2025. All rights reserved.

and the living mulch. In RC, the main crop gains a phenological advantage, thereby reducing the risk of competition from the living mulch, which is sown later (Lamichhane et al., 2023). For vegetable systems, instead, where crops are often more susceptible to weed competition, establishing PLMs before transplanting vegetable crops can improve the living mulch establishment and achieve better weed control compared with contemporary sowing (Bhaskar et al., 2021). Prior establishment of PLMs allows for mowing weeds before transplanting of vegetable crops: this would help for reducing weed dissemination. In both arable and vegetable systems, the living mulch persists in the field after the harvest of the main crop, covering efficiently the soil until the subsequent crop is sown (Leoni et al., 2024). At this stage, the subsequent crop can be established within an already-developed living mulch through direct seeding/transplanting or after strip tillage. Although this system can be effective for 2 or 3 crop cycles, studies on its long-term feasibility are limited. These systems often exhibit a significant increase in weed pressure after 2 or 3 main crops, especially for grasses and perennial weeds, prompting the necessity of soil tillage for the establishment of new living mulches (Brainard et al., 2012).

5.3 Long-term efficiency of permanent living mulches and their impact on weed community

It should be clarified that organic systems managed entirely without the use of tillage can be difficult to achieve, or rather, there is no evidence that this can be feasible for farmers, as there are no long-term experiments testing these systems. What can be affirmed at this moment is that the use of PLMs can be feasible and effective when considering a sequence of 2–3 subsequent crops, given the increase in weed pressure and the progressive reduction in mulch density due to self-competition mechanisms. Continuous conservation management tends to increase the weed seed bank in the first few centimetres of the soil favouring the presence of aggressive perennial and annual grasses with a large rate of seed production, such as *Lolium* spp., *Alopecurus myosuroides* Huds., and *Phalaris* spp. The control of some of these species can be very challenging without tillage since soil inversion is one of the most effective ways for their successful control. For dicotyledonous weeds, the impact of tillage systems depends on the species. For instance, on the one hand, living mulches tend to sensibly reduce some annual dicotyledonous, such as *Chenopodium* sp. (Westbrook et al., 2022) but, on the other hand, perennial weeds such as *Cirsium arvense* L., *Sonchus arvensis* L. and *Convolvulus arvensis* L., and annuals/biennials like *Conyza canadensis* L. can be predominant in organic conservation/regenerative systems. Given these considerations, it is recommended to temporarily discontinue the use of mulches and establish

© Burleigh Dodds Science Publishing Limited, 2025. All rights reserved.

them again after a crop sequence is managed according to standard organic operations in order to reduce the weed pressure. Furthermore, it is essential to examine any potential allelopathic effects from the previous living mulch on the new one. Unfortunately, current information on this matter is limited, and the duration required for establishing a new permanent living mulch system remains uncertain.

5.4 Choice of living mulches in organic arable and vegetable conservation/regenerative systems

Optimizing weed suppression without negative effects on crop yield can be accomplished by carefully choosing appropriate living mulches and implementing strategic management practices that capitalize on the morphological, physiological and developmental differences among main crops, living mulches and weeds (Verret et al., 2017). PLMs need to be tailored to each specific system, leading to differences in terms of selecting the most suitable species as well as their management. In contrast to conventional conservation agriculture where the use of vigorous cover crops such as vetch (*Vicia sativa* L.) or rye (*Secale cereale* L.) is recommended for maximizing weed control efficiency and allowing the establishment of effective dead mulches, in organic conservation/regenerative cropping systems the selection of suitable species for PLMs is even more important to maintain a satisfactory weed control and at the same time not having penalties on crop yield. This is because PLMs in such systems alternate between periods of co-cultivation with main crops and periods of cover cropping during fallow. Therefore, the first aspect that needs to be addressed is the choice of suitable species for living mulches.

The selection of suitable species for living mulches can be done based on plant morphological traits and growing characteristics. Species with short canopy (≤10 cm during the intercropping stage) and with a prostrate habitus are generally recommended for PLMs. Short canopy and prostrate habitus allow to contrast weed emergence and growth without interfering with the main crop and hindering its mechanical harvest. For example, prostrate and stoloniferous species such as white clover (*Trifolium repens* L.) are particularly interesting for temperate environments. Accurate cultivar selection is also crucial, as morphological characteristics can vary significantly among cultivars and ecotypes of the same species. For example, within white clover cultivars, there are dwarf and micro types (e.g. cv. Pipolina), with a smaller size compared with other cultivars, that can be particularly suitable for PLMs in vegetable systems. Notably, species with these characteristics often have a very superficial root system, making them less suitable for contexts with dry summers. In this regard, annual self-seeding species, such as subterranean clover (*Trifolium subterraneum* L.), can represent a valid alternative. Under

© Burleigh Dodds Science Publishing Limited, 2025. All rights reserved.

Mediterranean conditions, the growth cycle of annual self-seeding legumes terminates in late spring. After wheat harvest, biomass residues of these legumes are expected to form a dead mulch during the dry summer season and re-grow spontaneously from their seeds in the following autumn, to establish a cover crop until the subsequent spring. This characteristic makes these species interesting for systems in dry summer conditions. Subterranean clover is the most used self-seeding legume species, which has three subspecies (subsp. *subterraneum*, subsp. *yanninicum*, and subsp. *brachycalycinum*), each suitable for specific soil types. Other self-seeding species of interest include *Medicago polymorpha* L., *M. scutellata* (L.) Mill., *M. rotata* Boiss. and *M. arabica* (L.) Huds. However, as these species have undergone less genetic selection, they often exhibit seeds with high dormancy percentages, leading to regrowth that may not be sufficient to ensure good cover crop presence during the fallow period.

Another important trait of an effective living mulch against weeds is the growing rate for achieving total field cover. The faster the growth during the initial development stages, the greater the competitive effect against weeds. It has been observed that perennial species such as white clover, lucerne (*Medicago sativa* L.), or bird's-foot trefoil (*Lotus corniculatus* L.), exhibit slow growth during the early stages of development. On the contrary, self-seeding legume species often show a higher growth rate during the initial development phases, even though they may be less effective in weed control over the long term compared to perennial species. The most promising solution seems then to use mixtures composed of perennial legumes and self-seeding annuals, with the objective of achieving effective weed control both in the short and long term.

Considering that crop turnover in vegetable systems is much more frequent than in arable systems, it is important to find an effective living mulch all year round. Using a living mulch composed of a single species, even if it is perennial, will not guarantee continuous strong competition with weeds because, within a year, it will have periods of maximum biomass accumulation and other periods of vegetative stasis, corresponding to periods of good and poor competition with weeds, respectively. In this context, the use of a living mulch composed of a mixture of species can be interesting: species choice should be based on the temporal complementarity of periods of maximum biomass production, with the aim of keeping the living mulch always competitive with weeds (Leoni et al., 2020). For example, perennial legume species such as *Lotus corniculatus* and *Trifolium repens* show good soil cover in the summer months, while self-seeding annual species such as *Trifolium subterraneum* and *Medicago polymorpha* have good biomass production in the winter and spring periods. A mixture of perennial and self-seeding annual species can then help maintain constant soil cover within and across years.

© Burleigh Dodds Science Publishing Limited, 2025. All rights reserved.

5.5 Management options for perennial living mulches in organic arable and vegetable conservation and regenerative cropping systems

5.5.1 Arable systems

In arable, cereal-based cropping systems, during co-cultivation, the morphological traits and growth rate of living mulches can be modulated based on cropping system management (Leoni et al., 2022). In these systems, it is generally observed that the higher the competition from the main crop, the lower the growth capacity of the living mulch and its effects on weeds. For this reason, in low-productivity cropping systems where crop competitiveness is generally lower, the living mulch maximizes its functions with a higher impact on weed control. However, under these conditions, it is also more likely to have issues related to competition between the living mulch and the main crop. In such cases, it is suggested to adopt relay intercropping by establishing the living mulch with a postponed sowing date compared with that of the main crop, in order to provide a phenological advantage to the latter over the living mulch. A practical example, contextualized in a Mediterranean cereal system, would be the establishment of a winter cereal such as durum wheat after minimum tillage in autumn and then sowing the living mulch when wheat is at the tillering stage. The living mulch can be broadcast-sown with a subsequent pass of a light harrow to facilitate seed incorporation into the soil and allow mechanical control of weed seedlings. In high-productivity contexts or if highly competitive crops are used, the growth of living mulches can be significantly reduced due to competition from the main crop, sometimes leading to living mulch failure. In such situations, the use of self-seeding annual species should be avoided for two main reasons: (i) they would produce few seeds and biomass during the co-cultivation stage, compromising their regrowth and weed suppression ability during the following autumn, and (ii) they are susceptible to the shaded environment created by the more competitive crops. In these cases, a solution can be the use of a living mulch composed of shade-tolerant perennial species. In this respect, species such as lucerne, white clover or sulla (*Hedysarum coronarium* L., alias *Sulla coronaria* (L.) B.H.Choi & H.Ohash) can be a suitable choice in the Mediterranean environment.

Once the main crop is harvested, the living mulch persists in the field either as a living ground cover in the case of perennial species or as dead mulch when utilizing self-seeding species. In both scenarios, a continuous soil cover is ensured, reducing weed germination, growth and dissemination efficiency. Additionally, studies have shown an increase in seed predation by certain insects in the presence of a PLM (Schumacher and Gerhards, 2022). In a typical Mediterranean crop sequence, after a winter crop, a summer crop such as sorghum, corn, or sunflower is sown. In an inversion-tillage-based organic

© Burleigh Dodds Science Publishing Limited, 2025. All rights reserved.

system, cover crops may be used as green manure to maximize their positive and immediate impact on soil fertility. In a conservation/regenerative conventional system, the common management option involves the chemical termination of cover crops (e.g. by glyphosate) and then the direct seeding of the summer crop on dead mulch. Under organic management, the recommended approach is to maintain the living mulch and then use direct seeding methods for the summer crop. However, managing competition between the living mulch and the main crop, especially during the initial growth stages when crops are most vulnerable, is crucial. As such, reducing the vigour of the living mulch in this phase can be necessary to avoid competition. Strategies to reduce the vigour of the living mulch without the use of chemical methods often involve the use of strip tillage. This involves shallow cultivation in strips, typically 10–15 cm wide, where sowing of the summer crop then takes place. This practice helps the summer crop withstand the critical early growth phase without competition from living mulches. Depending on the species used, the living mulch colonises the strip-tilled area (e.g. with stolons, in the case of *Trifolium repens*), but only after the risk of competition for space and light has been avoided. After harvesting the summer crop, the system can theoretically proceed by sowing a new winter crop, taking into account potential challenges that may arise from the prolonged use of such a system, such as the increase in weed pressure over the years.

5.5.2 Vegetable systems

Permanent living mulches find application also in organic conservation/regenerative vegetable systems. Antichi et al. (2019) tested the performance of an innovative organic and conservation vegetable system in a three-year field experiment in a Savoy cabbage (*Brassica oleracea* L. var. *sabauda*)-spring lettuce (*Lactuca sativa* L.)-fennel (*Foeniculum vulgare* Mill.)-summer lettuce (*L. sativa*) crop rotation. Termination of cover crops, as well as the management of crop residues and a living mulch, was implemented by use of a roller crimper alternated with flaming. However, in this study, the organic conservation/regenerative management caused significant yield depletion for all crops and revealed suboptimal weed control and nitrogen availability. In accordance with the authors of this study, further insights into the management of living mulches are essential to enhance the suitability of these systems, especially regarding the management of the inter-row spaces. While within the rows weeds can be controlled by increasing the density of vegetable crops and prioritizing the use of more competitive species and cultivars, between rows the living mulch needs to be managed through periodic mowing. Vegetable crops are much more susceptible to living mulches competition compared to arable crops, so that even short-statured living mulches (ca. 10 cm) can already induce severe competition (Den Hollander et al., 2007).

© Burleigh Dodds Science Publishing Limited, 2025. All rights reserved.

Weeding robots can be used in organic conservation/regenerative vegetable systems for mowing weeds and managing living mulches between vegetable crop rows (Averill et al., 2022). Among them, Sportelli et al. (2022) tested an autonomous mower designed for turf mowing, adapted for organic conservation/regenerative vegetable systems, in a two-year rotation between cauliflower (*Brassica oleracea* L. var *botrytis*) and aubergine (*Solanum melongena* L.). This technique seems very promising since the system where the living mulch was managed with the autonomous mower produced an acceptable fresh marketable yield for both vegetable crops while guaranteeing good weed control. Notably, in the living mulch-based system energy consumption was lower than in the standard tillage-based system. However, the use of autonomous mowers needs to be further adapted to the vegetable systems, for example by implementing a GPS system that recognises the position of vegetable crop rows.

6 Opportunities, challenges and future perspectives

Organic conservation and regenerative systems have great potential to develop, especially if the present direction of agricultural policies – at least in Europe – is going to be confirmed. However, regardless of this, agricultural soil deterioration and its associated causes (e.g. desertification, land use change and habitat destruction, wind and water erosion, tilth destruction, overgrazing, overexploitation) have such a massive worldwide impact that it can easily be predicted they will be the focus of even stronger technical and policy action anywhere.

What are then the main hurdles to be overcome in order to foster organic conservation and regenerative agriculture? Part of the issue is related to the development of organic agriculture *per se*. Setting aside issues linked to produce prices and market accessibility, closing the yield gap between organic and conventional crop yields is believed to be a key objective to meet the goal of increasing the global acreage dedicated to organic production, as well as a major lever to convince part of the farmers still reluctant to embrace the conversion. The best global estimates regarding the organic yield gap report it to be around 19–20% (de Ponti et al., 2012; Ponisio et al., 2015): although this does not justify at all the fears of 'global food shortage' or even 'global famine' that are at times raised, it is enough to act as a psychological barrier for many farmers. Considering that a shift towards organic conservation/ regenerative systems would likely exacerbate this problem, at least initially, due to the suboptimal yields that may be obtained in the transition phase, the issue becomes even more critical. This is why considerable effort is being put into research aimed to provide novel technical solutions aimed to close the yield gap, without compromising the agroecological approach which should

© Burleigh Dodds Science Publishing Limited, 2025. All rights reserved.

be at the base of organic agriculture (see, e.g. the EU-funded OrganicYieldsUp project: https://cordis.europa.eu/project/id/101137068).

On the technical side, it is quite obvious that the two main critical issues to be solved in order to facilitate take-up of organic conservation and regenerative systems are (i) providing adequate nutrient supply to crops and (ii) ensuring good weed management. The prohibition to use chemical fertilisers and herbicides in organic agriculture makes the problem even more challenging but also calls for a paradigm shift in the management approach. Indeed, weed management in organic conservation/regenerative agriculture cannot rely on a mere substitution approach, because any non-chemical method could be at par with efficiency with the best chemical method. Therefore, effective weed management systems tailored to organic conservation/regenerative agriculture can only be envisaged through a system redesign approach, which is fully in line with the agroecological principles (Bàrberi, 2002). In this chapter, we have seen that cover crops, dead mulches, intercrops and living mulches are promising cultural practices in order to meet this goal. They are all important components of cropping system diversification, which is a key issue for ensuring agricultural sustainability, not just in organic conservation/regenerative agriculture. Interestingly, well-planned and optimally managed cover crops/dead mulches and intercrops/living mulches may also meet the other technical goal, i.e. adequate crop fertilisation.

7 Conclusion

It is clear that for optimising the management of cover crops, intercrops and living mulches, the best combination of practices should be adapted locally, on a case-by-case situation. This calls for the importance of having a good AKIS framework and developing participatory action research where solutions are co-designed, co-tested and co-evaluated in multi-actor teams including researchers, farmers, technicians and other relevant actors (Leclère et al., 2023). Besides this, important research avenues to be pursued to improve organic conservation/regenerative agriculture (and weed management therein) are: (i) unravelling the mechanisms of multispecific interactions (e.g. soil-plant-other biota) triggered in diversified cropping and how to optimise them to obtain the desired ecosystem services, also using digital and innovative sensoring technologies; (ii) trait-based genetic/genomic studies and breeding programmes to identify/develop improved genotypes to be specifically used as cover crops/intercrops/living mulches; (iii) development of improved machinery for sowing and terminating cover crops/living mulches under no-till or minimum-tilled regimes. System redesign coupled with new tailored technologies seems the best way to go to bridge the gap between science and practice in the perspective of fostering organic conservation/regenerative systems.

© Burleigh Dodds Science Publishing Limited, 2025. All rights reserved.

8 Acknowledgements

SC and FL have written paragraphs 4 and 5, respectively. PB has written all the other parts and done the final text editing.

9 Where to look for further information

- European Weed Research Society - Working Group on "Physical and Cultural Weed Control" - https://www.ewrs.org/en/Working-Groups/Physical-and-Cultural-Weed-Control/About
- Sustainable Agriculture Research and Education (SARE) Project "Weed Management in Organic Conservation Tillage/No Tillage" - https://projects.sare.org/sare_project/lnc04-240
- The European Alliance for Regenerative Agriculture (EARA) - https://eara.farm

10 References

Adeux, G., Munier-Jolain, N., Meunier, D. et al. (2019a). 'Diversified grain-based cropping systems provide long term weed control while limiting herbicide use and yield losses'. *Agron. Sustain. Dev.*, 39, 42. https://doi.org/10.1007/s13593-019-0587-x

Adeux, G., Vieren, E., Carlesi, S. et al. (2019b). 'Mitigating crop yield losses through weed diversity'. *Nature Sustain.*, 2, 1018–1026. https://doi.org/10.1038/s41893-019-0415-y

Adeux, G., Cordeau, S., Antichi, D. et al. (2021). 'Cover crops promote crop productivity but do not enhance weed management in tillage-based cropping systems'. *Eur. J. Agron.*, 123, 126221. https://doi.org/10.1016/j.eja.2020.126221

Antichi, D., Carlesi, S., Mazzoncini, M. et al. (2022). 'Targeted timing of hairy vetch cover crop termination with roller crimper can eliminate glyphosate requirements in no-till sunflower'. *Agron. Sustain. Dev.*, 42(5), 87. https://doi.org/10.1007/s13593-022-00815-2

Antichi, D., Sbrana, M., Martelloni, L. et al. (2019). 'Agronomic performances of organic field vegetables managed with conservation agriculture techniques: a study from central Italy'. *Agronomy*, 9(12), 810. https://doi.org/10.3390/agronomy9120810

Averill, K.M., Westbrook, A.S., Pineda-Bermudez, L., et al. (2022). 'Effects of Tertill® weeding robot on weed abundance and diversity'. *Agronomy*, 12(8), 1754. https://doi.org/10.3390/agronomy12081754

Baas, D., Mutch, D. and Ackroyd, V. (2011). 'A state-specific online cover crop decision tool for Midwest farmers'. *J. Extens.*, 49(6), 35. https://doi.org/10.34068/joe.49.06.35.

Bàrberi, P. (2002). 'Weed management in organic agriculture: are we addressing the right issues?' *Weed Res.*, 42, 176–193. http://dx.doi.org/10.1046/j.1365-3180.2002.00277.x

Bàrberi, P. (2006). 'Special Topic 4. Tillage: how bad is it in organic agriculture?' In P. Kristiansen, A. Taji, and J. Reganold (Eds.), *Organic Agriculture. A Global Perspective*. Collingwood (AU) and Wallingford (UK): CSIRO Publishing/CABI Publishing, pp. 295–303.

© Burleigh Dodds Science Publishing Limited, 2025. All rights reserved.

Bàrberi, P. (2019). 'Ecological weed management in sub-Saharan Africa: prospects and implications on other agroecosystem services'. *Adv. Agron.*, 156 (4), 219–264. https://doi.org/10.1016/bs.agron.2019.01.009

Barnes, J.P. and Putnam, A. (1983). 'Rye residues contribute weed suppression in no-tillage cropping systems'. *J. Chem. Ecol.*, 9(9), 1045–1057. https://doi.org/10.1007/BF00982210

Belde, M., Mattheis, A., Sprenger, B. et al. (2000). 'Long–term development of yield affecting weeds after the change from conventional to integrated and organic farming'. *Z. PflKrankh. PflSchutz,* Sonderh. XVII, 291–301.

Benbrook, C.M. (2016). 'Trends in glyphosate herbicide use in the United States and globally'. *Environ. Sci. Eur.*, 28(1), 1–15. https://doi.org/10.1186/s12302-016-0070-0

Bhaskar, V., Westbrook, A.S., Bellinder, R.R. et al. (2021). 'Integrated management of living mulches for weed control: a review'. *Weed Tech.*, 35(5), 856–868. https://doi.org/10.1017/wet.2021.52

Bilenky, M., Nair, A. and McDaniel, M. (2022). 'Effect of summer cover crops on cabbage yield, weed suppression, and N mineralization in a low input cropping system', *Front. Sustain. Food Syst.*, 6, 1021639. https://doi.org/10.3389/fsufs.2022.1021639

Bradshaw, L., and Lanini, W.T. (1995). 'Use of perennial cover crops to suppress weeds in Nicaraguan coffee orchards'. *Int. J. Pest Manag.*, 41(4), 185–194. https://doi.org/10.1080/09670879509371947

Brainard, D.C., Bakker, J., Noyes, D.C. et al. (2012). 'Rye living mulch effects on soil moisture and weeds in asparagus'. *HortScience*, 47, 58–63. https://doi.org/10.21273/HORTSCI.47.1.58

Brandsæter, L., Smeby, T., Tronsmo, A. et al. (2000). 'Winter annual legumes for use as cover crops in row crops in Northern Regions: II. Frost resistance study', *Crop Sci.*, 40, 175–181. https://doi.org/10.2135/CROPSCI2000.401175X

Canali, S., Diacono, M., Campanelli, G. et al. (2015). 'Organic no-till with roller crimpers: agro-ecosystem services and applications in organic Mediterranean vegetable productions'. *Sustain. Agric. Res.*, 4, 70. http://dx.doi.org/10.5539/sar.v4n3p70

Cheng, F. and Cheng, Z. (2015). 'Research progress on the use of plant allelopathy in agriculture and the physiological and ecological mechanisms of allelopathy'. *Front. Plant Sci.*, 6, 1020. https://doi.org/10.3389/fpls.2015.01020

Cooper J., Baranski M., de Lange M.N., et al. (2016). 'Shallow non-inversion tillage in organic farming maintains crop yields and increases soil C stocks: a meta-analysis'. *Agron. Sustain. Dev.*, 36, 22. https://doi.org/10.1007/s13593-016-0354-1

Creamer, N. and Baldwin, K.R. (2000). 'An evaluation of summer cover crops for use in vegetable production systems in North Carolina', *HortScience*, 35(4), 600–603. https://doi.org/10.21273/HORTSCI.35.4.600

Crossland, M., Fradgley, N., Creissen, H.E. et al. (2015). 'An online toolbox for cover crops and living mulches'. In *Getting the Most Out of Cover Crops. Aspects of Applied Biology,* 129, 1–5. https://pure.sruc.ac.uk/ws/portalfiles/portal/18232209/Crossland_et_al._An_online_toolbox_for_covercrops_and_living_mulches.pdf.

Deguine, J.-P., Aubertot, J.-N., Bellon, S. et al. (2023). 'Agroecological crop protection for sustainable agriculture'. *Adv. Agron.*, 178, 1–59. https://doi.org/10.1016/bs.agron.2022.11.002

© Burleigh Dodds Science Publishing Limited, 2025. All rights reserved.

Den Hollander, N.G., Bastiaans, L. and Kropff, M. J. (2007). 'Clover as a cover crop for weed suppression in an intercropping design: II. Competitive ability of several clover species'. *Eur. J. Agron.*, 26(2), 104–112. https://doi.org/10.1016/j.eja.2006.08.005

de Ponti, T., Rijk, B. and van Ittersum, M.K. (2012). 'The crop yield gap between organic and conventional agriculture'. *Agric. Syst.*, 108, 1–9. https://doi.org/10.1016/j.agsy.2011.12.004

European Commission. (2021). *Communication from the Commission to the European Parliament, The Council, The European Economic and Social Committee and The Committee of the Regions. EU Soil Strategy for 2030*, p. 26. https://eur-lex.europa.eu/legal-content/EN/TXT/PDF/?uri=CELEX:52021DC0699&from=EN.

European Commission. (2022). *European missions: A soil deal for Europe. 100 living labs and lighthouses to lead the transition towards healthy soils by 2030. Implementation Plan*, p. 77. https://research-and-innovation.ec.europa.eu/system/files/2021-09/soil_mission_implementation_plan_final_for_publication.pdf.

Frasconi, C., Martelloni, L., Antichi, D. et al. (2019). 'Combining roller crimpers and flaming for the termination of cover crops in herbicide-free no-till cropping systems', *PLoS ONE*, 14(2), e0211573. https://doi.org/10.1371/journal.pone.0211573

Grime, J.P. (1979). Primary strategies in plants. *Trans. Bot. Soc. Edinburgh*, 43(2), 151–160. https://doi.org/10.1080/03746607908685348

Gruber, H., Handel, K. and Broschewitz, B. (2000). 'Influence of farming system on weeds in thresh crops of a six-year crop rotation'. *Z. PflKrankh. PflSchutz*, Sonderh. XVII, 33–40.

Haramoto, E.R. and Gallandt, E. (2004). 'Brassica cover cropping for weed management: a review'. *Renew. Agric. Food Syst.* 19(3), 187–198. https://doi.org/10.1079/RAFS200490

Heap, I. (2024). *International herbicide-resistant database*. www.weedscience.org/Pages/Graphs/SOAGraph.aspx

Hobbs, P.R. (2007). 'Conservation agriculture: what is it and why is it important for future sustainable food production?'. *J. Agric. Sci.*, 145(2), 127–137. http://dx.doi.org/10.1017/S0021859607006892

Lamichhane, J.R., Alletto, L., Cong, W.F. et al. (2023). 'Relay cropping for sustainable intensification of agriculture across temperate regions: crop management challenges and future research priorities'. *Field Crops Res.*, 291, 108795. https://doi.org/10.1016/j.fcr.2022.108795

Latsch, R., Anken, T., Herzog, C. et al. (2017). 'Controlling *Rumex obtusifolius* by means of hot water'. *Weed Res.*, 57, 16–24. https://doi.org/10.1111/wre.12233

Lawson, A., Cogger, C., Bary, A. et al. (2015). 'Influence of seeding ratio, planting date, and termination date on rye-hairy vetch cover crop mixture performance under organic management'. *PLoS ONE*, 10(6), e0129597. https://doi.org/10.1371/journal.pone.0129597

Leclère, M., Loyce, C. and Jeuffroy, M.H. (2023). 'A participatory and multi-actor approach to locally support crop diversification based on the case study of camelina in northern France'. *Agron. Sustain. Dev.*, 43(1), 13. https://doi.org/10.1007/s13593-023-00871-2

Leoni, F., Lazzaro, M., Carlesi, S. et al. (2020). 'Legume ecotypes and commercial cultivars differ in performance and potential suitability for use as permanent living mulch

© Burleigh Dodds Science Publishing Limited, 2025. All rights reserved.

in Mediterranean vegetable systems'. *Agronomy*, 10(11), 1836. https://doi.org/10.3390/agronomy10111836

Leoni, F., Lazzaro, M., Carlesi, S. et al. (2024). 'Screening suitable legumes for living mulches to support nitrogen dynamics and weed control in a durum wheat-forage sorghum crop sequence'. *Field Crops Res.*, 307, 109246. https://doi.org/10.1016/j.fcr.2023.109246

Leoni, F., Lazzaro, M., Ruggeri, M. et al. (2022). 'Relay intercropping can efficiently support weed management in cereal-based cropping systems when appropriate legume species are chosen'. *Agron. Sustain. Dev.*, 42, 75. https://doi.org/10.1007/s13593-022-00787-3

Lorin, M., Jeuffroy, M., Butier, A. et al. (2015). 'Undersowing winter oilseed rape with frost-sensitive legume living mulches to improve weed control'. *Eur. J. Agron.*, 71, 96–105. https://doi.org/10.1016/j.eja.2015.09.001

McCloskey, M., Firbank, L.G., Watkinson, A.R. et al. (1996). 'The dynamics of experimental arable weed communities under different management practices'. *J. Veg. Sci.*, 7, 799–808. https://doi.org/10.2307/3236458

Mirsky, S., Curran, W., Mortensen, D. et al. (2009). 'Control of cereal rye with a roller/crimper as influenced by cover crop phenology'. *Agron. J.*, 1589–1596. https://doi.org/10.2134/agronj2009.0130

Mirsky, S., Ryan, M., Curran, W. et al. (2012). 'Conservation tillage issues: cover crop-based organic rotational no-till grain production in the mid-Atlantic region, USA'. *Renew. Agric. Food Syst.*, 27, 31–40. https://doi.org/10.1017/S1742170511000457

Mirsky, S.B., Ryan, M.R., Teasdale, J.R., et al. (2013). 'Overcoming weed management challenges in cover crop–based organic rotational no-till soybean production in the Eastern United States'. *Weed Tech.*, 27, 193–203. https://doi.org/10.1614/WT-D-12-00078.1

Nalewaja, J.D. (2003). 'Weeds and conservation agriculture'. In L. García-Torres, J. Benites, A. Martínez-Vilela, and A. Holgado-Cabrera (Eds.), *Conservation Agriculture*. Dordrecht: Springer. https://doi.org/10.1007/978–94-017–1143-2_25

Newton, P., Civita, N., Frankel-Goldwater, L. et al. (2020). 'What is regenerative agriculture? A review of scholar and practitioner definitions based on processes and outcomes'. *Front. Sustain. Food Syst.*, 194. https://doi.org/10.3389/fsufs.2020.577723

Nichols, V., Verhulst, N., Cox R. et al. (2015). 'Weed dynamics and conservation agriculture principles: a review'. *Field Crops Res.*, 183: 56–68. https://doi.org/10.1016/j.fcr.2015.07.012

Nosratti, I., Korres, N. E. and Cordeau, S. (2023). 'Knowledge of cover crop seed traits and treatments to enhance weed suppression: a narrative review'. *Agronomy*, 13(7), 1683. https://doi.org/10.3390/agronomy13071683

Peigné, J., Ball, B.C., Roger-Estrade, J. et al. (2007). 'Is conservation tillage suitable for organic farming? A review'. *Soil Use Manag.*, 23(2), 129–144. https://doi.org/10.1111/j.1475-2743.2006.00082.x

Perry, L., Blumenthal, D., Monaco, T. et al. (2010). 'Immobilizing nitrogen to control plant invasion'. *Oecologia*, 163, 13–24. https://doi.org/10.1007/s00442-010-1580-x

Pierce, K.W. and Nattrass, M.P. (2023). 'Rapid-screening bioassay assessing potential allelopathic influence on spinach by aqueous extract from fresh, whole-plant sorghum-sudangrass tissue'. *HortScience*, 58(10), 1183–1189. http://dx.doi.org/10.21273/HORTSCI17158-23

© Burleigh Dodds Science Publishing Limited, 2025. All rights reserved.

Ponisio, L.C., M'Gonigle, L.K., Mace, K.C. et al. (2015). 'Diversification practices reduce organic to conventional yield gap'. *Proc. Royal Soc. B*, 282, 20141396. https://doi.org/10.1098/rspb.2014.1396

Ranaldo, M., Carlesi, S., Costanzo, A. et al. (2020). 'Functional diversity of cover crop mixtures enhances biomass yield and weed suppression in a Mediterranean agroecosystem'. *Weed Res.*, 60(1), 96–108. https://doi.org/10.1111/wre.12388

Rasmussen, I.A., Askegaard, M. and Olesen, J.E. (1999). 'Weed control in organic crop rotations for grain production'. In Gut, D. (Ed.), *Proceedings 11th EWRS Symposium*, Basel. European Weed Research Society (EWRS), Wageningen, p. 98.

Reganold, J.P. and Wachter, J.M. (2016). 'Organic agriculture in the twenty-first century'. *Nature Plants*, 2, 15221. https://doi.org/10.1038/nplants.2015.221

Restuccia, A., Scavo, A., Lombardo, S. et al. (2020). 'Long-term effect of cover crops on species abundance and diversity of weed flora'. *Plants*, 9(11), 1506. https://doi.org/10.3390/plants9111506

Schreier, H., Bish, M. and Bradley, K.W. (2022). 'The impact of electrocution treatments on weed control and weed seed viability in soybean'. *Weed Tech.*, 36(4), 481–489. https://doi.org/10.1017/wet.2022.56

Schumacher, M. and Gerhards, R. (2022). 'Facilitation of weed seed predation by living mulch and cover crops'. *Weed Res.*, 62(5), 328–339. https://doi.org/10.1111/wre.12547

Sias, C., Wolters, B.R., Reiter, M.S. et al. (2021). 'Cover crops as a weed seed bank management tool: a soil down review'. Italian J. Agron., 16(4), 1852. https://doi.org/10.4081/ija.2021.1852

Smith, R.G., Warren, N.D. and Cordeau, S. (2020). 'Are cover crop mixtures better at suppressing weeds than cover crop monocultures?'. *Weed Sci.*, 68(2), 186–194. https://doi.org/10.1017/wsc.2020.12

Sportelli, M., Frasconi, C., Fontanelli, M. et al. (2022). 'Innovative living mulch management strategies for organic conservation field vegetables: evaluation of continuous mowing, flaming, and tillage performances'. *Agronomy*, 12(3), 622. https://doi.org/10.3390/agronomy12030622

Stagnari, F., Galieni, A., Speca, S. et al. (2014). 'Effects of straw mulch on growth and yield of durum wheat during transition to Conservation Agriculture in Mediterranean environment'. *Field Crops Res.*, 167, 51–63. https://doi.org/10.1016/j.fcr.2014.07.008

Sturm, D., Peteinatos, G. and Gerhards, R. (2018). 'Contribution of allelopathic effects to the overall weed suppression by different cover crops', *Weed Res.*, 58(5), 331–337. https://doi.org/10.1111/wre.12316

United Nations. (2015). *Resolution adopted by the General Assembly on 25 September 2015. 70/1. Transforming our world: the 2030 Agenda for Sustainable Development*, p. 35, https://documents-dds-ny.un.org/doc/UNDOC/GEN/N15/291/89/PDF/N1529189.pdf?OpenElement.

Verret, V., Gardarin, A., Pelzer, E. et al. (2017). 'Can legume companion plants control weeds without decreasing crop yield? A meta-analysis'. *Field Crops Res.*, 204, 158–168. https://doi.org/10.1016/J.FCR.2017.01.010

Vincent-Caboud, L., Vereecke, L., Silva, E.M. et al. (2019). 'Cover crop effectiveness varies in cover crop-based rotational tillage organic soybean systems depending on species and environment', *Agronomy*, 9(6), 319. https://doi.org/10.3390/agronomy9060319

© Burleigh Dodds Science Publishing Limited, 2025. All rights reserved.

Warren Raffa, D., Antichi, D., Carlesi, S. et al. (2021). 'Ground cover mulching in Mediterranean vineyards improves soil chemical, physical and biological health already in the short term'. *Agronomy*, 11, 787. https://www.mdpi.com/2073-4395/11/4/787

Wendling, M., Charles, R., Herrera, J. et al. (2019). 'Effect of species identity and diversity on biomass production and its stability in cover crop mixtures'. *Agric. Ecosyst. Environ.*, 281, 81–91. https://doi.org/10.1016/J.AGEE.2019.04.032

Westbrook, A. S., Bhaskar, V. and DiTommaso, A. (2022). 'Weed control and community composition in living mulch systems'. *Weed Res.*, 62(1), 12–23. https://doi.org/10.1111/wre.12511

Wezel, A., Bellon, S., Doré, T. et al. (2009). 'Agroecology as a science, a movement and a practice. A review'. *Agron. Sustain. Dev.*, 29, 503–515. https://doi.org/10.1051/agro/2009004

Zanin, G., Otto, S., Riello, L. et al. (1997). 'Ecological interpretation of weed flora dynamics under different tillage systems'. *Agric. Ecosyst. Environ.*, 66, 177–188. https://doi.org/10.1016/S0167-8809(97)00081-9

© Burleigh Dodds Science Publishing Limited, 2025. All rights reserved.

Chapter 11

Weed management in Conservation Agriculture-based production of annual crops

Ademir Calegari, Instituto Agronômico do Paraná (IAPAR), Brazil; Telmo Jorge Carneiro Amado, Federal University of Santa Maria, Brazil; Rolf Derpsch, Consultant (formerly GTZ), Germany; Ray Archuleta, USDA-ARS, USA; Martin Cubilla, FEPASIDAS, Paraguay; Frédéric Thomas, Farmer, France; Sergio Argenteri, Farmer, Italy; Fabricio Krzyzaniak, Agronomist Consultant, Paraguay; Kleso Silva Franco Junior, University CESEP, Brazil; Donizeti Aparecido Fornarolli, Consultant, Brazil; Benedito Noedi Rodrigues, Instituto Agronômico do Paraná (IAPAR), Brazil; Ronaldo Hissayuki Hojo, Regional Development Institute – Paraná, Brazil; Ken Moriya, FEPASIDAS/Agriculture Ministry, Paraguay; Cassio Egidio Cavenaghi Pretes, Londrina State University, Brazil; and Danilo Rheinheimer dos Santos, Federal University of Santa Maria, Brazil

1 Introduction

Defining, regenerating and maintaining healthy soils is seen as key to sustainable agricultural production (Reicosky et al., 2023). This principle underlies integrated management systems such as Conservation Agriculture (CA) which combine a range of techniques (minimal mechanical disturbance to the soil, permanent vegetation cover, and diversified crop rotations or other plant associations) to create a more sustainable agroecosystem (Derpsch et al., 2024). In developing this integrated management system, one of the biggest

http://dx.doi.org/10.19103/AS.2024.0145.17
© Burleigh Dodds Science Publishing Limited, 2025. All rights reserved.

challenges remains management of undesirable plants (traditionally labelled 'weeds') which threaten to unbalance the system.

The challenge of undesirable plants or weeds has encouraged the widespread use of herbicides for their control, even in CA systems. However, intensive use of herbicides such as glyphosate has accelerated the development of resistance, particularly in perennial and sub-perennial species. Persistent use of insecticides such as herbicides can also result in changes in and potentially an overall decrease in soil biology (macro, meso and micro-organisms), altering and reducing biodiversity in a way that can lead to an increase in the populations of some undesirable plant species, as well as reducing soil functions which promote crop growth. Broader soil management can also favor the concentration of nutrients and organic matter in the surface layers of the soil which can then accelerate the growth of undesirable plants (Calegari et al., 2008; Tiecher et al., 2020b). It is, therefore, desirable, wherever possible, to avoid using herbicides in the production system and, instead, to use other tools to manage and balance populations of undesirable plants. These include using soil cover and rotations in an integrated, no-till (NT) system such as CA. As a number of farmers have shown, it is possible to create balanced production systems that can control undesirable plants whilst achieving satisfactory (and even enhanced) yields and increasing farm net income in a sustainable and regenerative way. This chapter reviews the benefits and effects of service plants and crop rotation in NT cultivation systems such as CA in order to control undesirable plants in different annual crop production systems whilst minimizing the use of herbicides.

2 The evolution of a holistic approach to soil management

Conventional tillage-based soil management (including practices such as plowing and burning plant residues to clear soils for the next crop) has been used for the last 10,000 years and is still used to grow different crops in diverse agroecological conditions across the world (Lal, 2007; Reusser et al., 2015). Over time in some regions, these practices have resulted in significant soil losses due to wind and water erosion, as well as a decrease in the nutrient composition of soil, resulting in a loss of productive potential in these regions (Reicosky et al., 2021). For example, erosion rates from conventionally plowed agricultural fields have been found to be 1–2 orders of magnitude higher than rates of soil production (Lindstrom et al., 1990; Montgomery, 2007; Montanarella et al. 2016). According to Pimentel et al. (1995), e.g. about 430 million ha^{-1} (almost one-third of the global arable land area) has been lost to soil erosion.

These problems have been exacerbated by developments in weed management which have evolved from manual weeding to the use of

© Burleigh Dodds Science Publishing Limited, 2025. All rights reserved.

mechanized equipment like harrows, rotary hoes and brush cutters. However, like the growing use of (increasingly large) equipment for ploughing, these methods have also resulted in problems such as soil compaction or erosion. These developments in mechanical control have been accompanied by advances in chemical control through the use of herbicides which are easy to apply and can be very effective. However, as noted earlier, they have faced increasing problems from the development of resistant weed species as well as concerns about potential contamination of soil and water with herbicide residues and wider effects on biodiversity.

Other problems associated with intensive forms of agriculture include loss of soil organic carbon (SOC), soil contamination, soil acidification, soil salinization, soil biodiversity loss, soil compaction and poor soil nutrient and moisture status (Lal, 2015b). The disruption of agroecosystem functioning (e.g. through interventions such as tillage, monocropping and extensive use of synthetic fertilizers) has also been associated with proliferation of weed species which might normally have been kept in check, making weeds (predominantly early succession plants) a good indicator of ecosystem disturbance and a persistent problem in intensive systems (Knezevic et al., 2002; Lopes et al., 2013; Zimdahl, 2013). A growing global population, expected to reach nearly 10 billion by 2050 (United Nations, 2014; Sanderman et al., 2017), is exerting mounting pressure on the finite land area and resources for growing food. The combination with the cumulative environmental degradation associated with more intensive forms of agriculture jeopardizes our food security (Basche et al., 2016; Hatfield et al., 2017).

These concerns have driven the search for alternative approaches to conventional tillage. This lay behind the development of NT systems in the early 1960s in the USA and then in the early 1970s in Brazil and Latin America. Fundamental to these new NT systems was the need to minimize soil disturbance and to protect or restore the physical, chemical and biological properties of soil that could then reduce erosion risk, retain key nutrients and enable soil to deliver more sustainable crop production over the long term. A particular recent focus has been the way NT-based systems can protect and enhance the main groups of soil microorganisms such as bacteria, fungi, actinobacteria and algae, as well as macro fauna such as arthropods, which are particularly active in the rhizosphere (the area immediately around a plant's roots). This soil biological life is not just critical to nutrient cycling in soil and uptake by plants, but also in the delivery of other ecosystem services such as carbon sequestration (Tiecher et al., 2020a).

Interest in NT has led to the development of more holistic, integrated approaches, including Conservation Agriculture. (CA) is a broad term encompassing three main principles (Kassam and Friedrich, 2012; Lal, 2015a):

© Burleigh Dodds Science Publishing Limited, 2025. All rights reserved.

- continuous minimum soil disturbance (e.g. by abandoning plowing for alternatives such as direct drilling of seeds);
- continuous soil protection (e.g. with crop residue cover); and
- promotion of a diversity of crop species to optimize soil health (e.g. through diverse rotations to build soil organic matter, including integrating cover crops into a rotation).

By promoting soil health in these ways, CA aims to achieve a self-sustaining agroecosystem which is no longer reliant on costly and environmentally damaging external inputs such as synthetic fertilizers and pesticides. In this respect, cover crops in NT systems such as CA can play a broader role in optimizing the physical, chemical and biological properties of soil (Zibilske and Makus, 2009). Apart from their role in weed management (Hatcher and Melander, 2003), other benefits of well-managed cover crops include reduced erosion (Reicosky et al., 2021), improved soil organic-matter content (Poeplau and Don, 2015), promotion of good soil structure (Blanco-Canqui et al., 2011; Calegari et al., 2013), healthy microbial communities, and improved cycling and availability of nutrient sources such as nitrogen (N) (Schomberg et al., 2006; Tiecher et al., 2013, 2017; Reicosky et al., 2021). Maintaining soil cover has also been associated with control of insect pests (Peachey et al., 2002), pathogenic nematodes (DuPont et al., 2009) and pathogens (Conklin et al., 2002; Manici et al., 2004).

This broader role of cover crops in soil health can be seen, e.g. in the way cover crops help to retain soil moisture, an increasingly important issue given the increased risk of more extreme weather conditions (such as more intense periods of drought) linked to climate change. Cover crops can help keep undesirable plants under control by outcompeting them for key resources such as water. However, by providing soil cover and building organic matter, cover crops also both avoid water loss through evaporation and promote improved water infiltration/holding capacity in soil which benefits overall soil health and the entire production system (FAO, 2014).

3 Cover crops and weed control in Conservation Agriculture (CA) systems: mechanisms of action

The key to the control of undesirable plants is not to allow them to complete their life cycle (germination/emergence/growth), increase their populations and reseed at a level that may outcompete planted crops for resources. This requires combining a range of techniques, including an appropriate mix of cover crops/service plants which, by combining different species, can compete effectively with undesirable plants and maintain a high level of soil cover (typically at heights of 3–5 cm) (Creamer and Baldwin, 2020; Liebman et al., 2012; Teasdale

© Burleigh Dodds Science Publishing Limited, 2025. All rights reserved.

et al., 2007). Cover crops must integrate effectively with rotations of commercial crops both to prevent weed growth and to provide favorable conditions for commercial crop development with better performance. This mixture of cover crop species must be designed according to a range of criteria, including soil characteristics and an understanding of the characteristics and life cycles of native invasive plants with which they must compete.

The use of diverse cover crop species and mulches can reduce the germination and the development of weed seeds (Weston, 1996; Ohno et al., 2000; Lemessa and Wakjira, 2015) through allelopathic effects from plant exudates (Teasdale et al., 2007; Kruidhof et al., 2008b) which inhibit weed growth, and from competition between cover crops and weeds for resources such as light, water and nutrients (Crutchfield et al., 1985; Den Hollander et al., 2007; Kruidhof et al., 2008a). An additional potential benefit is the unique microclimate at the soil surface which can promote insect species predating weed seeds (Meiss et al., 2010; Petit et al., 2018).

One factor in weed suppression by cover crops is by interfering with red light transmittance through the cover crop canopy which reduces weed photosynthetic activity (Teasdale and Mohler, 1993; Kruk et al., 2006). Lawley et al. (2012) also attributed the weed-suppressive effect of a fall forage radish cover crop to interference with red light transmittance. These light transmittance effects are specific for weed seeds near the soil surface.

Other factors can also be involved. Mohler et al. (2012) found reduced emergence of several annual weeds, including velvet leaf (*Abutilon theophrasti*), lamb's quarters (*Chenopodium album*), Powell's amaranth (*Amaranthus powellii*), giant foxtail (*Setariafaberi*) and cockspur (*Echinochloa crus-galli*), following the incorporation of green manures consisting primarily of peas (*Pisum sativum*). They isolated the pathogens *Fusarium oxysporum* and *Fusarium chlamydosporum* from weed seeds collected from soils that had received green manures which provided a conducive environment for pathogens affecting weed seeds. Lou et al. (2016) also found weed suppression after a red clover cover crop. They concluded that the suppression was due to a combination of allelopathy and microbial infection of weed seeds and seedlings. They suggest a phased weed inhibition where allelopathic interactions are most important for a period after cover crop residue deposition. However, as some soil microbes break down allelochemicals, other members of the soil microbial community become more important and inhibit weeds through seed and/or seedling infection.

Allelopathic compounds are organic, often phytotoxic plant substances (Kimber, 1973; Liebl and Worsham, 1983). Although some have a stimulating effect, others have inhibitory effects on other plant species which can be exploited for weed suppression (Jabran et al., 2015). Despite first being described in the early twentieth century by Austrian botanist Hans Molisch, our

© Burleigh Dodds Science Publishing Limited, 2025. All rights reserved.

understanding of allelopathic interactions is still developing (Molisch, 1937; Kruidhof et al., 2008b). Allelopathic effects include inhibiting seed germination, respiration, photosynthesis, protein synthesis, cell division, nutrition and reproduction (Teasdale and Mohler, 1993; Teasdale et al., 2007).

Many cover crops have been reported to have allelopathic activity (Table 1). Many compounds with different action modes can be involved in allelopathic interactions. These compounds can be released into the soil in different ways (Khanh et al., 2005; Kelton et al., 2012). Some are released from living plants as root exudates or volatiles, while others are released from decomposing plant tissues (Lemessa and Wakjira, 2015). It is important to be aware that different cultivars of a plant species may have greater or lesser allelopathic potential, including cover crops such as rye (Reberg-Horton et al., 2005), wheat (Baghestani et al., 1999), sunflower (Leather, 1983), sorghum (Alsaadawi et al., 1986) and oats (Fay and Duke, 1977).

At the end of the cover crop cycle, the cover crop mulch can also continue to suppress weed emergence (Thiesen et al., 2000). The mulch or incorporated residues associated with terminated cover crops can impact weeds by physically or chemically inhibiting germination of weed seeds, releasing weed-inhibiting allelopathic compounds during decomposition, preventing weed emergence and immobilizing nitrogen and consequently its availability for weed growth (Weston, 1996; Ohno et al., 2000; Den Hollander et al., 2007; Teasdale et al., 2007). One issue to bear in mind is the way remaining mulches affect the efficacy of any subsequent use of herbicides, including the risk of leaching (Schmidt and Teschner, 2000; Mendes et al. 2019; Nunes and Vidal, 2016).

Table 1 Allelopathic and suppress effects of cover crops on different weeds

Cover crop species	Weeds controlled
Crotalaria juncea	*Brachiaria* sp., *Bidens pilosa, Galinsoga parviflora, Senchrus echinatus*, etc.
Avena strigose	*Brachiaria plantaginea*, other grass species
Mucuna pruriens, Crotalaria juncea, Canavalia ensiformis	*Cyperus rotundus*
Mucuna pruriens	*Bidens pilosa, Galinsoga parviflora, Senchrus echinatus*
Secale cereale and *Mucuna pruriens*	*Brachiaria plantaginea* can be controlled by many months
Vicia sativa	Sowed before maize promote efficient control of *Brachiaria plantaginea*
Lollium multiflorum	*Sida rhombifolia*
Wheat straw	*Cassia tora*
Tajetes patula	*Ipomoea* sp., *Amaranthus* sp., *Desmodium purpureum, Momordica charantia, Euphorbia heterophyla*

Source: Almeida and Rodrigues (1985), Calegari et al. (1993).

© Burleigh Dodds Science Publishing Limited, 2025. All rights reserved.

An important variable to bear in mind is that weed species composition and occurrence can evolve with the use of techniques such as cover crops (Gomee and Christoffoleti, 2008). This means that design principles for cover crop use may need to be defined for each plot or farm area and may need to evolve over time as weed dynamics change (Shahrtash and Brown, 2021). Where, occasionally, some invasive species are difficult to control and are localized in some plot areas, mechanical weeding, e.g. using hoes, may still be required.

4 CA-based weed management practices for annual crops

There is a huge volume of research on cover crops. This section seeks to review and synthesize some of the major studies in this area. It starts with a number of general overviews of the range of research. It then discusses work focusing on different types of cover crop before looking at examples of their use with particular crops.

In terms of assessing overall effects, the weed suppression capacity of cover crops has been investigated recently by Osipitan et al. (2019) in a meta-analysis of 53 studies published from 1990 to 2018 in different cropping systems. The results show that cover crop biomass production was inversely related to the amount of weed biomass (r^2 = 0.67) and weed density (r^2 = 0.64). In this study, the effectiveness of weed suppression was affected by management factors such as:

- choice of cover crop species;
- sowing dates within growing seasons;
- seeding rate;
- methods for cover crop establishment;
- termination date;
- delays in primary crop planting after cover crop termination; and
- integration of cover crops with other weed control techniques.

In general, grass cover crop species were more efficient in weed suppression than broadleaf species. Success in weed control relied particularly on sowing rate (with an increased sowing rate providing denser, more complete cover and thus better weed suppression). Cover crop effects are most significant in the first weeks after planting the following cash crop, highlighting the importance of timing in planting a cash crop. Devide et al. (2009) also found that the combination of cash and cover crops successfully controlled weed populations. Additionally, it decreased costs from herbicide weed control and provided additional income for the farmer, as well as better nutrient supply for crops from a more healthy soil with improved soil organic matter. Other

© Burleigh Dodds Science Publishing Limited, 2025. All rights reserved.

research has highlighted the potential cost savings from cover crop use for weed control. In a study of 58 farms in Paraná State in Brazil, Adegas (1998) found that, after 3 years with adequate weed management, weed control costs decreased on average by 35% with a reduction in herbicide use of 25%. Ruedell (1995), studying soybean, maize, wheat systems rotated with winter cover crop species, found a 42% reduction in weed control costs assuming farmers followed optimal weed management practices.

A particularly important and long-established area of research is the use of cover crop mixes. Derpsch and Calegari (1992) e.g. reviewed around 167 varieties in 51 countries in diverse cropping and farming systems and found the best results were achieved by using multiple species (4, 6, 8 or more) from different families of plants. As noted earlier, cover crops have a potentially multifunctional role, not just in weed suppression but in promoting the broader physical, chemical and biological health of soils, which can be best realized by using cover crop mixtures (Calegari et al., 2023; Reicosky et al., 2023). More recently, Thomas and Archambeaud (2016), working in different regions of France, have researched ways of combining different cover crop species in cropping systems such as wheat, barley, white oat, potatoes, canola and maize as well as perennial crops such as grapes and tree fruits. Other research has also shown the value of the strategic use of combining cover crop mixes and cash crops (Wiggins et al., 2016; Lamichhane et al., 2017; Marochi et al., 2018).

4.1 Types of cover crops

As suggested above, combinations of cereal and legume cover crops are widely used in many cropping systems (Isik et al., 2009). Generally, the use of legume cover crops is mainly due to their ability to fix nitrogen, while the use of cereals is mainly for superior cover and their consequent ability both to reduce nitrogen leaching while simultaneously suppressing weeds (Hooker et al., 2008; Bond and Grundy, 2001). Studying seed bank weed dynamics under different soil management systems, Voll et al. (1997, 2001), e.g. found that grasses provided particularly good soil cover in suppressing weed seed germination and emergence.

Results obtained in a cropping system developed over many years in Paraná State in South Brazil, including summer maize and soybean and winter cover crops, show both effective weed suppression and significant yield increases of cash crops (Calegari et al., 2008). Rodrigues and Almeida (2018) have shown that cover crops like black oats, oilseed radish and hairy vetch can be effective in reducing weed populations in NT systems. They found a linear correlation between the amount of biomass produced by a cover crop and its effectiveness in suppressing weeds. However, it is important to take into account environmental and other factors. In tropical environments,

© Burleigh Dodds Science Publishing Limited, 2025. All rights reserved.

where higher temperatures and precipitation can result in more rapid residue decomposition, *poacea* cover crops had a higher potential for weed control than legumes. However, legume cover crops remain critical to enhancing soil health (including biological activity). As noted, a consortium of legume and grass cover crops is an efficient tool to provide both weed control and improve soil health.

Different types of winter and summer cover crops have been shown to suppress weeds because of their fast pattern of growth (Calegari et al., 1993). For example, Favero et al. (2001) observed a 22–96% reduction of weed biomass using summer cover crops depending on species. Vasconcelos and Landers (1993) reported that planting grain crops into permanent cover crops, of which maize into siratro (*Macroptilium atropurpureum*) was the most successful, allowed a complete elimination of herbicide use. Fernandes et al. (1999) observed that *Crotalaria breviflora*, *Crotalaria spectabilis* and pigeon pea plots reduced densities of weeds, while Skora Neto (1993) also noted that pigeon pea intercropped with maize decreased weed infestation at and after harvesting.

Research by Meschede et al. (2007) found that sorghum had the highest dry cover and, together with millet and sunn hemp, showed a greater ability to suppress weeds due to a combination of good soil cover and allelopathic effects. In a cropping system including soybean and cotton, Ferreira and Lamas (2010) found the best results in weed suppression were *Brachiaria ruziziensis*, *Brachiaria ruziziensis* + *Crotalaria juncea*, *Brachiaria ruziziensis* + *Crotalaria spectabilis* and *Pennisetum glaucum* + *Raphanus sativus*, which all produced more than 6.8 t ha^{-1} of dry mass. However, straw residue of *Raphanus sativus* and *Avena strigosa* was found to inhibit early cotton growth and thus decrease cotton yields. However, studying maize intercropped with *Brachiaria ruziziensis*, Concenço at al. (2012) found less impact by the cover crop in weed suppression. In assessing different autumn/winter cover crops, Correia et al. (2006) found that weed emergence was lower using sorghum and brachiaria and ensuring higher levels of mulch, including preventing seedlings of *Bidens pilosa*, *Amaranthus* spp., *Commelina benghalensis* and *Leucas martinicensis* and grasses. Borges et al. (2014) demonstrated that *U. ruziziensis* and *S. sudanense* reduced weed infestation by more than 90% and maintained soil cover of more than 80% until the flowering of the following soybean crop, significantly decreasing the use of herbicides.

4.2 Examples of cover crop use with particular crops

There is a wealth of research on the use of different cover crops and their mixes with particular cash crops. A few examples are noted here. In Brazil, e.g. cover crops (mainly mixes of different species) can be grown to support maize,

© Burleigh Dodds Science Publishing Limited, 2025. All rights reserved.

sunflower, sorghum and cassava. CA practices have also been deployed in sugarcane cultivation, e.g. by effective soil cover which both controls weeds and improves overall soil fertility (Ferreira et al., 2010). Cassava crops, e.g. have slow initial growth, leaving soil potentially uncovered and facilitating the development of weeds. Azevedo et al. (2000) found that the use of cover crops helped both to establish the cassava crop and control weeds during the early stages of cassava growth.

Based on research from Paraguay, Florentin et al. (2010) have shown the successful combination of cover crops with cash crops such as cotton, maize, tobacco and cassava in decreasing weed populations. Similar results have been obtained in Brazil (Bolliger et al., 2006; Calegari et al., 2023; Reicosky et al., 2023). Calegari et al. (2006), Calegari et al. (2023) and Reicosky et al. (2023), working in southwestern Paraná State in Southern Brazil on clay soils, have investigated the use of different cover crops in a long-term experiment (currently running for more than 36 years). Studies of soybean and maize rotated with different winter/fall species have shown the strong effects of rye, black oat, hairy vetch, common vetch and mixes of these crops in weed control, including species such as *Brachiaria plantaginea*, *Bidens pilosa*, *Digitaria* spp., *Conyza* spp. and *Euphorbia* spp.

In soybean production, the use of *Urocloa ruziziensis* and *Sorghum sudanense* as cover crops to achieve soil cover of over 80% successfully suppressed weeds until the soybean flowering stage when the main crop was sufficiently established to outcompete potential weeds (Borges et al., 2014). The sorghum cover crop biomass promoted efficient suppression of weeds, since sorghum synthesizes allelopathic substances such as sorgoleone which inhibits photosynthesis and growth of competing plant species (Santos et. al, 2012). However, it is important to manage the cover crop carefully so it does not sprout and start to compete itself with the cash crop (Gomes et al., 2014). Gomes et al. (2014) also found that other potential cover crops, such as sunn hemp, had insufficient biomass to control weeds as well faster rates of decomposition in a tropical environment. Kliewer et al. (1998) have investigated the effects of different winter cover crop residues in NT soybeans cultivation (Table 2). Compared to fallow, crop residues of species such as oats (black and white), wheat and rye had strong effects in weed control.

In the case of coffee cultivation, Franco Junior et al. (2022) (Table 3) have shown that a mix of cover crops is an efficient tool in suppressing weeds, including fast-growing species such as *Bidens pilosa*. Moreira et al. (2013) have also shown that the longer-term intercropping of coffee trees with legumes was effective in weed control.

One example of the success in using cover crops is in the cultivation of oil seed rape (OSR). OSR is planted with a mix of companion plants, particularly legumes but also field beans, vetch, berseem clover, lentil, and fenugreek as

© Burleigh Dodds Science Publishing Limited, 2025. All rights reserved.

Table 2 Effects of winter species residue and fallow on weed dry mass in an NT soybean system in Paraguay

Winter species	Dry mass (kgha^{-1})
Fallow	7390 a
Radish	4258 b
White lupin	3735 bc
Field pea cv. Iapar-78	2263 cd
Triticale	1818 d
Sunflower	1653 de
Rye	760 de
White oat cv. IAC-7	723 de
Wheat	498 de
Black oat cv. Iapar 61	93 e

Kliewer et al. (1998). Averages followed by the same letter do not differ statistically by the Tukey test at 5%.

Table 3 Consortium of cover crops for weed control in coffee production

Undesirable plant species	Fallow	Mix summer legumes*	Undesirable plant species	Fallow	Mix**
	Plants m^{-2}			Plants m^{-2}	
Conyza spp.	12.3	0	*Conyza* spp.	14.8	18.2
Digitaria insularis	18.2	17.1	*Digitaria insularis*	18.3	6.3
Ipomoea spp.	7.14	0	*Ipomoea* spp.	29.9	10.0
Richardia brasiliensis	4.1	0			
Bidens pilosa	15.5	28.4			

*MIX summer legumes = (*Crotalaria spectabilis* + pigeon pea + cowpea) and Mix = (*Crotalaria spectabilis* + pigeon pea + cowpea + buckwheat). Adapted from Franco Junior et al. (2022).

well as smaller amounts of plants such as flax, buckwheat and sunflower (Hirel at al., 2011; Lorin et al., 2015). This extra early vegetation covers the soil faster and, as well as controlling weeds, improves nitrogen availability, supplies more biomass to soil and can confuse insect pests by making it harder to locate the target OSR crop. During the winter, most of these cover crops die off, leaving the OSR to grow in the following spring. Recent additions by networks of CA farmers in France include mix lucerne, clover or white clover. Those plants usually survive the winter and provide substantial cover as the OSR starts to grow in the Spring, both controlling weeds and adding N to the system.

A specific example of using mixes of cover crops is that provided by Sergio Argenteri, farming 20 ha in the Piemonte Region of Italy (Sergio Argenteri,

© Burleigh Dodds Science Publishing Limited, 2025. All rights reserved.

Personal communication). Farming 20 ha, he has developed a CA system over the past 6 years, with a cropping system comprising mainly maize, wheat, soybean, sunflower and pea, rotated with summer/spring and autumn/winter service plants. He uses mixes of cover crops with 16 winter species such as *Avena strigosa, secale cereale, Raphanus sativus, Pisum sativum, Vicia sativa, Vicia villosa, Facelia tanacetifolia, Trifolium* spp, *Vicia faba, Linum usitatissimum* and *Lupinus angustifolius*. Summer species include *Pennisetum glaucum, Crotalaria* spp., *Vigna unguiculata, Fagopirum esculentum, Helianthus annuus, Sorghum* spp., among others. Results have been promising with soil cover both successfully controlling weeds and improving soil health (including indicators such as higher amounts of soil organic matter and earthworm numbers) which have also improved crop yields.

These examples suggest that selection of the right cover crop mix can be optimized by good soil diagnostics to identify soil structural and nutritional deficiencies that point to species with particular root characteristics or ability to fix specific nutrients such as N. Measuring the activity of enzymes such as Beta glycosidase and Arilsulfatase has been found to provide a good general indicator of soil health (Mendes et al., 2019).

Once soil health requirements have been identified, experience of using CA in Brazil e.g. has shown that cover crops can be successfully combined with other soil improvement practices. These include adding lime (calcium carbonate and magnesium carbonate) to neutralize potential acidity, in particular to precipitate exchangeable aluminum (Kaminski et al., 2002; Gatiboni et al., 2003). A synergistic effect has been found by combining lime and cover crops that helps deepen the root system and improve crop nutrient intake (Rheinheimer et al., 2024a, 2024b). Combining practices such as cover crops and lime to restore soil health sufficiently to allow a second summer crop in some fields. Other amendments that can be combined with cover crops include biostimulants. In soybean cultivation, this includes applying a mixture of *Bradyzrizobium japonicum*, *Azospirillum* spp. and different *Bacillus* spp. to increase biological nitrogen fixation.

These practices can also be combined with increased use of precision and other technologies agriculture which enable precision planting and precision fertilizer application to meet the 4 Rs (right source, right rate, right time and right place) and, where necessary, more targeted use of herbicides, which help to optimize resource use whilst decreasing infield spatial yield variability (MacRae et al., 1990; Riemens et al., 2002). In terms of effects on soil structure, light and median compacted soil layers can be effectively loosened by a robust cover crop root system. It is, however, also possible to combine cover cropping with occasional soil chiseling, a shallow tillage technique that involves loosening the soil using a chisel plow or cultivator, particularly to alleviate significantly compacted soils.

© Burleigh Dodds Science Publishing Limited, 2025. All rights reserved.

5 Conclusion

As this chapter has shown, cover crops are an essential tool in NT-based systems such as CA as well as an integrated weed management program. The broad range of potential cover crop species allows their use in highly diverse agroecosystems around the world with a wide range of cash crops. Optimizing their use requires balancing a range of factors, including:

- identifying key objectives for their use, given that cover crops have a multifunctional role beyond weed suppression, particularly in improving soil health with consequent benefits for cash crop yields;
- understanding weed life cycles and the ways these can be disrupted by cover crops;
- understanding mechanisms of action in the way cover crops interact with weeds and with cash crops; and
- developing a tailored solution best fitted to a specific local context to meet a particular mix of objectives.

In many cases, the solution is to use a cover crop mix comprising 4, 6, 8, 10 or more species. This remains challenging given the number of variables which need to be accounted for. This is particularly so since the use of cover crops for weed control does not happen in a vacuum but is, rather, a crucial component of a holistic ecological cropping system design. Specific areas where research is needed include the development of highly allelopathic cover crop species and varieties, further development of regionally suitable species and varieties, and refining of the timing of cover crop planting and termination for maximum weed suppression during critical periods for crop growth.

Ongoing concern about the negative impacts and viability of continued herbicide use mean that both producers and consumers have a growing interest in more sustainably managed food crops, including certified organic and regenerative cropping systems. This means that interest in cover crops for weed control will continue to grow. This interest will sharpen the need for more research in this area, as well as taking advantage of the many innovative strategies farmers have developed for themselves in looking for better solutions to facilitate effective weed control and in developing more sustainable and resilient agroecosystems.

6 Where to look for further information

- Calegari, A., et al. (2008), '*Impact of Long-Term No-Tillage and Cropping System Management on Soil Organic Carbon in an Oxisol: A Model for Sustainability*', *Agron. J.*, 100, 1013–1019.

© Burleigh Dodds Science Publishing Limited, 2025. All rights reserved.

- Rodrigues, B.N. and Almeida, F.S. (2018). *Guia de herbicidas*, 7ª. Edição, Produção Independente, 764p.
- Reicosky, D.C., et al. (2023), Cover Crop Mixes for Diversity, Carbon, and Conservation Agriculture, In: Islam, R. and Sherman, B. (eds.), *Cover Crops and Sustainable Agriculture*, 1 ed. CRC Press: Boca Raton, Florida, 2023. p.170–209.
- Montgomery, D.R. (2007), *'Soil erosion and agricultural sustainability'*, *Proc. Nat. Acad. Sci. USA*, 104(13), 268–272.
- Teasdale, J.R., et al. (2007), Cover crops and weed management, In: Updadhyaya, M.K. and Blackshaw, R.E. (eds.), *Non-chemical Weed Management: Principles, Concepts and Technology*, CABI Publishing, Wallingford, UK. pp. 49–64.
- Florentin, M.A., et al. (2010), *'Geen manure/cover crops and crop rotation in conservation agriculture on small farmers'*, *Integr. Crop Manag.*, 12, 11–Plant Production and Protection Division Food and Agriculture Organization of the United Nations Rome FAO) ISBN 978-92-5-106839-7 ISSN 1020-4555. 109p.

7 References

Adegas, F.S. (1998), 'Manejo integrado de plantas daninhas em plantio direto no Paraná', In *Seminário nacional sobre manejo e controle de plantas daninhas em plantio direto*, 1, Palestras, pp. 17–26. Aldeia Norte, Passo Fundo, RS, Brazil.

Almeida, F.S. and Rodrigues, B.N. (1985), *Guia de herbicidas; contribuicao para o uso adequado em plantio direto e convencional*, Londrina, Iapar.

Alsaadawi, I.S., et al. (1986), 'Allelopathic suppression of weed and nitrification by selected cultivars of Sorghum bicolor (L.) Moench', *J. Chem. Ecol.*, 12, 209–219.

Azevêdo, C.L.L., et al. (2000), 'Levantamento de plantas daninhas na cultura da mandioca, em um ecossistema semiárido do Estado da Bahia', *Magistra*, 12, 41–49.

Baghestani, A., et al. (1999), 'Determination of allelochemicals in spring cereal cultivars of different competitiveness', *Weed Sci.*, 47, 498–504.

Basche, A.D., et al. (2016), 'Simulating long-term impacts of cover crops and climate change on crop production and environmental outcomes in the Midwestern United States'. *Agric., Ecosys. Environ.*, 218, 95–106.

Blanco-Canqui, H., et al. (2011), 'Addition of cover crops enhances no-till potential for improving soil physical properties', *Soil Sci. Soc. Amer. J.*, 75, 1471–1482.

Bolliger, A., et al. (2006), 'Taking stock of the Brazilian "Zero till revolution": a review of landmark research and farmers' practice', *Adv. Agron.*, 91, 47–110.

Bond, W. and Grundy, A.C., (2001), 'Non-Chemical Weed Management in Organic Farming Systems', *Weed Research*, 41, 383–405.

Borges, W.L.B., et al. (2014), 'Supressão de plantas daninhas utilizando plantas de cobertura do solo', *Planta Daninha*, 32(4), 755–763.

Calegari, A. (2006), *Sequestro de carbono, atributos físicos e químicos em diferentes sistemas de manejo em um latossolo argiloso do sul do Brasil*, Tese Doutorado, Universidade Estadual de Londrina, 191p.

© Burleigh Dodds Science Publishing Limited, 2025. All rights reserved.

Calegari, A., et al. (1993), *Adubação verde no sul do Brasil*, 2nd, Ed. Assessoria e Serviços a Projetos em Agricultura Alternativa (AS-PTA), Rio de Janeiro.

Calegari, A., et al. (2008), 'Impact of long-term no-tillage and cropping system management on soil organic carbon in an oxisol: a model for sustainability', *Agron. J.*, 100, 1013–1019.

Calegari, A., et al. (2013), 'Influence of soil management and crop rotation on physical properties in a long-term experiment in Paraná, Brazil', *Commun. Soil Sci. Plant Analysis*, 44(13), 2019–2031.

Calegari, A., et al. (2023), 'Evaluating C trends in clayey Cerrado Oxisols using a four-quadrant model based on specific arylsulfatase and β-glucosidase activities', *Appl. Soil Ecol.*, 183, 104742.

Concenço, G., et al. (2012), 'Phytosociology in agricultural areas submitted to distinct winter cropping management', *Planta Daninha*, 30(2), 297–304.

Conklin, A.E., et al. (2002), 'Effects of red clover (Trifolium pratense) green manure and compost soil amendments on wild mustard (Brassica kaber) growth and incidence of disease', *Plant and Soil*, 238(2), 245–256.

Correia, N.M., et al. (2006), 'Influência do tipo e da quantidade de resíduos vegetais na emergência de plantas daninhas', *Planta Daninha*, 24(2), 245–253.

Creamer, N.G. and Baldwin, K.R. (2000), 'An evaluation of summer cover crops for use in vegetable production systems in North Carolina', *Hort Sci.*, 35, 600–603.

Crutchfield, D.A., et al. (1985), 'Effect of winter wheat (Triticum aestivum) straw mulch level on weeds control', *Weed Sci.*, 34, 110–114.

Den Hollander, N.G., et al., (2007), 'Clover as a cover crop for weed suppression in an intercropping design II. Competitive ability of several clover species', Eur J Agron, 26, 104–112.

Derpsch, R. and Calegari, A. (1992), *Adubação verde de inverno*, Londrina, IAPAR, 2a. edição, 80p. (IAPAR, Circular, 73).

Derpsch, R., et al. (2024), "Nature's laws of declining soil productivity and conservation agriculture', *Soil Security*, 14, 100127.

Devide, A.C.P., et al. (2009), 'Produtividade de raízes de mandioca consorciada com milho e caupi em sistema orgânico', *Bragantia*, 68(1), 145–153.

DuPont, S.T., et al. (2009), 'Effects of cover crop quality and quantity on nematode-based soil food webs and nutrient cycling', *Applied Soil Ecology*, 41(2), 157–167.

FAO. (2014), *The State of Food Insecurity in the World 2014: Strengthening the Enabling Environment for Food Security and Nutrition*, FAO, Rome. http://www.fao.org/soils-portal/soil-degradation-restoration/global-soilhealth-indicators-and-assessment/global-soil-health/en/

Favero, C., et al. (2001), 'Modificações na população de plantas espontâneas na presença de adubos verdes', *Pesq. Agropec. Bras.*, 36, 1355–1362.

Fay, P.K. and Duke, W.B. (1977), 'An assessment of allelopathic potential in Avena germ plasm', *Weed Science*, 25(3), 224–228.

Fernandes, M.F., et al. (1999), 'Fitomassa de adubos verdes e controle de plantas daninhas em diferentes densidades populacionais de leguminosas', *Pesq. Agropec. Bras.*, 34, 1593–1600.

Ferreira, A.C.B. and Lamas, F.M. (2010), 'Espécies vegetais para cobertura do solo: influência sobre plantas daninhas e a produtividade do algodoeiro em sistema plantio direto', *Rev. Ceres*, 57(6), 778–786.

© Burleigh Dodds Science Publishing Limited, 2025. All rights reserved.

Ferreira, E.A., et al. (2010), 'Manejo de plantas daninhas em cana-crua', *Planta Daninha*, 28(4), 915–925.

Florentin, M.A., et al. (2010), 'Geen manure/cover crops and crop rotation in conservation agriculture on small farmers', *Integr. Crop Manag.*, 12, 11–Plant Production and Protection Division Food and Agriculture Organization of the United Nations Rome FAO) ISBN 978–92-5-106839-7 ISSN 1020-4555. 109p.

Franco Júnior, K.S., et al. (2022), Sistema produtivo do café em plantio direto/cultivo mínimo/conservacionista', In: Sistemaplantio direto no Brasil, Editora Aldeia Norte, Passo Fundo, capítulo 10, 218–244.

Gatiboni, L.C., et al. (2003). 'Alterações nos atributos químicos de solo arenoso pela calagem superficial no sistema plantio direto consolidado', *Ci. Rural*, 33, 283–290.

Gomes, D.S., et al. (2014), 'Supressão de plantas daninhas pelo uso de cobertura vegetal', *Rev. Bras. Agroec.*, 9(2), 206–213.

Gomee, J.R.F.G. and Christoffoleti, P.J. (2008), 'Biologia e manejo de planas daninhas em áreas de plantio direto', *Planta Daninha*, 26(4), 789–798.

Hatcher, P.E. and Melander, B. (2003), 'Combining physical, cultural and biological methods: Prospects for integrated non-chemical weed management strategies' *Weed Research*, 43(5), 303–322.

Hatfield, J.L., et al. (2017), 'Soil: the forgotten piece of the water, food, energy nexus', *Adv. Agron.*, 143, 1–46.

Hirel, B., et al. (2011). 'Improving nitrogen use efficiency for sustainable agriculture', *Sustainability*, 3(9), 1452–1485.

Hooker KV, et al. (2008), 'Evaluation of cover crop and reduced cultivation for reducing nitrate leaching in Ireland', *J Environ Qual,* 37(1), 138–145.

Isik D, et al., (2009), 'Weed suppression in organic pepper (Capsicum annuum L.) with winter cover crops', *Crop Prot.,* 28, 356–63.

Jabran, K., et al. (2015), 'Allelopathy for weed control in agricultural systems', *Crop Prot.*, 72, 57–65.

Kaminski, J., et al. (2002), 'Estimativa da acidez potencial em solo e sua implicação no cálculo da necessidade de calagem', *Rev. Bras. Ci. Solo*, 26(4), 1107–1113.

Kassam, A. and Friedrich, T. (2012), *An ecologically sustainable approach to agricultural production intensification: Global perspectives and developments*, Field Actions Science Reports, Reconciling Poverty Eradication and Protection of the Environment, Special Issue 6 (2012). Institute Veolia Environment, France.

Kelton, J., et al. (2012), 'Allelopathic weed suppression through the use of cover crops', in: Price, A.J. (ed.), *Weed Control*, Intech Press, Rijeka, Croatia. pp. 115–130.

Khanh, T.D., et al. (2005), 'Weed Control The exploitation of crop allelopathy in sustainable agricultural production', *J. Agron. Crop Sci.*, 191, 172–184.

Kimber, R.W.L. (1973), 'Phytotoxicity from plant residues. II. The effect of time of rotting of straw from some grass and legumes on the growth of wheat seedlings', *Plant and Soil*, 38, 347–361.

Kliewer, I., et al. (1998), 'Feasibility of reducing herbicide use and non-plant control costs in wheat and soybean cultures in direct planting system, through the introduction of short-term green manure', In: *I National Seminar on Management and Control of Plants in Straight Planting. Summary of Palestras*, North Village. Passo Fundo, Brazil, 1998. pp. 120–123.

Knezevic, S.Z., et al. (2002), 'Critical period for weed control: the concept and data analysis', *Weed Sci.*, 50, 773–786.

© Burleigh Dodds Science Publishing Limited, 2025. All rights reserved.

Kruidhof, H.M., (2008a), 'Cover crop residue management for optimizing weed control', *Plant and Soil*, 318(1–2), 169–184.
Kruidhof, H.M. (2008b), 'Ecological weed management by cover cropping: effects on weed growth in autumn and weed establishment in spring', *Weed Research*, 48(6), 492–502.
Kruk, B., et al. (2006), 'Light and thermal environments as modified by a wheat crop: effects on weed seed germination', *J. Appl. Ecol.*, 43, 227–236.
Lal, R. (2007), 'Carbon sequestration', *Phil. Trans. R. Soc.*, 363, 815–830.
Lal, R. (2015a), 'A system approach to conservation agriculture', *J. Soil Water Conser.*, 70, 82–88.
Lal, R. (2015b), 'Cover cropping and the "4 per thousand" proposals', *J. Soil Water Conser.*, 70, 141A.
Lamichhane, J.R., et al. (2017), 'Integrated weed management systems with herbicide-tolerant crops in the European Union: lessons learnt from home and abroad', *Crit. Rev. Biotech.*, 37, 459–475.
Lawley, Y.E., et al. (2012), 'The mechanism for weed suppression by a forage radish cover crop', *Agron. J.*, 104, 205–214.
Leather, G.R. (1983), 'Sunflowers (*Helianthus annuus*) are allelopathic to weeds', *Weed Sci.*, 31, 37–42.
Lemessa, F. and Wakjira, M. (2015), 'Cover crops as a means of ecological weed management in agro-ecosystems', *J. Crop Sci. Biotech.*, 18(2), 133–145.
Liebl, R.A. and Worsham, A.D. (1983), 'Inhibition of pitted morning glory (Ipomoea lacunosa L.) and certain other weed species by phytotoxic components of wheat (Triticum aestivum L.) straw', *J. Chem. Ecol.*, 9, 1027–1043.
Liebman, M., et al. (2012), 'Use of legume green manures as nitrogen sources for corn production', *Renewable Agric. Food Sys.*, 27, 180–191.
Lindstrom, M.J., et al. (1990), 'Soil movement by tillage as affected by slope', *Soil Till. Res.*, 17(3–4), 255–264.
Lopes, A.A.C., et al. (2013), 'Interpretation of microbial soil indicators as a function of crop yield and organic carbon', *Soil Sci. Soc. Am. J.*, 77, 461–472.
Lorin, M., et al. (2015), 'Undersowing winter oilseed rape with frost-sensitive legume living mulches to improve weed control', *Euro. J. Agron.*, 71, 96–105.
Lou, Y., et al. (2016), 'Interactions between allelochemicals and the microbial community affect weed suppression following cover crop residue incorporation into soil', *Plant Soil*, 399, 357–371.
MacRae, R.J., et al., (1990), 'Farm-scale agronomic and economic conversion from conventional to sustainable agriculture', *Adv. Agron.*, 43, 155–198.
Manici, L.M., et al. (2004), 'Effect of green manure on Pythium spp. Population and microbial communities in intensive cropping systems', *Plant Soil*, 263, 133–142.
Marochi, A., et al. (2018), 'Managing glyphosate resistant weeds with cover crop associated with herbicide rotation and mixture', *Ci. Agrot.*, 42(4), 381–394.
Mendes, K.F., et al. (2019), 'Transport of atrazine via leaching in agricultural soil with mineral oil addition', *Planta Daninha*, 37, 1–7.
Meiss, H., et al. (2010), '*Weed seed predation increases with vegetation cover in perennial forage crops*', *Agric. Ecosys. Environ.*, 138(1), 10–16.
Meschede, D.K., et al. (2007), 'Avaliação de diferentes coberturas na supressão de plantas daninhas no cerrado', *Planta Daninha*, 25, 465–471.
Mohler, C.L., et al. (2012), 'Reduction in weed seedling emergence by pathogens following the incorporation of green crop residue', *Weed Research.*, 52, 467–477.

© Burleigh Dodds Science Publishing Limited, 2025. All rights reserved.

Molisch, H. (1937), *Der EinflusseinerPflanze auf die andere-Allelopathie*, Fischer, Jena.
Montanarella, L., et al. (2016), 'World's soils are under threat', *Soil*, 2, 79–82.
Montgomery, D.R. (2007), 'Soil erosion and agricultural sustainability', *Proc. Nat. Acad. Sci. USA*, 104(13), 268–272.
Moreira, G.M., et al. (2013), 'Fitossociologia de plantas daninhas do cafezal consorciado com leguminosas', *Planta Daninha*, 31(2), 329–340.
Nunes, A.L. and Vidal, R.A. (2016), 'Leaching of the S-metolachlor herbicide associated with paraquat or glyphosate in a no-tillage system', *Pesq. Agrop. Tropical*, 46, 254–258.
Ohno, T., et al. (2000), 'Phytotoxic effects of red clover amended soils on wild mustard seedling growth', *Agric. Ecosyst. Evoron.*, 78, 187–192.
Osipitan, O.A., et al. (2019), 'Impact of cover crop management on level of weed suppression: a meta-analysis', *Crop Sci.*, 59, 833–842.
Peachey, R.E., et al. (2002), 'Effect of cover crop and tillage systems on symphylan (Simphlya: Scutigerella immaculate, Newport) and Pergamasus quisquiliarum Canestrini (Acari: Mesostigmata) populations, and other soil organisms in agricultural soils', *Appl. Soil Ecol.*, 21, 59–70.
Petit, S., et al. (2018), 'Biodiversity-based options for arable wheat management: a review', *Agron. Sust. Dev.*, 38(5), article 48.
Pimentel, D., et al. (1995), 'Environmental and economic cost of soil erosion and conservation benefits', *Science*, 267, 1117–1123.
Poeplau, C. and Don, A. (2015), 'Carbon sequestration in agricultural soils via cultivation of cover crops–A meta-analysis', *Agric. Ecosyst. Environ.*, 200, 33–41.
Reberg-Horton, S.C., et al. (2005), 'Changes over time in the allelochemical content of ten cultivars of rye (Secale cereale l.)', *J. Chem. Ecol.*, 31(1), 179–193.
Reicosky, D.C., et al. (2021), 'Challenging balance between productivity and environmental quality: tillage impacts', In: Hatfield. J.L. and Sauer, T.J. (eds.), *Soil Management: Building a Stable Base for Agriculture*, American Society of Agronomy and Soil Science Society of America, Madison, WI, pp. 13–38.
Reicosky, D.C., et al. (2023), 'Cover crop mixes for diversity, carbon, and conservation agriculture', In: Islam, R. and Sherman, B. (eds.), *Cover Crops and Sustainable Agriculture*, 1 ed. CRC Press: Boca Raton, Florida, pp. 170–209.
Reusser, P., et al. (2015), 'Quantifying human impacts on rates of erosion and sediment transport at a landscape scale', *Geology*, 43(2), 171–174.
Rheinheimer, D.S., et al. (2024a), 'Kinetics of soil re-acidification under no-tillage system', *Soil Till. Res.*, 242(1), 106137.
Rheinheimer, D.S., et al. (2024b), 'Soil aluminum saturation threshold for subtropical crops in no-tillage system', *Soil Res.*, 62(3), SR23174.
Riemens, M., et al., (2002), 'An Integrated Weed Management framework: A pan-European perspective, Eur. J. Agron., 133, 126443.
Rodrigues, B.N. and Almeida, F.S. (2018). *Guia de herbicidas*, 7ª. Edição, Produção Independente, 764p.
Ruedell, J. (1995), *Plantio direto na região de Cruz Alta*, FUNDACEP/BASF, Cruz Alta. 134p.
Sanderman, J., et al. (2017), 'Soil carbon debt of 12,000 years of human land use', *PNAS*, 114(36), 9575–9580.
Santos I.L.V.L., et al. (2012), 'Sorgoleone: Benzoquinona lipídica de sorgo com efeitos alelopáticos na agricultura como herbicida', *Arq.Inst. Biol.*, 79, 135–144.

© Burleigh Dodds Science Publishing Limited, 2025. All rights reserved.

Schomberg, H.H., et al. (2006). *Influence of cover crops on potential nitrogen availability to succeeding crops in a Southern Piedmont soil*. Springer, Berlin / Heidelberg. *Biology and Fertility of Soils*, 42(4), 299–307.

Schmidt, M. and Teschner, K. (2000), "Integration of Rainwater Science: Results of Researchfor the Potsdamer Platz Project - Part 1: Chemical Constituent Retention in Extensive Vegetative (green) Roofs" ("Kombination vonRegenwasserbewirtschaf tungsmassnahmen: Ergebnisse der Voruntersuchungen fuer das Project Potsdamer Platz – Teil 1: Stoffrueckhalt extensiver Dachgegruenung,") GWF Wasser Abwasser 670–675.

Shahrtash, M. and Brown, S.P. (2021), 'A path forward: promoting microbial-based methods in the control of invasive plant species', *Plants*, 10(5), 943–956.

Skora Neto, F. (1993), 'Controle de plantas daninhas através de coberturas verdes consorciadas com milho', *Pesq. Agropec. Bras.*, 28, 1165–1171.

Teasdale, J.R. and Mohler, C.L. (1993), 'Light transmittance, soil temperature, and soil moisture under residue of hairy vetch and rye', *Agron. J.*, 85, 673–680.

Teasdale, J.R., et al. (2007), 'Cover crops and weed management', In: Updadhyaya, M.K. and Blackshaw, R.E. (eds.), *Non-Chemical Weed Management: Principles, Concepts and Technology*, CABI Publishing, Wallingford, UK. pp. 49–64.

Thiesen, G., et al. (2000), 'Reduction of Brachiaria plantaginea infestation on soybean crop with black oat residues', *Pesq. Agropec. Bras.*, 35, 753–756.

Thomas, F. and Archambaud, M. (2016), *Les couverts végétaux: Gestion pratique de l'interculture*, 2ed. Éditions France Agricole, Paris. 302p.

Tiecher, T., et al. (2013), 'Long-term effect of different soil management systems and winter crops on soil acidity and vertical distribution of nutrients in a Brazilian Oxisol', *Soil Tillage Res.*, 133, 32–39.

Tiecher, T., et al. (2017), 'Soil fertility and nutrient budget after 23-years of different soil tillage systems and winter cover crops in a subtropical Oxisol', *Geoderma*, 308, 78–85.

Tiecher, T., et al. (2020a), 'The role and management of soil mulch and cover crops in conservation agriculture systems', In: *Burleigh Dodds Series in Agricultural Science*. 1st ed. Burleigh Dodds Science Publishing, Cambridge. pp. 179–248.

Tiecher, T., et al. (2020b), 'Effect of 26-years of soil tillage systems and winter cover crops on C and N stocks in a Southern Brazilian Oxisol', *Rev. Bras. Ci. Solo*, 44, e0200029.

United Nations. (2014), *Concise Report on the World Population Situation*, United Nations Department of Economic and Social Affairs, Washington, DC.

Vasconcelos, H. and Landers, J.N. (1993), 'Projeto Morrinhos: Relato de uma experiência de Manejo de resíduos no cerrado', In: Arantes, N.E. and Souza, P.I.M. (eds.), *Cultura de soja nos cerrados*, Potafos, Piracicaba, SP, Brazil. pp. 523–535.

Voll, E., et al. (1997), 'Dinâmica de populações de trapoeraba (Comelina benghalensis L.) sob manejos de solo e de herbicidas', *Pesq. Agropec. Bras.*, 32(6), 571–578.

Voll, E., et al. (2001), 'Dinâmica do banco de sementes de plantas daninhas sob diferentes sistemas de manejo de solo', *Planta Daninha*, 19(2), 171–178.

Weston, L.A. (1996), 'Utilization of allelopathy for weed management in agroecosystems', *Agron. J.*, 88, 860–866.

Wiggins, M.S., et al. (2016), 'Evaluating cover crops and herbicides for glyphosate-resistant Palmer amaranth (Amaranthus palmeri) control in cotton', *Weed Tech.*, 30, 415–422.

© Burleigh Dodds Science Publishing Limited, 2025. All rights reserved.

Zibilske, L.M. and Makus, D.J. (2009), 'Black oat cover crop management effects on soil temperature and biological properties on a Mollisol in Texas, USA', *Geoderma*, 149(3–4), 379–385.

Zimdahl, R.L. (2013), *Fundamentals of Weed Science*, 4th ed., Academic Press, San Diego, CA, USA.

© Burleigh Dodds Science Publishing Limited, 2025. All rights reserved.

Chapter 12

Weed management in Conservation Agriculture systems: farmers' testimonies

Gottlieb Basch, Mediterranean Institute for Agriculture, Environment and Development (MED) – Universidade de Évora, Portugal; Søren Ilsøe, Knudstrupgaard Farm, Denmark; Miguel Barnuevo-Rocko, Finca Munibañez s/n Chinchilla, Spain; Anderson Schmitz, Schmitz Family Farm, Brazil; Marcelo Zanella, Agricultural Research and Rural Extension Company of Santa Catarina (EPAGRI), Brazil; Benjamin Dias Osorio Filho, Chalé do Seival/State University of Rio Grande do Sul, Brazil; Luiz Antônio Pradella, Pradella Group, Brazil; Valmor dos Santos, Inovação Agrícola, Brazil; Corey Loessin, Aidra Farms Ltd, Canada; and Marie Luise Carolina Bartz, Federal University of Santa Catarina, Brazil/University of Coimbra, Portugal

http://dx.doi.org/10.19103/AS.2024.0145.20
© Burleigh Dodds Science Publishing Limited, 2025. All rights reserved.

1 Introduction

Since the dawn of agriculture several millennia ago, farmers have tried to manage unwanted, spontaneous vegetation to give their crops an advantage over such vegetation and produce more of the desired yield in the crops they grow. After the shift by humans from a hunter-gatherer existence to producing their own food through agriculture, a variety of methods to manage weeds have been developed based on the knowledge and technologies available at a given time. Apart from hand weeding to reduce the pressure of weeds on crops, the use of soil tillage has been the preferred method over centuries to suppress weeds, whether selectively or non-selectively, through uprooting, burial or creating unfavourable conditions for unwanted vegetation.

Although the first organic chemical herbicide was developed at the end of the nineteenth century, the broad development and use of both organic and synthetic compounds as weed killers started in the middle of the last century. There is no doubt that chemical weed control was revolutionary, making weed management more effective, faster and cheaper. It also allowed for reducing soil tillage intensity, as weed control was no longer solely dependent on tillage. At the same time, interest in reduced tillage started to grow in countries such as the United States in response to the Dust Bowl conditions in the 1930s. In the decades that followed, more and better herbicides became available. This helped to stimulate more interest in no-till planting which, in turn, helped no-till seeding equipment to develop and improve.

The history of no-till development is well documented (Phillips and Young, 1973; Phillips and Phillips, 1984; Derpsch, 1998; Baker at al., 2007). However, the definition of a new concept known as Conservation Agriculture (CA), which integrated minimum soil disturbance (ideally no-till) into a more complex system of sustainable soil management, was launched at the turn of the last century. The additional components which define CA, permanent organic soil cover and crop diversity, provide opportunities for integrated weed management in CA systems.

As detailed in the previous chapters, there are many approaches, whether stand-alone or combined with one another, that can contribute to weed management in CA systems. The adequate combination of approaches will depend on several factors such as agroecological conditions, crops and cropping systems, history of land use, and soil management.

Conventional, tillage-based crop establishment in combination with herbicides, whether applied in pre-seeding, pre-emergence, or post-emergence, has developed over decades and adapted to emerging challenges such as the appearance of new weeds or the development of herbicide resistance. However, the relatively recent and sudden shift from tillage to CA-based crop production has confronted farmers with a new reality from the

© Burleigh Dodds Science Publishing Limited, 2025. All rights reserved.

day of transition. The chapters in this book have discussed the many aspects of this transition and changes in weed management, whether in different cropping systems or through the use of different cultural, physical, chemical, or biological approaches. In this concluding chapter, several testimonies by farmers about their experiences of weed management after the shift to CA are presented to emphasize the importance of practitioners' views when profound changes in farming methods are needed.

The testimonies in this chapter comprise different agroecologies and farming systems:

- two in Europe, one in the Atlantic Central region (Denmark) and the other in the Mediterranean South (Spain), both working in annual cropping systems;
- three in Brazil, one in the South, in the Santa Catarina state, with a temperate but hot summer climate, specialized in horticultural crops; another in the Cerrado biome of Bahia state in the Northeast, dedicated to annual broadacre crops; and a third one practicing organic no-till farming in Rio Grande do Sul; and
- finally, a farmer from the cold semi-arid Canadian prairies, cropping cereals, oilseed rape, and pulses.

Each is listed by country and farmer.

2 Farmer experiences of weed management under Conservation Agriculture systems: Denmark (Søren Ilsøe)

2.1 Farm and farming system description

Location: Knudstrupgaard, Denmark (55.41775 11.62764).

Cropping area: 295 ha.

Minimum-tillage since the year 2000 and a transition to direct seeding following the principles of CA since 2011.

Several different seeders have been tested on the farm over the years, but we have ended up with a disc drill as the best solution. We use a Weaving GD 6-meter machine with very little soil disturbance.

Over the years, the crop rotation has become simpler with fewer crops due to problems with roe deer, which cause major damage to broad beans. In addition, summer drought has become more widespread, resulting in significant yield losses in broad beans. Also, the price is simply too low for this crop to be profitable.

Winter rapeseed is the only broadleaf crop, and its position in the rotation is important because grass weeds can be effectively controlled in this crop.

© Burleigh Dodds Science Publishing Limited, 2025. All rights reserved.

Winter wheat is grown on the largest proportion of the farm area. Currently, there is a transition to bread wheat varieties to meet the desire to grow higher-quality bread-making wheat instead of feed wheat, in part, because the demand for wheat for human nutrition is expected to increase in the coming years.

Spring barley is grown focussing on malting varieties. It is a challenging crop to grow throughout Denmark due to the short growing season. Farmers practicing direct seeding, in particular, face problems in achieving satisfactory yields as a result of the slower germination of spring barley in the no-till system.

However, on our farm, over the last 12 years under CA, average yields compared to the last 10 years with traditional ploughing (1990–1999) have increased by 6.9% in winter wheat, while spring barley yields have grown by 26.6%.

Fuel and time consumption have decreased, and the average fuel consumption for all field work is now 42 L/ha.

2.2 Challenges and solutions in weed management: grass weed species occurrence and dynamics

At the end of the 'ploughing period' in the late 90s, there were major problems with Couch Grass (*Elymus repens* L.). However, within a few years after switching to no-till cultivation, this grass weed started to disappear and now, under pure no-till, this problem weed is completely forgotten; it simply does not occur anymore.

In the years 2000–2010, a new grassy weed appeared all over the country. This was Barren Brome (*Bromus sterilis* L.), likely due to the phasing out of Isoproturon, a very effective grass weed herbicide. Although a very common weed on conventionally cultivated fields, it has never been a major problem on our farm and nowadays is almost non-existent. It is easy to control. The use of glyphosate applied before sowing seems to contribute decisively to keeping this species under control.

Soft Brome (*Bromus hordeaceaus* L.) is almost absent today on our no-till cultivated fields. However, it can be found in roadside embankments and fences in some places.

In contrast, Italian Ryegrass (*Lolium multiflorum* Lam.) is becoming an increasing problem in some parts of Denmark, and resistance has developed in certain areas. Luckily, it has never been detected on our farm.

Black-grass (*Alopecurus myosuroides* Huds.) has become a huge problem on farms with intensive winter crop rotations. It is a major issue but, again, this species has never been detected on our farm.

Annual Meadow-grass (*Poa annua* L.) can be found in the fields and is almost exclusively seen in trampled areas. So far, it is not a problem as its

© Burleigh Dodds Science Publishing Limited, 2025. All rights reserved.

occurrence is still very rare. In some years with more favourable conditions for the species, it may be seen to a slightly larger extent.

Rats's Tail Fescue (*Vulvipa myuros* L.), on the other hand, is a problem (Fig. 1). This genus was first detected in 2006 in a field with Red Fescue (*Festuca rubra* L.). It is not known whether it originated from seed contamination or was introduced with a foreign straw baler. This new species has escaped regular weed control and spread to the other fields with the combined harvester. Additionally, the problem was underestimated at first since the species was relatively unknown.

Since then, it has been the biggest challenge in annual cropping systems (Fig. 1, left), and it is responsible for higher glyphosate consumption than if this species were not present. Controlling large plants requires 1000–1100 grams of glyphosate per hectare. To ensure an effective treatment, we acidify the spray solution and add ammonium sulphate. Without this grass weed, glyphosate usage could be halved!

In spring barley, the species cannot be controlled, so field desiccation is crucial before sowing (Fig. 1, right).

In winter wheat, there is a moderate effect of the herbicide mix of Mateno Duo + Boxer (a.i.: Aclonifen, Diflufenican + Prosulfocarb) at growth stage 10–11. Atlantis (a.i.: Mesosulfuron-methyl, Propoxycarbazone, Mefenpyr-diethyl (Safener)) also has an effect. In early spring, Broadway, a mix of Florasulam, Pyroxsulam, and Cloquintocet (a.i.) has an effect that is highly dependent on weather conditions and growth.

In winter rapeseed, Kerb 400 SC (Propyzamid) is an important solution that is nearly 100% effective against *V. myuros* L. if applied correctly. This means that having winter rapeseed crop in the rotation is very important. It is probably impossible to completely eradicate this weed but, as long as it is kept under control and only occurs at acceptable densities, yield impacts are tolerable.

Figure 1 Rats's Tail Fescue (*Vulvipa myuros* L.); Infested field (left) (Photo: Muhammad Javaid Akhter), sprayed-off before seeding (right).

© Burleigh Dodds Science Publishing Limited, 2025. All rights reserved.

2.3 Broadleaved weed occurrence and dynamics

The species composition of dicotyledonous weeds has changed significantly since we stopped ploughing. Previously, common chickweed (*Stellaria media* L.) was widespread but is now very rare. Mayweed (*Tripleurospermum inodorum* L.) has also been greatly reduced. This species is typically a problem in winter oilseed rape but is now only seen at the edges of fields. However, it must be controlled as weed plants can grow very large and cause problems at harvest, as well as produce a lot of new seeds, thereby increasing the seed bank.

Shepherd's purse (*Capsella bursa-pastoris* L.) is commonly found but easily controlled. Sow-thistle (*Sonchus* spp.) is also common and easily controlled. In years with lower plant densities and rainy weather, it can sprout very late in the season and grow rapidly.

Field poppy (*Papaver rhoeas* L.) occurs in all fields but is easily controlled. Common field-speedwell (*Veronica persica* Poir.) is widespread in all fields and easily controlled with standard techniques. Cranesbill (*Geranium* spp.) is increasing in prevalence. It did not exist 10 years ago, but now it is found in almost all fields. It can be quite difficult to control with low doses of glyphosate, but it is easily controlled with other common herbicides.

Cleavers (*Galium aparine* L.) has become more widespread and is the dominant weed in most fields. It has a tremendous ability to reproduce, especially in winter oilseed rape, where it can sprout late in the season and produce seeds. It is easily controlled in cereal crops using 2-*methyl*-4-*chlorophenoxyaceticacid (*MCPA*)* but persists longer in winter oilseed rape before harvest.

Creeping thistle (*Cirsium arvense* L.) is the new major weed problem! Commonly used herbicides have very little effect on this species, and it requires a targeted application of MCPA herbicide, which increases overall herbicide use. The species invades from areas where fallow is legally required (farmers are required by the Danish government to leave some land fallow on environmental grounds). Seeds are blown from these areas far into cultivated areas. Mechanical mowing is not allowed in these areas from May to September and, during this period, the species disperses many seeds over the fields via the wind.

2.4 Changes in pesticide use after going ploughless (1999) and converting to Conservation Agriculture in 2011

When looking at the overall use of pesticides on the farm, it naturally varies greatly from year to year due to the weather conditions and, especially, the amount of rainfall.

Insecticides have been completely phased out and have not been used for the past 10 years.

© Burleigh Dodds Science Publishing Limited, 2025. All rights reserved.

Fungicides appear to have reduced in dosage generally, but more years of experience are needed to confirm this trend. In recent years on our farm, the application of around 50% of the dosage used under conventional conditions has proven to be sufficient.

There can be some slug attacks in the field, but the promotion of aboveground biodiversity under CA conditions, especially birds, helps to control this problem. Figure 2 shows seagulls hunting. They visit fields in the mornings to find slugs but, unfortunately, earthworms as well.

When we have dense and well-developed crops, the strong competition they provide against weeds requires reduced amounts of herbicide. We can now skip glyphosate application before the establishment of winter wheat after spring barley (Fig. 3) since a good barley stand helps to suppress weeds, as

Figure 2 Seagulls invading a no-till cereal field feeding on slugs, earthworms, and other small animals.

Figure 3 Field (left) and detailed (right) view of dense residue cover allowing for avoiding pre-seeding herbicide treatment.

© Burleigh Dodds Science Publishing Limited, 2025. All rights reserved.

well as due to the absence of grass weeds that are successfully controlled in previous crops.

Using the web-based application, Plant Protection Online allows for a more targeted, need-based dosage and helps in the selection of products, providing great help in finding good solutions and individual application strategies at the field level. When cover crops are successfully established and dense, they can effectively compete with weeds (Fig. 4). Unfortunately, it is not always possible to establish them at the desirably early stage after the main crop.

A test with planting pure summer vetch at a high plant density has proven to be very effective in suppressing weeds, possibly due to allelopathy. More focus should be placed on this approach in the coming years. For the past 2 years, winter rapeseed has been grown without the use of glyphosate before sowing (Fig. 5). This method works very well when the previous crop has been weed-free, and it will likely become common practice in the future.

The use of new technologies for digital weed recognition will definitely become widely adopted, and new machines for non-chemical control are also gaining popularity. Direct seeding with disc openers (Fig. 6) is also very effective in avoiding the germination of new weeds from the existing seed pool in the soil as almost no soil disturbance takes place. After several years of using disc openers, the number of weeds has significantly reduced. Weed seeds on the soil surface are consumed by beetles and many are largely destroyed by fungi and birds, so it is very important to disturb the soil as little as possible to optimize these biological methods of control.

Figure 4 Crop establishment into a dense cover crop.

© Burleigh Dodds Science Publishing Limited, 2025. All rights reserved.

Figure 5 Rapeseed established into cereal residues without pre-seeding weed control.

Figure 6 No-till with strongly inclined disc openers for minimum soil disturbance.

© Burleigh Dodds Science Publishing Limited, 2025. All rights reserved.

3 Farmer experiences of weed management under Conservation Agriculture systems: Spain (Miguel Barnuevo-Rocko)

3.1 Farm and farming system description

Munibañez Farm where I started using no-tillage is located in the south-central part of Spain, in Chinchilla in the province of Albacete. We have two types of soils: some are deep, basic, with low organic matter content and loamy and clay loam textures (Calciorthid). Other soils are shallow with a calcareous crust at shallow depths, with little organic matter and light-to-loamy and sandy loam textures.

The average annual rainfall is 375 mm, and the climatic classification is Continental Mediterranean. Rainfall is irregularly distributed in autumn, winter and spring, with long dry periods, especially in summer. The frost-free period occurs from May to September. Due to climate change, in recent years, there has been a tendency for the spring to start much earlier and for autumns to lengthen: the frost-free period is widening from April to October, but there are still heavy frosts in winter. The average temperature has increased by more than 1.5°C, and, more seriously, high temperatures begin to occur in the spring (March and April), significantly affecting the phenology of both herbaceous and woody crops. In 2023, the duration of high temperatures was so severe that, together with a prolonged drought since December 2022, it led to a complete failure of rainfed crops in a large part of the province. Irrigated cereal crops suffered yield losses of more than 35%.

At the farm level, there are two types of managements:

- Rainfed farming on an area of 300 ha, with barley, oats, triticale, wheat, and legumes, mainly peas and vetches, all sown in autumn. Spring-sown summer crops are very risky due to erratic rainfall in spring. Sunflower has been tried sporadically on the deeper soils, but frequently results in very low yields.
- On the irrigated part of the farm, crops can be grown throughout the year, choosing species that are adapted to the climate. Rotations used consist mainly of alfalfa, maize, barley, wheat, oats, and ryegrass for fodder, oilseed rape, and sunflower, all sown using the no-till system on an area of around 100 ha.

In addition to cultivating our own farm, we provide services for no-till crop establishment on other farms representing around 1000 ha/year (500 ha of winter cereals and leguminous crops and 500 ha of mainly maize and sunflower).

© Burleigh Dodds Science Publishing Limited, 2025. All rights reserved.

3.2 Challenges and solutions in weed management

In rainfed farming, when we started in 1993 with no-till, the usual cropping practice was to have one crop every second year. This meant that 1 year a crop was planted (e.g. barley, wheat, oats, triticale) and the following year the land was kept fallow. Weed problems were relatively low, with wild oats (*Avena sterilis* L.) and poppies (*P. rhoeas* L.) being the main problems. During the fallow period, wild oats were controlled using glyphosate and poppies with a selective 2,4-D or similar treatments within the cereal crop. Erosion rates were quite high.

During the initial years of no-till, a tine-coulter seeder was used without incorporating fertilizer. Fertilizer was broadcast at the surface and applied at the same rates as in the conventionally tilled plots. During the first years, we used the same doses in both systems (tillage and no-tillage).

After 3 years of cereal monoculture, it became clear that we had to rotate with leguminous crops every third year to ensure sufficient nutrients in the soil. We also concluded that 'continuous' cultivation (every year) was much better than fallow in between two main crops. This was because the fields that were not cultivated for 1 year resulted in the development of spontaneous vegetation (Fig. 7), consisting mainly of volunteers from the previous crop and weeds. To deal with this vegetation, glyphosate was applied at the beginning of spring (April and May) before weed seed setting. Occasionally, there was a

Figure 7 Spontaneous vegetation developing in years under fallow.

© Burleigh Dodds Science Publishing Limited, 2025. All rights reserved.

need to have another spray-off treatment at the end of winter to keep some grass weeds such as brome grass (*B. sterilis* L.), wild oats (*A. sterilis* L.), annual ryegrass (*Lolium rigidum* Gaudin), and fescues (*Vulpia* sp.) under control.

In the early 2000s, we started to notice the proliferation of some weeds such as Veronica (*Veronica* sp.), 'verruguera' or European heliotrope (*Heliotropium europaeum* L.), and shepherd's purse (*C. bursa-pastoris* L.). However, we did not find it necessary to eliminate these as they did not compete with our crops. We started to apply the principle of tolerating some plants which contributed to forming a protective cover for the soil.

It was also during these years that we began to have problems with prickly saltwort (*Salsola kali* L.) (Fig. 8, left). In some summers, when it rained in July–August after harvest, there were significant infestations of this weed. These had to be eliminated, as otherwise, they greatly hindered subsequent sowing in autumn with tine-coulter seeders. The control of this weed was expensive and complicated, as eradication was difficult. The solution came within a few years, with the use of disc seeders of several brands (Semeato, Tatu Marchesan, Kuhn, etc.) which all had almost no problem in 'cutting' prickly saltwort and sowing on top of them (Fig. 8, right).

Conyza sp., mainly hairy fleabane/horseweed (*Conyza bonariensis* L.), also began to appear, but their presence did not compete with crops as they developed in August and September, which is an off-season cropping period under rainfed farming conditions in Spain.

During this period, we also tested different glyphosate doses in the autumn treatments to see whether it was possible to reduce the application rates. The conclusion was that glyphosate doses could be reduced to 0.75 L/ha (360 g/L). However, over the years, the experience taught us that it was not worth 'fine-tuning' too much, as low doses could not be 100% effective, thus risking generating resistance, especially in brome grass and annual ryegrass.

Another lesson we learnt, especially in irrigated plots, was that the earlier treatment (15–20 days before sowing) was always better for sowing quality than treating volunteers of the previous crop and weeds just before sowing.

Figure 8 Prickly saltwort (*Salsola kali* L.) infestation after unusual summer rainfall.

© Burleigh Dodds Science Publishing Limited, 2025. All rights reserved.

In 2000, we started to experiment with sowing on irrigated fields with large amounts of residues from the previous crop. Such conditions are prone to become a big problem if the seed drills are not able to cut the large amount of plant remains and place the seed at the correct depth. To overcome this potential problem, we started to use sweepers to clean the sowing line (Fig. 9). At the same time, we started to carry out post-emergence herbicide treatments as pre-emergence treatments were less effective due to the large quantity of residues and organic matter on the surface of the soil.

At this time we also started to apply fertilizers with the seed drill. The improvement in yields, especially in maize, was noticeable.

In recent years, we have had some problems due to autumn droughts. The lack of rain in September–October, and in some years even November, causes a delay in the germination of the first weed 'wave'. This means that the spray-off before autumn sowing is not very effective in protecting a crop sown in almost dry soil against the weeds mainly germinating after sowing together with the crop. This has made subsequent treatments with other active ingredients than glyphosate necessary, and thus more expensive. It obliges us to have spraying equipment of sufficient width as possible and in good condition to make the most of the short time available to carry out treatments in the autumn–winter period. The price of spraying equipment, as well as seed drills, has risen enormously in recent years.

One interesting aspect is that, in certain years, either because of a strong infestation of a certain weed species or because of a low expectation of grain yield mowing the cereal crop for fodder was preferable, thus avoiding the seed-setting of weeds. The problem of controlling weeds by mowing is that, in the end, other creeping weeds that are difficult to control by mowing can become established. Weed control on irrigated land has turned out to be much easier, especially with alfalfa in the crop rotation.

Figure 9 Row-cleaners used for crop establishment into residue-rich no-till seedbed.

© Burleigh Dodds Science Publishing Limited, 2025. All rights reserved.

As for insecticides after the shift to no-till, we still used them preventively until 2005–2006, applying them together with the seeding operation. Subsequently, we observed that there were no problems, so we have not used them since then as a preventive measure.

4 Farmer's experiences of weed management under Conservation Agriculture systems: Brazil (Anderson Schmitz and Marcello Zanella)

4.1 Farm and farming system description

My name is Anderson Schmitz, and my wife is Milena Jasper Schmitz. We are farmers in the municipality of Aguas Mornas, Santa Catarina, Brazil. My family has been growing vegetables for at least 30 years. My parents, Afonso Schmitz and Aurea Sebold Schmitz, and I farm a total area of 20 ha. The area used for cultivation is 3 ha, where we grow tomatoes, eggplant, cauliflower, bell peppers, zucchini, string beans, and Japanese cucumbers. The property is located in a mountainous region with a predominance of Quartz Neosol soils and a humid subtropical climate, with hot summers and rainfall of approximately 1500 mm/year.

Our family has always grown vegetables using the conventional system, which involves a lot of soil preparation and intensive use of herbicides, fungicides, bactericides, and insecticides. We have frequently discussed as a family how we could improve our management to reduce production costs, soil losses from constant tillage in hilly terrain, and the direct and indirect damage caused by herbicides used to control weeds between rows of crops.

In 2016, I participated in a course for rural youth organized by the Agricultural Research and Rural Extension Company of Santa Catarina (EPAGRI). I learned about a new system for working with vegetables. Through the classes and technical visits to properties in the vegetable-producing region of Florianópolis, I discovered new ways of cultivating and managing the cultivation systems based on the No-Till System for Vegetables (SPDH). The principles of this system are based on promoting plant health, a better understanding of how plants grow, their relationships with each other and with microorganisms, their capacity to rebuild and improve the cultivation environment, and their ability to co-exist with spontaneous plant growth throughout their life cycle. This approach improved soil quality and, in many cases, eliminated problems with soil fungi and bacteria that were harming the plants.

The first step in changing our property's management was acquiring equipment such as a roller crimper (Fig. 10, left) and a brush cutter to manage cover crops and spontaneous plants and a two-wheel walking tractor equipped with a roller, cutting disk, and straight rotary device to open rows for crop establishment with minimal soil disturbance (Fig. 10, right).

© Burleigh Dodds Science Publishing Limited, 2025. All rights reserved.

Figure 10 Roller crimper to terminate the cover crop consisting of pearl millet (left); two-wheel walking tractor, equipped with a roller, cutting disk and straight rotary device as row opener (right).

4.2 Challenges and solutions in weed management

One of the main challenges of this new approach was overcoming the fear of allowing coexistence between commercial/cash crops and spontaneous plant/weed growth. Rethinking crop management, including changing fertilization, was also a significant challenge, mainly because we were accustomed to conventional management, where the presence of spontaneous plants during the growing season of the main crop was considered unthinkable.

After 6 years of managing cropping areas without turning over the soil, we have gathered new knowledge that has allowed a considerable reduction in the use of chemical inputs. We have already reduced the use of pesticide products to control fungal and bacterial diseases by 60%, insecticides by 50%, and herbicides by 90%. Additionally, we have reduced the use of fertilizers by 40%. Previously, we used more inputs than needed, which ended up harming the plants. This reduction in inputs has also led to a reduction in labour and production costs, along with an increase in productivity and better-quality food.

In the conventional system, we used to harvest approximately 75 000 kg of tomatoes per hectare. Today, we harvest up to 140 000 kg/ha. In the case of cauliflower, where we used to achieve a harvest index of 60%, we now can harvest more than 90% of the plants. We have also eliminated erosion that occurred in areas under soil tillage, reduced the use of water for irrigation, and increased soil organic matter content. The use of cover crops, ensuring permanent soil cover throughout the year, has facilitated management of spontaneously growing plants/weeds.

In long-cycle crops such as tomatoes, the residues left by the cover crop suppress spontaneous plants during the initial phase of the main crop. At a later stage, spontaneous vegetation coexists with the commercial crop and is managed mechanically, thus maintaining enhanced biodiversity in the system (Fig. 11, left). This approach is also effective for bell pepper and eggplant crops.

© Burleigh Dodds Science Publishing Limited, 2025. All rights reserved.

Figure 11 Tomato crop cultivated under the 'No-Till System for Vegetables (SPDH)' established into oats residues with emerging spontaneous vegetation controlled by cutting (left); weed control in a cauliflower crop solely based on the residues of a pearl millet cover crop (right).

In string bean, cauliflower, and Italian zucchini crops, residues of cover crops are sufficient to suppress spontaneous vegetation throughout the whole cycle without any need for additional management (Fig. 11, right).

In general, the population of spontaneous plants remains constant. However, with the adoption of management practices under the SPDH, we noticed that in the hottest period (summer), grass species such as *Brachiarias*, *Digitarias*, and *Commelina* spp. predominate. In winter, *Sonchus* spp. and *Galinsoga* sp. are more common. As the system evolves and soil quality improves, some spontaneous plants no longer establish themselves. Since we mechanically manage spontaneous plants coexisting with vegetables, we only perform chemical control when climbing species such as Ipomea appear in order to prevent them from climbing on the plants. After harvesting vegetable crops, we immediately sow cover crops so that they establish themselves as quickly as possible, avoiding the appearance of new spontaneous plants.

We work with winter and summer cover crops. In the summer, we mainly use Millet, and in the winter Black Oats intercropped with Vetch. The largest volumes and consequently the best coverage always occur with summer plants because Millet produces a large amount of biomass, covering the soil for a long period (Fig. 11, right). The winter cover crops produce less biomass but, when rolled without using herbicides, they sufficiently cover the soil to suppress weeds effectively during a cropping cycle of up to 80 days, which is enough for commercial crops such as cauliflower. We realized that, when we used herbicides to terminate cover crops, the process of residue degradation is accelerated, thus promoting faster soil exposure. When the cover crops do not produce enough biomass to fully cover the soil, the weeds that germinate between the rows are managed mechanically with a brush cutter to prevent them from shading the main crop. The objective is to manage but not destroy them altogether so as to maintain biodiversity in the system.

© Burleigh Dodds Science Publishing Limited, 2025. All rights reserved.

The establishment of cover crops is done by broadcasting seeds superficially. We still do not have the equipment to perform direct no-till seeding, as many producers already do in the region, which would be the preferable and optimal option. Our system of cover crop establishment requires some adaptations to guarantee sufficient germination. We broadcast the cover crop seeds into standing spontaneous vegetation and later brush cut this vegetation so that the seeds are protected under the spontaneous vegetation residues. If soil moisture is insufficient, we irrigate twice in a period of 1 week to ensure uniform germination, obtaining a crop stand necessary for good biomass production and comprehensive soil cover.

In the SPDH that we practice, crop rotation is fundamental. We always use intercropping with a cash crop and later with cover crops. This way, the soil will always have plants and will always be covered with some kind of vegetation. This management has facilitated coexistence with spontaneous plants without harming the commercial crop. Through crop rotation, we have areas in production and under cultivation throughout the year, providing continuous income for the family. Our experience is that with adequate management of the system, the use of appropriate machinery, and planning of commercial crops and cover crops, we can reduce and even eliminate the use of herbicides in our system of horticultural crop production.

5 Farmer's experiences of weed management under Conservation Agriculture systems: Brazil (Luiz Antônio Pradella and Valmor dos Santos)

5.1 Farm and farming system description

In our testimony we would like to respectfully address all the food producers in this immense and diverse Brazil, diverse both in terms of soil and climate and in cultural traditions inherited from our ancestors, including descendants of Europeans and others. Our main objective is to provide some brief historical background, offer a diagnosis of current challenges, and offer some solutions. Our main aim is to stimulate interest and independent thinking which does not necessarily have to align with ours. The important thing is to think and reflect as a basis for productive change.

It is important to emphasize the agroclimatic conditions of our 'Cerrado' region which, unlike the northern hemisphere, experiences high temperatures practically year-round, severely penalizing any agricultural mismanagement. It is also important to note that in our region, land tenure is characterized by large farms that are highly mechanized and grow mainly cotton, soybean, and corn in intensive forms of production. These types of production have had a number of negative consequences, including on soil health. We have learned that a concerted action of all involved parties is needed to understand problems

© Burleigh Dodds Science Publishing Limited, 2025. All rights reserved.

that have emerged, take the right corrective actions, and change established behaviours to overcome problems.

One example of emerging problems was the push in the early 2000s to authorize the use of modern biotechnologies such as genetic modification (GM). The development of new GM pesticide-resistant crop varieties made it easier to use pesticides to manage pests, diseases, and invasive weed species. However, it also had some negative impacts, particularly on fibre quality, increased susceptibility of new varieties to some diseases, and greater demands on soil fertility. With conventional soybean production, producers struggled to control invasive plants, but there were practically no reports of plants resistant to glyphosate and/or graminicides. Issues with white mold (stem rot) (*Sclerotinia sclerotiorum* (Lib.) de Bary) or soybean cyst nematode (*Heterodera glycines* Ichinohe) were also minimal.

However, the use of new GM varieties has been associated with the development of corn stunt disease (*Spiroplasma kunkelii* Whitcomb) transmitted by the corn leafhopper (*Dalbulus maidis* DeLong) and, more recently, by the fall armyworm (*Spodoptera frugiperda* Smith) in maize cultivation. Both seem to have been 'selected for' by more susceptible GM varieties and have become serious pests, far more virulent than the corn earworm (*Helicoverpa armigera* Hübner). This has raised legitimate questions about placing too much reliance on biotechnology. It is fair to say that, on most farms, the weed problem is growing. Perhaps it is time to consider alternative approaches to control invasive plants more intelligently. By shifting the focus from anthropic to syntropic solutions, we can use nature to act in our favour.

5.2 Challenges and solutions in weed management

This section describes in more detail some of the invasive plant problems that exist in our region (Cerrado of Western Bahia) which, we believe, are not very different from other regions. Given that invasive plants can be defined as those that are in an undesirable place at a given time, we first need to think of cultivated plants, so-called volunteers, that can become invasive in following crops. This is the case for all our main crops: cotton, soybean, and corn, which are all based on genetically modified Roundup-Ready varieties which can therefore become invasive and resistant to control. Secondly, there are several indigenous plants such as crowfoot grass (*Eleusine indica* L.), shrubby false buttonweed (*Spermacoce verticillata* L.), hairy spurge (*Euphorbia hirta* L.), all highly tolerant to glyphosate, and sour grass (*Digitaria insularis* L.) and hairy fleabane/horseweed (*C. bonariensis* L.), which are resistant to glyphosate. The development of resistance has been slowed, but not halted, by the development of new herbicides such as Enlist and Xtend.

© Burleigh Dodds Science Publishing Limited, 2025. All rights reserved.

In the next section, we will look at solutions for managing both volunteers and indigenous 'weed' species with a particular focus on combining chemical control and intercropping. An important point to note that modern concepts of integrated weed management no longer aim to completely eliminate a weed but rather to adopt a more balanced approach by limiting weed populations to levels that do not significantly affect cash crop yields

Cotton seeds that survive harvesting are dormant. When chemical control is used, these potential volunteer plants (known as 'tiguera' cotton) remain on the soil surface and gradually become inviable. Mechanical destruction, e.g. via tillage can both damage soil structure and mean that cotton seeds are incorporated into the soil where they invariably germinate and emerge in different waves, making them extremely difficult to control. Avoiding soil disturbance also favours rainwater infiltration into the soil profile, promoting aquifer recharge. To illustrate this point, Fig. 12 shows the result of chemical management of cotton stalks, while Fig. 13 shows the result of mechanical management of stalks with significant resprouting.

The chemical approach to destroying cotton bolls has continuously evolved. Since the late 1990s, chemical control has been combined intercropping the subsequent maize crop with a grass species, *Brachiaria* sp., based on the No-Till System (NTS) (Fig. 14) which involves no-till crop establishment into plant/crop residues. This raises the question of how to avoid competition between *Brachiaria* sp. and maize as the main crop. Experience shows that, if well managed, the two can coexist, contributing to the integrated control of invasive plants.

In this case, the main problem is the cotton itself, either through volunteer seeds or shoot regrowth (stalks), since the following maize crop on its own

Figure 12 Chemical control of cotton stalks (no resprouting).

© Burleigh Dodds Science Publishing Limited, 2025. All rights reserved.

Figure 13 Mechanical management of stalks (a lot of resprouting, not allowed by law).

Figure 14 Maize under the No-Till System intercropped with *Brachiaria* sp.

provides excellent conditions for hosting volunteers. However, an intercropped species can smother cotton volunteers and resprouts. Figure 14 shows the excellent control of cotton resprouting and volunteers in this system. By incorporating *Brachiaria* sp. into the maize planting, the system benefits from the aggressive growth habit of *Brachiaria* sp., which helps suppress the growth of cotton volunteers. The key is to manage *Brachiaria* sp. in a way that does not compete excessively with the maize crop. This can be done through timely mowing or herbicide application that targets *Brachiaria* sp. without harming maize, allowing both crops to coexist while suppressing unwanted cotton volunteers. Residues from the maize and *Brachiaria* sp. also contribute to soil

© Burleigh Dodds Science Publishing Limited, 2025. All rights reserved.

organic matter, improve water infiltration, and reduce erosion, creating a more sustainable cropping system. In western Bahia and other regions, soybeans can also be used as a living cover which avoids the development of invasive plants as well as delivering other benefits such as improving water infiltration and storage in the soil (Fig. 15).

Figures 16–19 showcase one of the most advanced cotton NTS in the world, developed in western Bahia through a Technical Cooperation (TC) between Regional Consultants (CRs) and Rural Producers (PRs). Only in a well planned and executed NTS, is it possible to effectively control weeds in a crop, avoiding such problems as over-reliance on herbicides (which promotes resistance), heavy cultivation (which both damages soil structure and can even promote some invasive plants), insufficient cover (which then fails to prevent volunteer or invasive plant emergence) or competition between cover and cash crop

Figure 15 No-Till System of soybean in rotation with cotton – excellent control of stalk resprout and volunteers.

Figure 16 Cotton under the No-Till System in rotation with maize + *Brachiaria* sp.

© Burleigh Dodds Science Publishing Limited, 2025. All rights reserved.

Figure 17 Cotton under a high-quality No-Till System in rotation with maize + *Brachiaria* sp.

Figure 18 Maize under a high-quality No-Till System into residues of *Panicum Maximum* cv. Mombaça, intercropped with *Brachiaria brizantha* cv. Piatã.

which may then affect cash crop yields). Complementing the use of *Brachiaria* sp. is intercropping of maize with another grass species, *Crotalaria spectabilis* Roth, a practice developed approximately 10 years ago. Implementing this technology in the field requires a lot of experience as operations are more complex. Figure 20 illustrates the performance of this intercropping species in maize.

Figure 21 (left) shows soybean and *Brachiaria* planted simultaneously at a small plot scale. With the remaining soil moisture retained in the system, there is exponential growth of the grass shortly after the soybean harvest. This requires careful management, e.g. with selected applications of herbicide to

© Burleigh Dodds Science Publishing Limited, 2025. All rights reserved.

Figure 19 Maize under a high-quality No-Till System, intercropped with *Brachiaria ruziziensis*.

Figure 20 Maize under a high-quality No-Till System, intercropped with *Crotalaria spectabilis*.

ensure the resulting cover is properly controlled. Figure 21 (right) shows this approach at field scale.

Figures 22 and 23 show the key role that cover or 'second' crops can play in suppressing weeds. Figure 22 shows a field with (T1) and without (T2) Brachiaria as a second crop after soybean, where T2 shows a mix of invasive species. In Fig. 23, without Brachiaria (T2), hairy spurge (*Euphorbia hirta* L.), which is tolerant to glyphosate, finds an excellent opportunity to develop.

Figures 24 and 25 show pearl millet (*Cenchrus americanus* L.) and sorghum (*Sorghum bicolor* L.) after soybean, intercropped with *Brachiaria brizantha*

© Burleigh Dodds Science Publishing Limited, 2025. All rights reserved.

Figure 21 Soybean intercropped with *Brachiaria*, at plot scale (left), at field scale (right).

Figure 22 Field after soybean with (T1) and without (T2) *Brachiaria brizantha* cv. Marandu.

Figure 23 Field after soybean with (T1) and without (T2) *Brachiaria brizantha* cv. Marandu. T2 infested with hairy spurge.

© Burleigh Dodds Science Publishing Limited, 2025. All rights reserved.

Figure 24 Pearl millet (*Cenchrus americanus* L.) intercropped with *Brachiaria brizantha* cv. Piatã.

Figure 25 Rainfed sorghum (*Sorghum bicolor* L.) intercropped with *Brachiaria ruziziensis*.

cv. Piatã and *B. ruziziensis*, respectively. No soybean volunteers are visible, indicating full control of this potential invasive plant. After the harvest of the cereal crop (corn, millet, or sorghum), *Brachiaria* sp. develop rapidly (Fig. 26) providing enough biomass for excellent soil cover for the following soybean crop (Figs. 27 and 28).

Some farmers argue that it is only feasible to use *B. ruziziensis* Germ. & C.M. Evrard on their farms, claiming that other grass species such as *Brachiaria brizantha* cv. Piatã, cv. Marandu, and *Megathyrsus maximus* B.K. Simon & S.W.L. Jacobs, cv. Mombaça require higher doses of glyphosate to control. While *B.*

© Burleigh Dodds Science Publishing Limited, 2025. All rights reserved.

Figure 26 Pearl millet (*Cenchrus americanus* L.) intercropped with *Brachiaria brizantha* cv. Piatã, right after harvest (left), 10 days after harvest (right). Source: authors.

Figure 27 No-till (NT) of soybean in soil covered with guinea grass (*Megathyrsus maximus* B.K.Simon & S.W.L.Jacobs, cv. Mombaça). Source: authors.

ruziziensis is an excellent choice, the other species mentioned can accelerate the improvement of a NTS without relying on higher doses of glyphosate.

As suggested by the discussion above, achieving a high-level NTS requires co-operation between different parties, exchange of ideas and a period of experimentation and leaning from practical experience. We are fully aware that achieving a high-level NTS is hard work and requires teamwork, planning, and patience from everyone involved. A fundamental objective of ours to call upon those all those involved to develop agriculture further together. We are facing increasing climatic adversities and greater risks for agricultural activities. Only through the development of appropriate soil and crop management technologies we can mitigate these impacts and increase the resilience of production systems.

© Burleigh Dodds Science Publishing Limited, 2025. All rights reserved.

Figure 28 No-till (NT) of soybean in rotation with corn into soil covered with *Brachiaria ruziziensis*. Source: authors.

6 Farmer experiences of weed management under CA systems: Brazil (Benjamin Dias Osorio Filho)

6.1 Farm and farming system description

Organic farming has always attracted my attention. While studying agronomy, I became interested in agroecology and soil management. I chose the path of teaching, and since then, I've been a university lecturer involved in research and extension work in regenerative and organic agriculture. Meanwhile, my family's property was leased for conventional soybean and rice production, which created a paradox that bothered me. In 2020, we started an organic grain production project on the property, focusing on soybean, maize, and beans. We began with just over 20 ha and have since expanded to around 70 ha. Along with producing organically, I have always been concerned with soil conservation and carbon sequestration. This laid the foundation for a fundamental shift to organic no-till farming.

© Burleigh Dodds Science Publishing Limited, 2025. All rights reserved.

In the first year of my Agronomy degree, back in 1999, I had the privilege of attending a lecture by Dr. Ana Maria Primavesi, a prominent figure in Agroecology in Brazil. According to her, organic no-till farming is viable as long as a large amount of mulch is provided, and cover crops are properly managed near maturity. These observations stuck with me and ultimately guided the implementation of our project more than 20 years later. We chose to use winter cover crops based on black oats (*Avena strigosa* Schreb) and Persian (*Trifolium resupinatum* L.) and white (*T. repens* L.) clovers. In May 2020, we sowed this mixture but due to the presence of a considerable seed bank of ryegrass (*L. multiflorum* L.), ryegrass emerged alongside the other crops. Through allelopathy, it dominated the area, hindering the growth of oats and clovers. Consequently, we were forced to rely on ryegrass cover for direct sowing of our summer crops. The pleasant surprise was that, thanks to the allelopathic effect of the ryegrass, the incidence of spontaneous plant growth was strongly inhibited, and ryegrass became an excellent cover crop for our production system, especially for soybean and beans. The ryegrass biomass is flattened with a knife roller when the summer crops are sown (Fig. 29).

For maize, however, while ryegrass cover suppresses spontaneous plants due to its higher C/N ratio, it also causes nitrogen immobilization, requiring higher doses of nitrogen fertilizer. In our organic maize crop, chicken litter is

Figure 29 Sowing soybeans on ryegrass straw managed with a knife roller.

© Burleigh Dodds Science Publishing Limited, 2025. All rights reserved.

Figure 30 Growing organic soybeans on ryegrass straw.

our main nitrogen source, but it is costly because it must be sourced from some distance away. One alternative we have not yet successfully implemented, due to the allelopathic properties of ryegrass, is to grow winter legumes before maize. In the case of soybeans and beans, ryegrass cover has proven to be an effective ally in suppressing spontaneous plant growth (Fig. 30).

6.2 Challenges and solutions in weed management

In the first harvest (2020/2021), weeds were less of a problem due to the previous years of conventional herbicide use, which reduced the seed bank and perhaps left some residual effects. However, in subsequent cropping seasons, spontaneous plants such as fanpetals (*Sida* spp.), rice grass (*Echinochloa* spp.) (Fig. 31), millet (*Digitaria* spp.), and corriola (*Ipomoea* spp.) were quite prevalent. In case of deficient crop establishment, weeds easily occupy the space where the cover crop was cut to allow the main crop to emerge (Fig. 32). It is important to note that these weeds emerge late, thanks to the ryegrass cover. During the critical period for preventing soybean interference, the area remains relatively weed free. However, at the end of the crop cycle, when the soybean leaves begin to fall, weeds become dominant, affecting grain filling and making mechanical harvesting difficult.

Spontaneous plant management in our system is based on the allelopathic effect of ryegrass cover crops and the use of mechanical control between the

© Burleigh Dodds Science Publishing Limited, 2025. All rights reserved.

Figure 31 Rice grass (Echinochloa spp.) infesting the space between rows of organic soybeans.

rows. Ryegrass has proven to be an excellent ally in weed suppression and is a low-cost cover crop. Thanks to the seed bank established over the past 20 years, we do not need to sow ryegrass annually. When the summer crops finish in the autumn, ryegrass begins to emerge spontaneously. To increase its biomass and, consequently, its weed-suppressing effect, we apply fertilizers, especially poultry waste, earlier during summer crop growth. This nitrogen also enhances ryegrass growth. Another strategy we have employed is rolling the ryegrass during the milky grain stage while it is still green and combining it with earlier sowing of soybeans or beans. At this stage, there is still some regrowth, but the suppressive effect lasts longer.

In our pursuit of organic no-till grain production, we have found allies along the way. One such partner is the company Gebana Brasil®, which buys our organic grains and is researching technologies to support such initiatives. The company has provided a prototype of an inter-row brush cutter (Fig. 33). The machine, attached to a tractor, cuts the aerial parts of weeds without disturbing the cultivated rows. It has proven effective in controlling dicotyledonous weeds, which do not regrow after being cut. However, grass weeds tend to regrow a few days after mowing, requiring improvements to the prototype. One proposed solution is using winged rods that cut the roots with minimal soil disturbance, preserving the straw cover.

© Burleigh Dodds Science Publishing Limited, 2025. All rights reserved.

Figure 32 Failures of soybeans' emergence allowing spontaneous oat to emerge where the disc cut through the straw cover.

In addition to summer crops, we plan to grow winter grains such as white oats, wheat, or peas. However, ryegrass, which is crucial as a cover crop in our system and supports summer no-till farming, becomes an obstacle for winter planting. To address this, we aim to introduce turnip rape (*Raphanus sativus* L.) immediately after the summer crops to inhibit ryegrass growth, making no-till planting of winter crops feasible. Each situation, depending on its environmental conditions – such as soil type, climate, seed bank size, and commercial crop preferences – requires specific spontaneous plant management strategies. As Ana Primavesi suggested, organic no-till farming is viable, provided there is abundant biomass and proper timing for managing cover crops. Observation and continuous adjustment are key.

© Burleigh Dodds Science Publishing Limited, 2025. All rights reserved.

Figure 33 Spontaneous plant control using an inter-row brush cutter developed by Gebana Brasil®.

In summary, there is no question that direct seeding/zero till production has stabilized and improved production, nearly eliminated any soil erosion and has resulted in soil organic matter increase. The use of diesel fuel is also much lower per unit output than would be required if tillage were still a significant weed control means. There are a few weed species that are increasing, and multi-group herbicide resistant populations are a cause for concern. This will require new modes of action, and/or a more diverse set of weed control tools. Modern production is currently fairly heavily reliant on herbicide weed control, and this needs to change and become more diverse if the same level of effective control is to be maintained. Farms here are very large with a low labour component, so to be successful, additional methods will need to be able to be incorporated into production systems without dramatic increases if any in labour requirements. Robotic and autonomous machinery employed for weed control may well be part of the future system.

There is not a single farmer I know or have ever met that would reverse the adoption of zero till production. Everyone recognizes the benefits to the soil and to the farm business. Challenges must be overcome using methods and technology that allows zero till to continue, as it is too important from a soil conservation and soil health perspective to deviate from it.

© Burleigh Dodds Science Publishing Limited, 2025. All rights reserved.

7 Farmer experiences of weed management under CA systems: Canada (Corey Loessin)

7.1 Farm and farming system description

Owner: Corey Loessin, Aidra Farms Ltd.

Radisson, Saskatchewan, Canada (52.5°N, 107.5°W).

Our farm is in north-central province of Saskatchewan in the heart of the Canadian Prairies. We have thin black soils (Chernozem), primarily loam texture with clay subsoil. This region would be classed 'semi-arid' and receives about 300 mm of moisture annually including snowfall, although precipitation is highly variable with both extended dry and wet spells being common. The frost-free season is about 100–120 days from May to September so there is only time for one crop per year. Our cold winters help to lessen soil carbon losses, reduce disease pressures, improve soil structure among other benefits.

We currently farm 3500 acres (1400 ha) in a continuous cropping system (no fallow) with a crop rotation that alternates a broadleaf crop with a monocot crop. A typical 4-year rotation is spring wheat – canola – wheat – pulse crop, then back to wheat (Fig. 34). Spring barley or oats can replace wheat in the rotation, and the pulse can be either lentils or peas. Canola is typically the most profitable so it is always part of the crop cycle.

We have been direct seeding since 1996 so approaching 30 years with the only soil disturbance occurring during seeding. We have always used a Bourgault air drill with 3/4 inch (1.9 cm) knife style openers and the fertilizer blend is applied with mid-row banding discs (Fig. 35).

Before and into the 1980s, farms in this area would crop 2/3 of their land base each year and fallow 1/3. Fallow was performed by several tillage passes with a disc or cultivator through the growing season to control weeds and try to save moisture for the next crop (dust mulch). However, soil erosion from

Figure 34 Wheat (left), lentils (centre), and canola (right) as key crops in the rotation under Conservation Agriculture.

© Burleigh Dodds Science Publishing Limited, 2025. All rights reserved.

Figure 35 Planting equipment used on the farm with knife opener for seed placement and discs for fertilizer banding.

wind, primarily, was catastrophic. When the top soil got dry, and had little or no cover, frequent prairie winds would move tonnes of topsoil to the field edge and beyond. Lighter texture soils lost so much soil that some fields became uneconomic to farm.

During the 1980s, several farms switched to continuous cropping. Some equipment improvements made this easier. I began farming in 1991 and we continuously cropped our land at that time. Seeding was done with an air seeder using large sweep openers with maximum disturbance. Depth control was poor, and packing following seeding was a separate operation. Fertilizer application was also done in a separate pass, either early in spring or late in previous fall. Soil erosion was reduced but not eliminated.

7.2 Challenges and opportunities in weed management

Weed control was also very challenging during this period of transition to continuous cropping. Glyphosate was not widely used as it was too expensive. Quack Grass (*E. repens* L.) was seriously out of control and the seeding operation dragged the rhizomes all over the field where they promptly grew new plants in the soft, moist soil. Canada Thistle (*Cirsium arvense* L.) was also very prevalent in fields as the elimination of the fallow part of the cycle was the only real control mechanism. Both of these perennial weeds were major problems and caused a lot of crop loss. Several annual or winter annual weeds were also common problems, such as Narrow-Leaved Hawksbeard (*Crepis tectorum* L.), Shepherd's Purse (*C. bursa-pastoris* L.), Stinkweed (*Thlaspi arvense* L.), and Wild Buckwheat (*Polygonum convolvulus* L.).

Two changes occurred that completely changed the direction of this production system – glyphosate became more economical to use, and the

© Burleigh Dodds Science Publishing Limited, 2025. All rights reserved.

practice of a pre-seed burn-off application became common. Typically applied at 0.5 L/acre (1.25 L/ha), virtually all weeds present before seeding were controlled. Fall (post-harvest) spraying also provided excellent control of the two biggest weed issues – Quack Grass and Canada Thistle. The other major change was the development of narrow opener direct seeding machine capable of seeding and fertilizing in a single field pass and disturbing the soil very little.

Another development of great significance was the introduction of herbicide-tolerant canola. Systems with either glyphosate-tolerant or glufosinate-tolerant canola became available, and enabled the control of a wide spectrum of weeds previously difficult or impossible to control. Cleavers (*G. aparine* L.) was a common weed issue in canola crops as the seed is inseparable from canola seed. These new systems controlled it. Stinkweed was also very common in canola crops and could now be controlled.

Introduction of pulse crops (first field peas and then lentils) was another helpful development that enabled the use of some different herbicides during that year. Both crops are rather weak competitors and require robust weed control programs. Perennial weeds must be well controlled before the pulse crop in the crop rotation.

Generally, planting of annual crops occurs earlier in the season in a zero till system than what used to be done using a conventional tillage-based system. Obviously, the less preparation work that needs to be done ahead of zero till planting partly enables this. Additionally though, the changes in the soil that improve water infiltration from multi-year zero till allow field operations to begin earlier. The earlier start helps crops to begin growth ahead of weeds and increases the likelihood of full maturity being reached before a damaging fall frost.

About 10 years following the introduction of direct seeding, Quack Grass had virtually disappeared from the fields. Canada Thistle was also much better controlled as well. A pre-seed application of glyphosate, followed by an in-crop herbicide application, then a pre-harvest or post-harvest application of glyphosate dramatically changed the weed spectrum and weed population density on fields. Overall, fields now have the fewest weeds that they have ever had during my farming career. Some annual weeds such as Wild Oats (*Avena fatua* L.), Shepherd's Purse, Cleavers, and Wild Buckwheat are still present in small, well-controlled populations. Many winter-annual weeds that used to be very prevalent such as Narrow Leaved Hawksbeard and Stinkweed are now very rare. We have not seen a Quack Grass plant on any of our fields in 20 years. Canada Thistle (wind blown seeds) will still be present particularly if a fall control application is not done for 2 years in a row. Fields with very low weed populations can now occasionally forego the spring burn-off application with no economic penalty.

© Burleigh Dodds Science Publishing Limited, 2025. All rights reserved.

A few weeds are increasing in zero till and are cause for some concern. One small area weed is Field Horsetail or Scouring Rush (*Equisetum arvense* L.). It typically grows in small patches that do not spread readily, but there is no herbicide that can control it. This weed was very minor back when tillage was a frequent part of the cropping system, so it clearly thrives under no-tillage.

Another problematic weed is Kochia (*Bassia scoparia* L.). It has been increasing rather dramatically and has also developed resistant populations to several groups of herbicides. Weakly competitive pulse crops enable Kochia to become established and then its control becomes challenging. The extensive use of Roundup Ready canola is also becoming a problem, as glyphosate-tolerant kochia populations are also rising rapidly. To a lesser extent, Foxtail Barley (*Hordeum jubatum* L.) has increased. Later fall glyphosate applications still seem to control it fairly well. Some annual weeds, such as Stinkweed and Shepherds Purse, may begin to increase, as they are largely resistant to herbicides used in pulse crops. However, several other herbicide groups still provide good control, thus underlining the positive impact of diverse crop rotations.

Herbicide application technology has undergone considerable advances. GPS-controlled machines have been common for some time. Nearly all sprayers now have many sections along the boom with automatic control. Our latest unit has individual nozzle control and turn compensation, so application rate is very precise and overlap is virtually eliminated. The first 'see-and-spray' technology is being trialled by some. It seems green on brown spraying (where computer vision technology is used to identify weeds in fallow fields) works fairly well, although the economics of the equipment is not yet favorable, given its current cost. Green on green spraying (which seeks to identify weeds in a growing crop) is being developed but is not yet working to the point of commercial availability.

Almost all farms have their own sprayer. Spray timing is critical, and farms can best time the operation with their own unit. Custom applicators are quite rare, with the exception of some aerial applications (plane and helicopter) more common for crop desiccation or insect control.

Other aspects of weed control are starting to be implemented. Weed seed destructors on combine harvesters are starting to be used. This technology may become widespread over the next few years. Other non-herbicide control strategies – timely mowing of problem patches, for example, or strategic patch tillage operations, are being incorporated by some.

Most farmers are paying attention and rotating herbicide groups through fields as best they can (various published and online tools are available). That said, one of the most worrisome issues on the horizon are increased populations of herbicide-resistant weeds.

© Burleigh Dodds Science Publishing Limited, 2025. All rights reserved.

Costs of weed control have been relatively stable for some time. Some product costs have decreased typically as patent protection runs out. However, the cost of application equipment has increased dramatically. Our system remains reliant on herbicides as the most cost-effective weed control method.

Virtually all modern combine harvesters have highly effective crop residue chopping and spreading capability. While crop residue chopping was once a limitation, it no longer is. Furthermore, widespread adoption of semi-dwarf cereal crop varieties means there is less residue produced which further enables successful direct seeding. Some operations include a harrow post harvest to help spread residue and provide some further straw breakdown.

The principal reason farmers adopted no-till (approximately 75% of the cropped acres in western Canada) was to reduce wind erosion of top soil. The system also reduced diesel fuel consumption (ours is now about 18 L/acre/year; 44 L/ha/year). It saves topsoil moisture which enables better crop establishment. Continuous cropping with a diverse crop rotation also improves the soil. On our home quarter, e.g. soil organic matter has increased from 3% to 4.5% during the 33 years we have farmed it. Soil tilth has visibly improved, as has moisture-holding capacity and infiltration rate. Earthworms are very prevalent throughout the fields now while they were almost non-existent when we started.

Insecticides are used in zero tillage systems relatively rarely. Insecticide seed treatment on canola seed is 100% employed for the control of flea beetles, and its effectiveness reduces the need for foliar applications dramatically. Occasionally, other insect issues do arise – currently grasshopper is a significant pest in the dry cycle and in the driest areas. However, overall, insecticide use is quite low.

Fungicide use is also occasional. Wet cycles necessitate some disease control measures but many years there is simply not enough moisture to create an environment that is conducive to serious disease.

Nearly 100% of farms practice responsible and 'closed-loop' systems with respect to pesticide containers. Common and higher rate products are now almost exclusively handled in returnable and reuseable totes (typically 1000 L). Smaller use rate products in plastic jugs are emptied, rinsed, and then returned to collection depots at all retailers in large bags for collection and recycling. This program is run by CleanFarms – an initiative funded by manufacturers, and participated in by virtually all users of their products – works exceptionally well (https://cleanfarms.ca/).

Although our farm and many others have been in zero till systems for a long time, and it is difficult to make comparisons, it does seem like the amount of pesticide use has actually fallen as a ratio of productivity. Some studies conducted by the University of Saskatchewan quantify this. Many weeds, as well

© Burleigh Dodds Science Publishing Limited, 2025. All rights reserved.

as some insects and diseases that used to be very problematic, are either not present or not present in economically damaging levels. The zero till system itself is partly responsible for this. Improvement in genetics of every crop type has made a major contribution (disease resistant varieties, for example). And again, a more diverse crop rotation has improved soil health, reduces disease pressures in crops and enables farms to optimize use of resources.

Cover crops are not used at all in this region. Generally, there is hardly enough moisture for one crop, let alone more. Fall planted crops are also used little, although they can work satisfactorily some years. They would be a good addition from a weed control/weed life cycle perspective.

8 Conclusions and the future of weed management in Conservation Agriculture systems

The testimonies of farmers who have transitioned from conventional-to-CA-based farming systems many years ago provide valuable insights into what must be taken into account regarding weed management after making this change:

- There is no 'one-size-fits-all' solution. The adaptation of weed management strategies, alongside the significant change in soil management, needs to be tailored to existing conditions in terms of crops, weeds, soils, climate, and agricultural management practices and their interactions.
- Changes in the composition and incidence of weed communities will occur, requiring farmers to have a better knowledge of prevalent weed species and their ecology, their sensitivity to herbicides, available herbicides (particularly their mode of action and efficacy at different crop residue levels and weed development stages), and the period of weed interference on crop yield. A key issue is keeping the seedbank at a low level (even with diverse composition), preventing the prevalence of a single or small group of weeds well adapted to the crops and management practices. This requires crop diversification and diverse management practices.
- Under CA, successful weed control strategies are accomplished by the combined use of several direct and indirect methods and management practices, both preventive (e.g. cover crops, mulch layers) and cultural methods (e.g. intercropping, crop varieties, sowing time). In other words, the full implementation of all CA principles is key for keeping the weed population manageable, thus avoiding negative impacts on yields, especially in the long term, and even allowing for a reduction in herbicide use over time.
- Biotechnological advancements, such as the development and use of single or multi-herbicide tolerant crop varieties, could facilitate weed

© Burleigh Dodds Science Publishing Limited, 2025. All rights reserved.

management (not only under CA) in the short term. However, they might not be sustainable in the medium/long term due to the adaptation of weed communities, the reliance on chemical weed control, and the likely disincentive for alternative weed management strategies offered by the CA approach.

- A continuous attitude of observation and improvement, willingness to take risks, and eagerness to innovate are key ingredients for making weed management under CA successful and less reliant on herbicides over time.

The testimonies of the pioneer farmers presented in this chapter, though from different corners of the globe, are certainly not exhaustive regarding the vast variety of conditions and farming systems. However, these examples demonstrate that it is possible to adapt to and cope with site- and cropping system-specific challenges after a shift towards CA. They can serve as case studies providing potential approaches for different agroecological and farming conditions.

The main objectives of soil tillage have always been to ease the placement of seeds, plants, or tubers into the soil and to reduce competition from unwanted vegetation with the established or soon-to-be-established crop. While the challenge of the first objective has gradually been overcome through the development of increasingly sophisticated machinery, often equipped with precision technologies able to adapt to local conditions, the challenge of the second objective remains for both tillage-based and no-till cropping systems. Due to their effectiveness, low cost, ease of use, and high return, chemical weed control is widely used among farmers in both conventional and CA systems.

In some regions more than others, existing concerns among consumers and society as a whole, along with more restrictive regulations on the use and environmental exposure to pesticides, are increasingly limiting the use of agrochemicals, especially pesticides. Based on these concerns, the benefits and necessity of using pest and disease control products for crop production and global food security are often overlooked. When it comes to herbicides, the benefits of replacing mechanical, tillage-based weed control with chemical weed control are almost completely ignored. These benefits largely derive from the avoidance of tillage-induced soil disturbance, thus reducing soil erosion to a minimum, as well as the mineralization and loss of soil organic matter, not to mention the positive impact on soil life.

Moreover, the widespread assumption that no-till farming is more reliant on herbicides is not true. While tillage may terminate existing weeds, it also creates loose, bare ground, perfect conditions for weed seeds to germinate. In essence, tillage can increase weed germination and growth. Conversely, no-till farming, in combination with other elements of CA, hosts a wide range of

© Burleigh Dodds Science Publishing Limited, 2025. All rights reserved.

mechanisms and processes that, over time, lead to reduced weed pressure and a decrease in overall herbicide use. Most of these mechanisms and processes are well known and described (Eslami, 2014; Singh et al., 2015; Mhlanga et al., 2016; Basch et al., 2020), while others require further investigation and field studies (Rueda-Ayala et al., 2015).

9 Acknowledgements

The lead author of this chapter deeply appreciates the contributions of the farmers who dedicated a great deal of their valuable time to share their testimonies and respond to countless questions throughout the process of compiling this chapter. Their empirical experience is as valuable as long-term scientific experiments, and I owe them a heartfelt thank you for all their efforts.

10 References

Baker, C.J., et al. (2007), *No-tillage seeding in conservation agriculture* (2nd ed.) CABI and FAO. Cromwe Press, Trowbridge, p. 352.

Basch, G., et al. (2020), 'Weed management practices and benefits in CA systems'. In A. Kassam (Ed.), *Advances in conservation agriculture, volume 2: practice and benefits*, Cambridge, UK, Burleigh Dodds Science Publishing, pp. 105–141. https://doi.org/10.19103/AS.2019.0049.04. (ISBN: 978-1-78676-2689).

Derpsch, R. (1998), 'Historical review of no-tillage cultivation of crops', *Proc. 1st JIRCAS Seminar on Soybean Research. No-tillage Cultivation and Future Research Needs.* 1–18.

Eslami, S.V. (2014), Weed management in conservation agriculture systems. In B.S. Chauhan and G. Mahajan (Eds.), *Recent advances in weed management*, Springer, New York, NY, USA, pp. 87–124.

Mhlanga, B., et al. (2016), 'Weed management in maize using crop competition: a review'. *Crop Prot.*, 88, 28–36.

Phillips, E.R. and Phillips, S.H. (1984), *No-tillage agriculture, principles and practices*, Van Nostrand Reinhold Co., New York, p. 319.

Phillips, S.H. and Young, H.M. (1973), *No-tillage farming*, Reiman Associates, Milwaukee, Wisconsin, p. 224.

Rueda-Ayala, V., et al. (2015), 'Investigation of biochemical and competitive effects of cover crops on crops and weeds', *Crop Prot.*, 72, 79–87.

Singh, V.P., et al. (2015), Weed management in conservation agriculture systems'. In M. Farooq and K.H.M. Siddique (Eds.), *Conservation agriculture*, Springer International Publishing, Cham, pp. 39–78. ISBN 978-3-319-11620-4.

© Burleigh Dodds Science Publishing Limited, 2025. All rights reserved.

Index

© Burleigh Dodds Science Publishing Limited, 2025. All rights reserved.

© Burleigh Dodds Science Publishing Limited, 2025. All rights reserved.

© Burleigh Dodds Science Publishing Limited, 2025. All rights reserved.

© Burleigh Dodds Science Publishing Limited, 2025. All rights reserved.

© Burleigh Dodds Science Publishing Limited, 2025. All rights reserved.

© Burleigh Dodds Science Publishing Limited, 2025. All rights reserved.

www.ingramcontent.com/pod-product-compliance
Ingram Content Group UK Ltd.
Pitfield, Milton Keynes, MK11 3LW, UK
UKHW021826100926
8556IPUK00003B/176

9 781801 46784